ORGANIC CHEMISTRY ONLINE 2.0 WORKBOOK
FOR MCMURRY'S

Organic Chemistry
FIFTH EDITION

Paul R. Young
University of Illinois, Chicago

Australia • Canada • Mexico • Singapore • Spain • United Kingdom • United States

Assistant Editor: Melissa Henderson
Marketing Manager: Steve Catalano
Marketing Assistant: Christina DeVeto
Editorial Assistant: Dena Dowsett-Jones
Production Coordinator: Stephanie Andersen
Cover Design: Vernon Boes
Cover Photo: J. Jamsen, Natural Selection
Print Buyer: Nancy Panziera
Printing and Binding: Globus Printing Company

COPYRIGHT © 2000 by Brooks/Cole
A division of Thomson Learning
The Thomson Learning logo is a trademark used herein under license.

For more information, contact:
BROOKS/COLE
511 Forest Lodge Road
Pacific Grove, CA 93950 USA
www.brookscole.com

All rights reserved. No part of this work may be reproduced, transcribed or used in any form or by any means—graphic, electronic, or mechanical, including photocopying, recording, taping, Web distribution, or information storage and/or retrieval systems—without the prior written permission of the publisher.

For permission to use material from this work, contact us by
Web: www.thomsonrights.com
fax: 1-800-730-2215
phone: 1-800-730-2214

Printed in the United States of America

5   4   3   2   1

ISBN: 0-534-37191-4

CONTENTS

Preface
Structure & Bonding 1
    Conversions to Line Drawings 2
    Conversions from Line Drawings 4
    Unshared Pairs of Electrons 6
    Drawing Resonance Structures 8
Alkanes, Cycloalkanes & Isomerism 9
    Identifying Isomers 10
    Identifying Structural Elements 11
    Nomenclature 12
    Conformational Analysis 15
Alkenes and Cycloalkenes: Nomenclature and Ionic Addition of Halogen Acids ... 16
    Nomenclature 17
    Degrees of Unsaturation 18
    Carbocation Rearrangements 19
    Addition of Halogen Acids 20
Alkenes: Additions and Oxidation Reactions 21
    An Overview of Addition Reactions 23
    An Overview of Oxidation Reactions 24
    Exercises in Regiochemistry 24
    Addition Reactions—I 25
    Addition Reactions—II 26
    Addition Reactions—III 27
    Addition and Oxidation Reactions 28
    Synthesis 29
Alkynes: Addition and Oxidation Reactions 30
    An Overview of Alkyne Reactions 31
    Reactions 32
Stereochemistry 33
    Recognizing Symmetry 35
    Identifying Chiral Centers 36
    Conversion to Fisher Projections 37
    Identifying Isomerism 38
    Assigning Absolute Configuration 40
    Optical Activity 43
    Reactions Generating Stereocenters 44
Alkyl Halides: Preparation, Substitution & Elimination Reactions 46
    Trends in S_N2 and E2 Reactivity 47
    Trends in S_N1 and E1 Reactivity 48
    Preparation of Alkyl Halides: An Overview 49
    An Overview of S_N2 Reactions 50
    Nomenclature 51
    Substitution Reactions 52
    Elimination Reactions 53
    More Reactions 54
Spectroscopy 55
    Infrared Spectroscopy Problem Set 56
    Nuclear Magnetic Resonance 60
    ^{1}H NMR Spectroscopy Problem Set 62
    ^{13}C NMR Spectroscopy Problem Set 67
    Mass Spectrometry—Background 69
    Mass Spectrometry Problem Set 72
    Integrated Spectroscopy Problem Sets 73
Conjugated Dienes 83
    Cycloaddition Reactions 86
    Diels-Alder Reactions—Synthesis 87
    Diels-Alder Reactions—Synthesis II 88
    Diels-Alder Reactions—Synthesis III 89
Arenes 90
    Benzene Derivatives: Nomenclature 94
    Aromaticity 96
    An Overview of Electrophilic Aromatic Substitution 97

CONTENTS - *Continued*

Arenes—*Continued*
- Electrophilic Aromatic Substitutions 99
- Benzene Derivatives: *Other Reactions* 100
- Benzene Derivatives: *Synthesis* 101

Alcohols, Ethers & Phenols 103
- An Overview of Alcohol Reactions 107
- An Overview of Reactions that Yield Alcohols 108
- Alcohols and Thiols: Nomenclature 109
- Preparation of Alcohols by Reduction of Carbonyl Groups 110
- Preparation by the Grignard Reaction 111
- Alcohols: Reactions 112
- Alcohols: Synthesis—I 113
- Alcohols: Synthesis—II 114
- Ethers: Preparation and Reactions 115
- Ethers: Synthesis Problems 116

Aldehydes & Ketones 117
- Reactions of Aldehydes & Ketones 122
- Nomenclature 123
- Selected Reactions—I 124
- Selected Reactions—II 125
- Synthesis—I 126
- Synthesis—II 127

Carboxylic Acids 128
- Nomenclature 131
- Reactions that Yield Carboxylic Acids 132
- Selected Reactions of Carboxylic Acids 133
- Synthesis 134

Acyl Compounds (Carboxylic Acid Derivatives) 135
- Reactions of Acid Halides, Anhydrides & Esters 139
- Nomenclature 141
- Tetrahedral Intermediates 142
- Reactions that Yield Acyl Derivatives 143
- Reactions of Acyl Compounds 144
- Synthesis 145

Carbonyl α–Substitution Reactions 146
- Reactions—I 153
- Reactions—II 154
- The Malonic Acid Synthesis 155
- The Acetoacetate Ester Synthesis 156
- Synthesis 157

Carbonyl Condensation Reactions 158
- Aldol-Type Condensation Reactions 165
- Claisen-Type Reactions 166
- Conjugate Additions: Reactions 167
- Carbonyl Compounds: *Reactions in Sequence* 168

Aliphatic and Arylamines 169
- Reactions of Amines 175
- Reactions of Arylamines & Diazonium Salts 176
- Reactions of Phenols 177
- Amines & Arylamines: Nomenclature 178
- Amines: Reactions 179
- Aryl Amines: Reactions of Diazonium Salts 180
- Aliphatic Aryl Amines: Synthesis 181

"Pushing Electrons": Representing Reaction Mechanisms 182
- Reaction Mechanisms Problem Set 185

Practice Examinations 187
- Practice Examination #1: *McMurry*, Chapters 1-6 187
- Practice Examination #2: *McMurry*, Chapters 7-12 191
- Practice Examination #3: *McMurry*, Chapters 13-16 196
- Practice Examination #4: *McMurry*, Chapters 17-24 201
- Practice Examination #5: Alkanes & Cycloalkanes 205
- Practice Examination #6: Spectroscopy 209
- Cumulative Examination #1: *McMurry*, Chapters 1-24 214
- Cumulative Examination #2: *McMurry*, Chapters 1-24 223

Structure & Bonding

The exercises in this section focus on the proper rendering of organic structures and the interconversion of different structural formats. Further, exercises are included which require unshared pairs of electrons and atomic charges to be drawn in structures containing heteroatoms (within organic chemistry, "heteroatoms" generally refers to anything that is not carbon or hydrogen), and exercises which require resonance forms to be drawn for a variety of simple structures.

The structural interconversions largely cover the relationship between "condensed" structure, "line-bond" drawings, in which all atoms and all bonds are shown, and the more common "line" or "structural" drawings which are used in most examples. In strict "line drawings", hydrogens are not shown and the carbon skeleton is shown as a series of short connecting lines. In these structures, a **vertex** represents a CH_2 group, and a **truncated line** represents a CH_3 group. Heteroatoms are shown, and the structure is generally written with bond angles of 109-120° to simulate the shape of the actual backbone. An example of these types of structural interconversions is shown below:

The "purity" of line drawings, however, is often compromised in day-to-day use and "hybrid" structural representations are commonly used in which terminal methyl groups or side chains may be shown in condensed format and hydrogen atoms may be shown to clarify stereochemistry, such as shown below:

The exercises on unshared pairs of electrons and atomic charges calls upon those skills you learned in Freshman Chemistry and can easily be extended to Organic Chemistry by simple analogy; i.e., divalent oxygen is neutral (H_2O), monovalent oxygen is anionic (HO^-), trivalent oxygen is cationic (H_3O^+), etc. Resonance, however, is a topic which is generally new to organic chemistry, but is easily understood if the following simple set of rules is followed:

- move only electrons, not atoms,
- electrons can move only between adjacent sp and sp^2 centers, and between heteroatoms attached to these centers, and finally,
- always remember that resonance forms are electronic **limits**; and just like a mathematical limit, you never get there! The actual molecule is the **hybrid**, it *never* looks like any of the limits, but is always something in between.

Drawing resonance forms is an important skill to master early in Organic Chemistry, because the principles of resonance stabilization and charge delocalization are central to a mechanistic understanding of organic reactions. Resonance stabilization will be a reoccurring theme throughout the course, and a clear understanding of what resonance structures mean (and what they don't mean) will be important to your success as a student of Organic Chemistry.

Structure & Bonding: *Conversions to Line Drawings*

Redraw the following compounds using "skeletal" or "line" drawings.

a.
```
       Br
        |
  H₃C—C⋯CH₃
        |
        CH₃
```

b.
```
  H₃C     CH₂CH₂CH₃
     \   /
      CH
      |
      CH₃
```

c. (cyclohexane drawn with all H's shown)

d. Newman projection:
```
   H    H    CH₃
    \   |   /
     (front C)
     /     \
    H      CH₃
       |
     CH₂CH₃
```

e.
```
             CH₃
              |
   H₃C       C⋯CH₃
       \   /
        CH₂
              |
              CH₃
```

f.
```
   CH₃CH₂    H
         \  /
          C
         / \
        H   CH₃
        |
        C
       / \
      H   CH₃
```

g.
```
         Br     CH₃
          |      |
    H—C═C—C═C—CH₃
       |      |
       H      H
```

h.
```
    H  H  H  H  H
    |  |  |  |  |
  H—C—C—C—C—C—CH₂
    |  |  |  |  |
    H  H  H  H  CH₃
```

i.
```
       H   H
        \ /
    H₂C—C—CH₂
        |   |
    H₂C—C—CH₂
        / \
       H   H
```

Structure & Bonding: *Conversions to Line Drawings, continued*

Redraw the following compounds using "skeletal" or "line" drawings.

j.

k.

l.

m.

n.

o.

p.

q.

Structure & Bonding: *Conversions from Line Drawings*

Redraw the following compounds using "Kekulé" (line-bond) structures.

a. 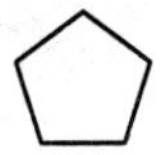 cyclopentane

b. 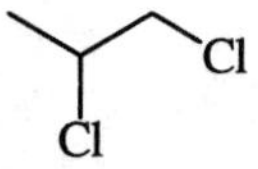 1,2-dichloro-2-methylpropane (CH₃)₂C(Cl)CH₂Cl — structure shown: (CH₃)₂CH-CHCl-CH₂Cl... (line drawing with two Cl substituents)

c.  cyclopropane

d. 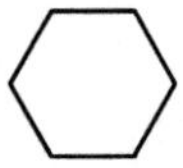 cyclohexane

e.  cyclopropanone (three-membered ring with =O)

f. cyclobutane

g. 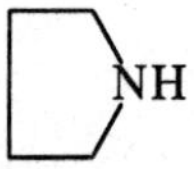 pyrrolidine (five-membered ring with NH)

h. 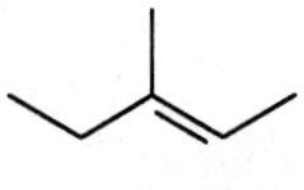 2-methyl-2-butene

i. 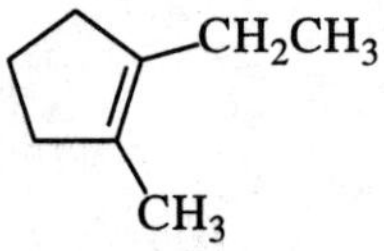 1,2-disubstituted cyclopentene with CH₂CH₃ and CH₃ groups

Structure & Bonding: *Conversions from Line Drawings, continued*

Redraw the following compounds using "Kekulé" structures.

j.

k.

l.

m.

n.

o.

p.

q.

r.

Structure & Bonding: *Unshared Pairs of Electrons*

For each of the following structural drawings, fill in any unshared pairs of electrons, paying particular attention to atomic charges.

a.

b.

c.

d.

e.

f.

g.

h.

i.

Structure & Bonding: *Unshared Pairs of Electrons, continued*

For each of the following structural drawings, fill in any unshared pairs of electrons, paying particular attention to atomic charges.

j.

k.

l.

m.

n.

o.

p.

q. H—O—Cl

r.

Resonance: *Drawing Resonance Structures*

For each of the molecules shown below, draw all of the major resonance forms. Be sure to clearly show the movement of electrons using "curved arrows".

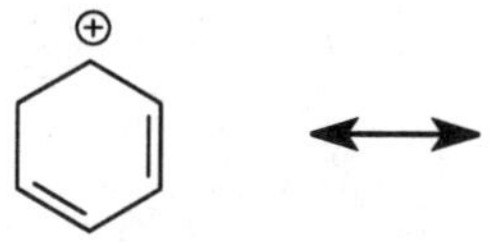

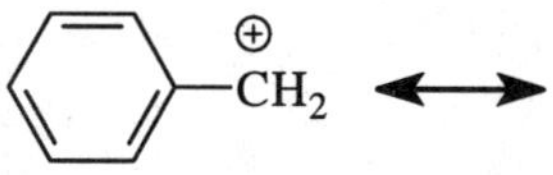

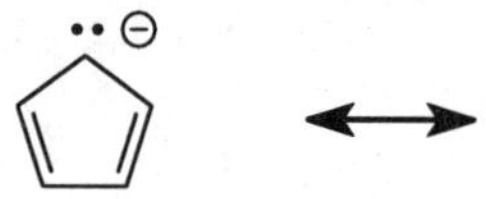

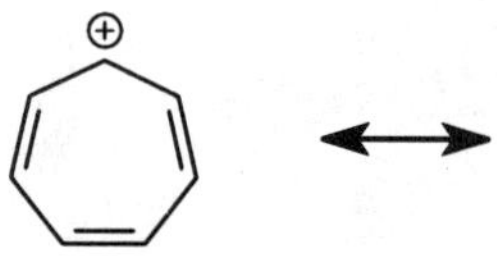

Alkanes, Cycloalkanes & Isomerism

The exercises in this section focus on identifying isomerism and recognizing structural elements within molecules, nomenclature of alkanes and cycloalkanes, and structural features and interconversions within substituted cyclohexanes. Isomerism, within these chapters, is limited to simple **constitutional** and *cis-trans* isomers. Constitutional isomers are defined as molecules having different "constitutions", that is, they have the same number and types of atoms, but they are attached is a different numerical sequence to each other. *Cis-trans* isomers, our first example of **stereoisomers**, differ from each other not in the numerical sequence of attachment, but in the actual three-dimensional spacial relationship between groups in the molecules. Simple examples of constitutional and cis-trans isomers are shown below:

...constitutional isomers   ...*cis-trans* isomers

For simple unbranched alkanes, the name is composed of a prefix to indicate the number of carbons in the main chain, followed by the suffix *ane* to indicate the functional group is an alkane. For branched alkanes:

1. **The longest continuous chain** determines the base name for the alkane.

2. Number the atoms in the chain, beginning at the end nearer the first substituent on the chain. If there are substituents occurring at equal distances from both ends of the chain, begin numbering at the end providing **the lowest number sequence at the first point of difference**. When two equal chains compete for selection as base chain, choose the one with the greatest number of substituents.

3. Substituents on the chain are named as **alkyl radicals** and are numbered using the numbering system from (2). When two or more substituents are identical, use the multipliers: di-, tri-, tetra-, etc. **Names are to be arranged alphabetically without regard for these multiplier prefixes.** Write the name as a single word, using hyphens to separate prefixes and commas to separate numbers.

4. Complex substituents are numbered from the **point of attachment** to the main chain and are included in parenthesis.

5. **Cycloalkanes** are named by adding the prefix *cyclo*.

6. Substituents on cycloalkanes are numbered to give **the lowest possible numbers, or lowest possible number at the first point of difference**. If more than one type of substituent is present, begin numbering alphabetically.

Some simple examples of these rules are given below:

2-methyl-4-propylheptane

2,2,6-trimethyl-4-propyloctane
not 3,7,7-trimethyl-5-propyloctane

1-ethyl-2-methyl-3-isopropylcyclohexane

Alkanes, Cycloalkanes & Isomerism: *Identifying Isomers*

For each of the following sets of structures, indicate if the compounds are:
  a) identical   b) isomers   c) different compounds and not isomeric

a. ____ $H_3C-CH_2-CH_2-CH_2-CH_3$ and   (hexane skeletal structure)

b. ____ CH_3CH_2\C=C/H with CH$_3$ and H   and   (2-methyl-2-pentene skeletal)

c. ____ $CH_3CH_2-CH(Cl)-CH_2CHClCH_3$ and   (2,4-dichlorohexane skeletal)

d. ____   H-C(H)(H)-C(H)(H)-C(CH$_3$)(Cl)-C(H)(H)-C(H)(H)-Cl   and $CH_3CH_2(CClCH_3)CH_2CH_2Cl$

e. ____ $H_3C-C(CH_2CH_3)(CH_3)-O-CH_3$ and   (tert-structure with OCH$_3$)

f. ____   H-C≡C-CH$_2$Cl   and $ClCH_2-C≡C-H$

g. ____ $H_3C-CH(CH_3)-C(=O)-NHCH_3$ and   (H-N(CH$_3$)-C(=O)-CH$_2$CH$_2$CH$_3$)

h. ____   (cyclohexanone with CH$_3$ at position 3)   and   (3-methylcyclohexanone)

i. ____ H_3C-CH_2-C(CH$_3$)(CN)(H) with wedge   and   (skeletal with CN wedge)

Alkanes, Cycloalkanes & Isomerism: *Identifying Structural Elements*

For each of the compounds shown below, answer each of the following questions ("yes" or "no").

Does the molecule:

a. have the empirical formula (CH_2)?
b. contain a secondary alkyl halide center?
c. represent an isomer of 1-bromo-2-chloropentane?
d. contain at least one sp^2 center?

a.

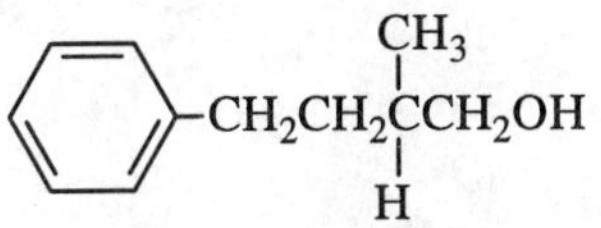

b.

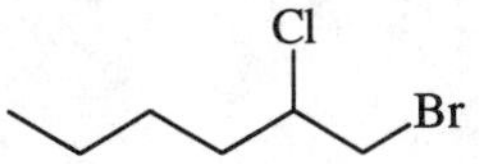

c.

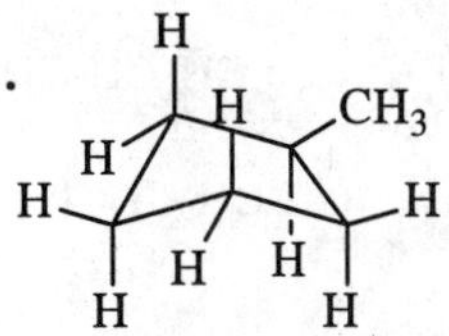

d.

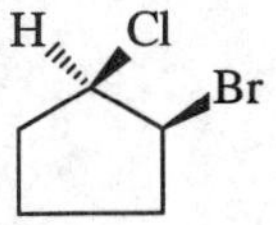

e.

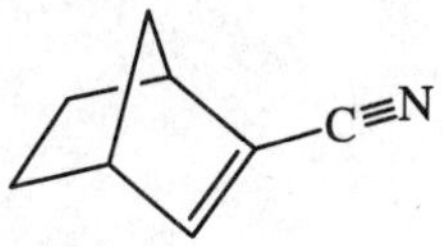

f.

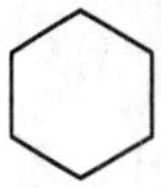

g.

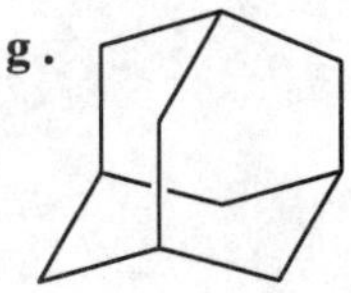

h.

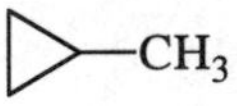

i.

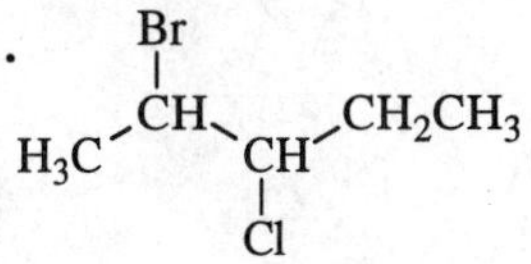

j.

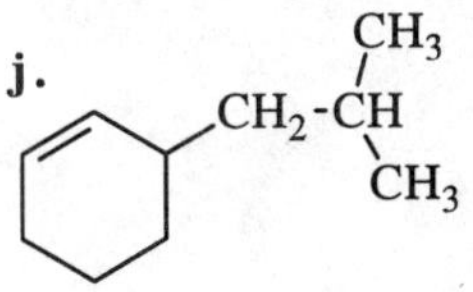

k.

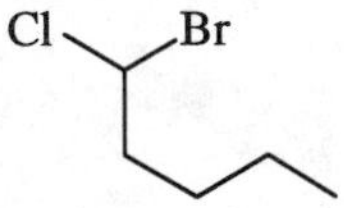

l.

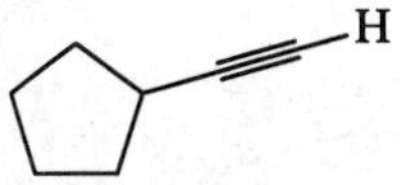

m.

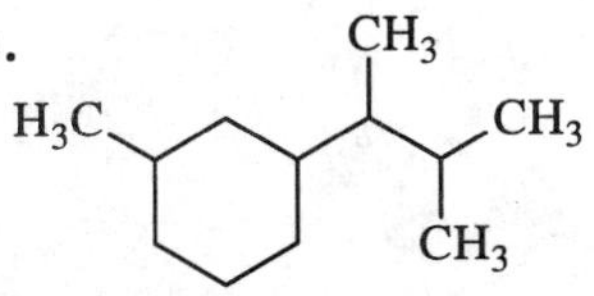

Alkanes & Cycloakanes: *Nomenclature*

Give the proper IUPAC name for each of the following compounds.

a.

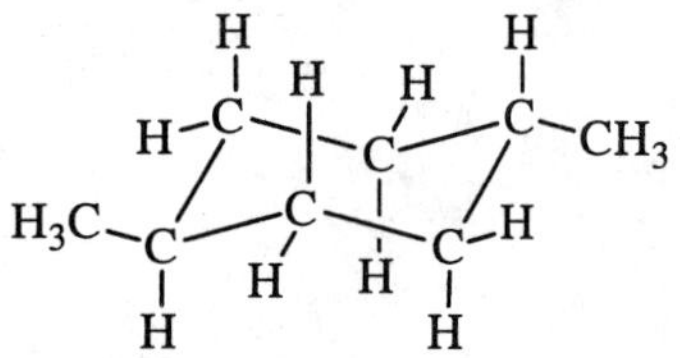

b.

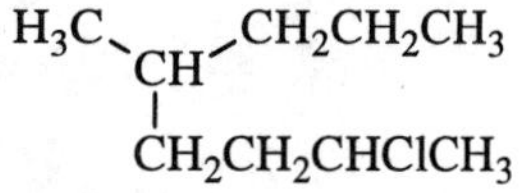

c.

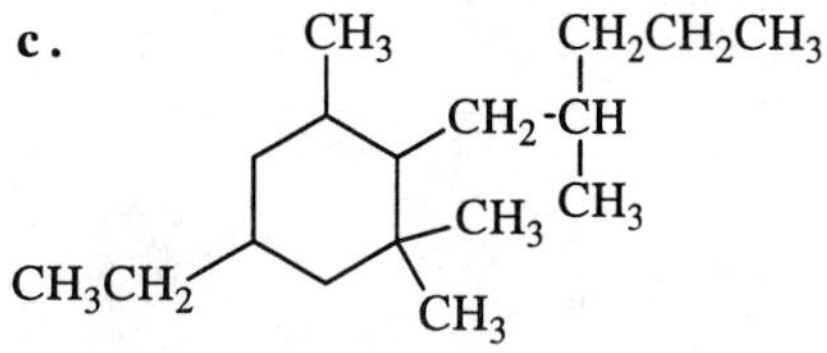

d.

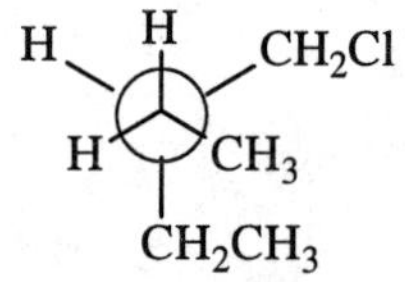

e.

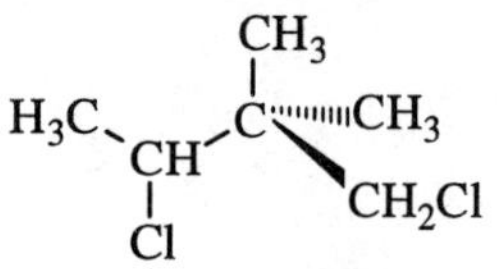

f.

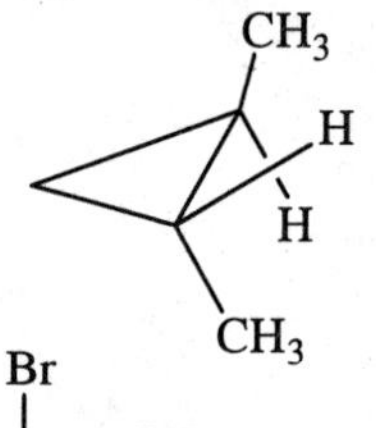

g.

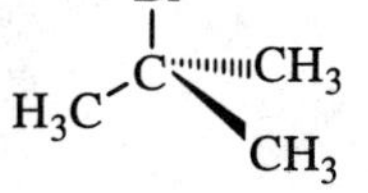

h.

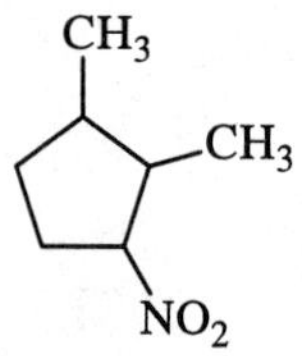

i.

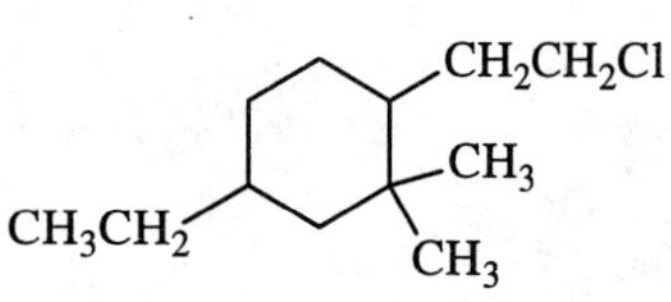

Alkanes & Cycloalkanes: *Nomenclature, continued*

Give the proper IUPAC name for each of the following compounds.

j.

k.

l.

m.

n.

o.

p.

q.

r.

Alkanes & Cycloakanes: *Nomenclature*

Draw the structure corresponding to the following IUPAC names.

    a. *trans*-1-3-dibromocyclopentane

    b. *cis*-1-isopropyl-3-methylcyclohexane

    c. (cyclopentylmethyl)cyclopentane

    d. 2,2,5-trimethylheptane

    e. *trans*-1-chloro-2-methylcyclopropane

    f. *trans*-1,3-dimethylcyclobutane

    g. hexylcyclooctane

    h. *cis*-1-methyl-3-nitrocyclopentane

    i. *trans*-1,2-dibromocyclohexane

Alkanes & Cycloakanes: *Nomenclature*

Alkanes & Cycloakanes: *Conformational Analysis*

Draw each of the compounds shown below in its **most stable conformation.**

a.

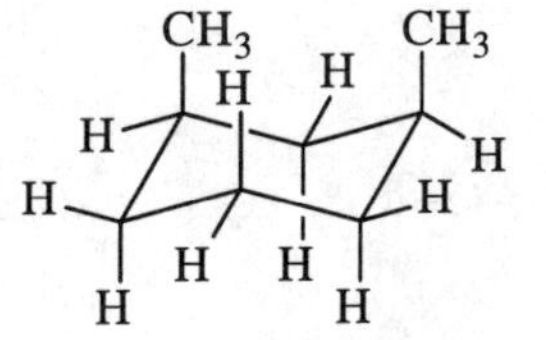

b.

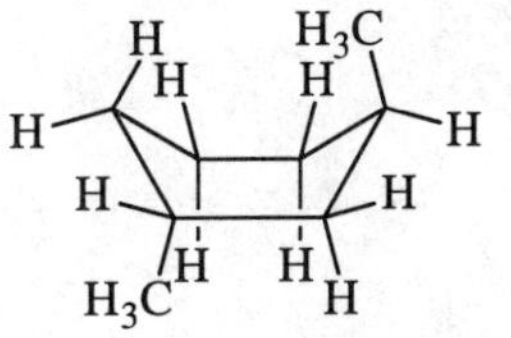

c.

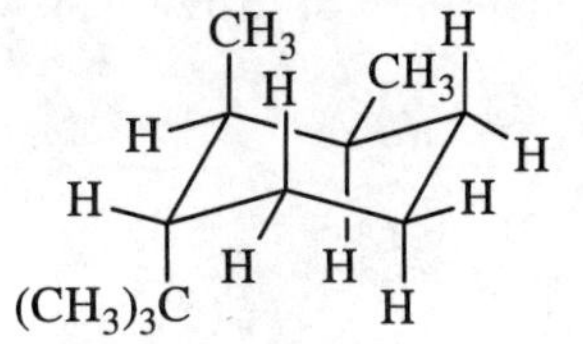

d.

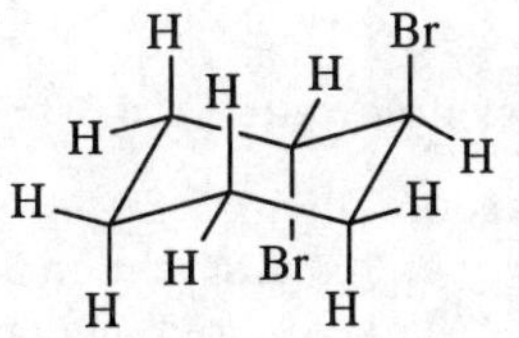

Redraw each of the compounds shown below using "chair" and "boat" drawings.

a.

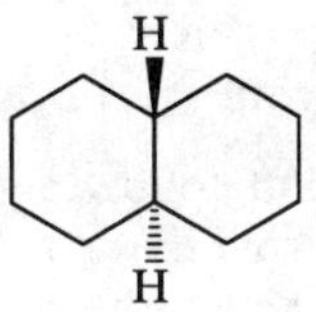

b.

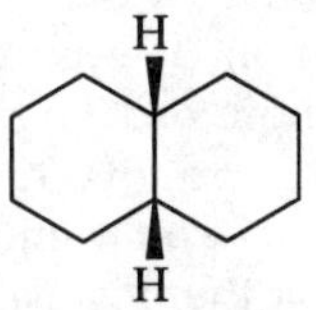

c.

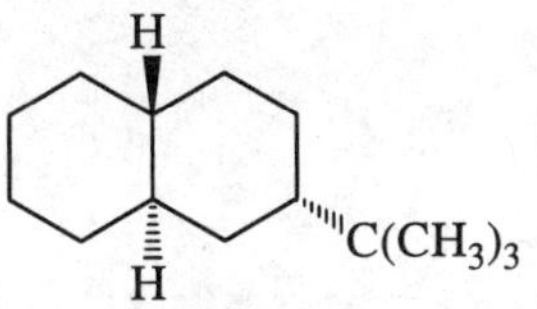

d.

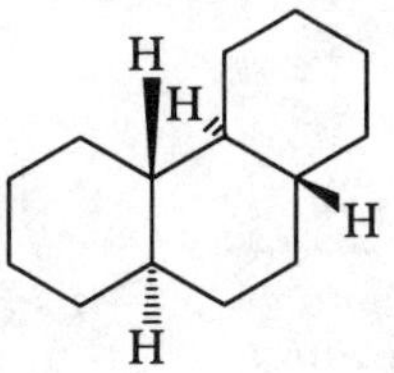

Alkenes and Cycloalkenes: Nomenclature and Ionic Addition of Halogen Acids

The exercises in this section focus on nomenclature of simple alkenes and cycloalkenes, and on the mechanism of the simple ionic addition of HCl and HBr to alkenes to give alkyl halides, with the possibility of structural rearrangement.

For simple alkenes, the name is constructed using the same basic set of rules we have used previously for alkanes, with the following exceptions:

1. Find the **longest chain** *containing the alkene*.

2. Number the chain, giving the **double bond** *the lowest possible number*.

3. Number substituents according to their positions on the chain; *use -diene, -triene*, etc. for multiple double bonds, as appropriate.

4. for cycloalkenes, **begin numbering at the double bond and proceed** *through* **the double bond in the direction to generate the lowest number at the first point of difference.**

Some simple examples of these rules are given below:

2-ethyl-1-pentene   6,6-dimethyl-3-heptene   3,3,6-trimethylcyclohexene   4,6,6-trimethylcyclooctene

The addition of halogen acids to alkenes is a multistep reaction in which the double bond is first protonated to give an intermediate carbocation. The **regiochemistry** of this protonation is in the direction to form the most stable initial carbocation (the driving force for "Markovnikov" regiochemistry). The stability of carbocations can generally be predicted using a simple set of rules:

1. Carbocations **adjacent to heteroatoms** with unshared pairs of electrons are the most stable,

2. carbocations in which the positive charge can be **delocalized by resonance** over two or more centers are more stable than charge-localized carbocations, and,

3. simple alkyl carbocations **decrease** in stability in the order, **tertiary, secondary, primary, methyl**.

Once protonation has occurred to give the most stable initial carbocation, following the guidelines shown above (or following Markovnikov's rule, and all of the *exceptions* to Markovnikov's rule), the initially formed carbocation has the opportunity to **rearrange** to form another, more stable carbocation. Predicting whether this will happen simply requires examination of the molecule to determine if a site exists which would form a more stable carbocation center. That is, if the initial carbocation is **secondary**, is there a **tertiary** carbon in the molecule? If there is, the potential exists for one or more rearrangement steps to occur, involving transfer of hydride or alkyl anions between adjacent centers to give this more stable carbocation structure. Once the *most stable* carbocation has been attained, halide anion simple attacks the carbocation center (from either face, since the carbocation is sp² and planar) to give the alkyl halide. Examples of these steps are shown below:

Alkenes: *Nomenclature*

Give the proper IUPAC name for each of the following compounds.

a. [structure]

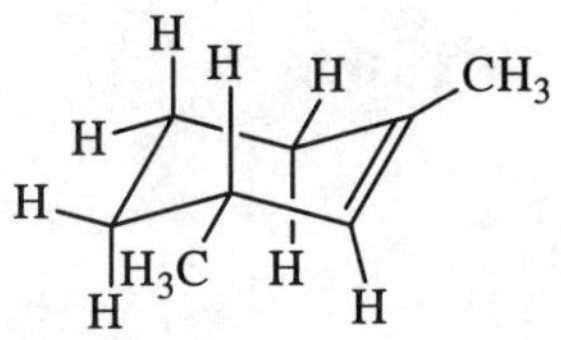

b. [structure]

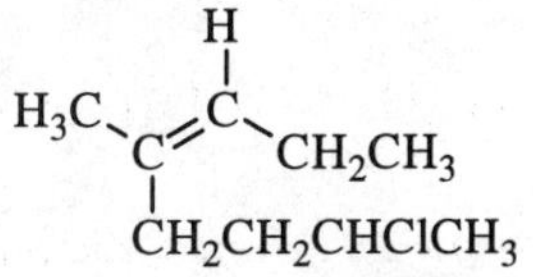

c. [structure]

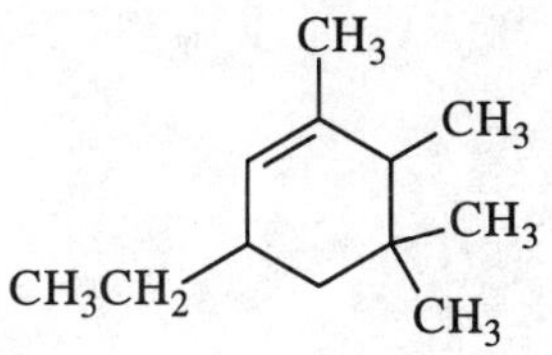

d. [structure]

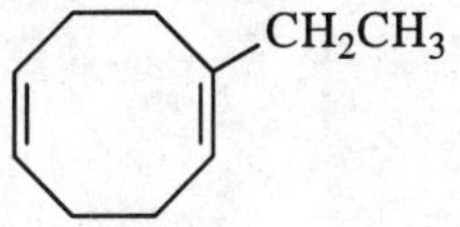

e. [structure]

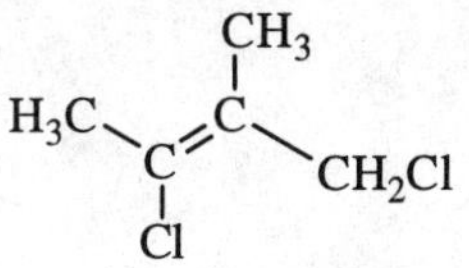

f. [structure]

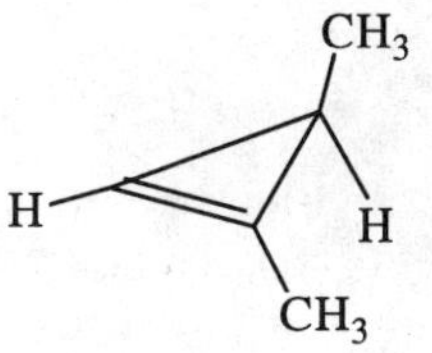

g. [structure]

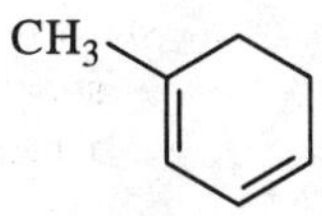

h. [structure]

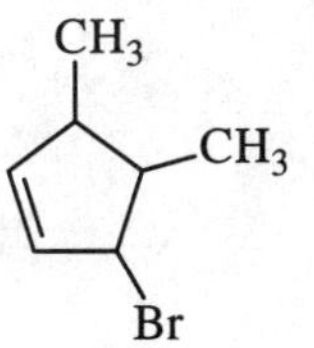

i. [structure]

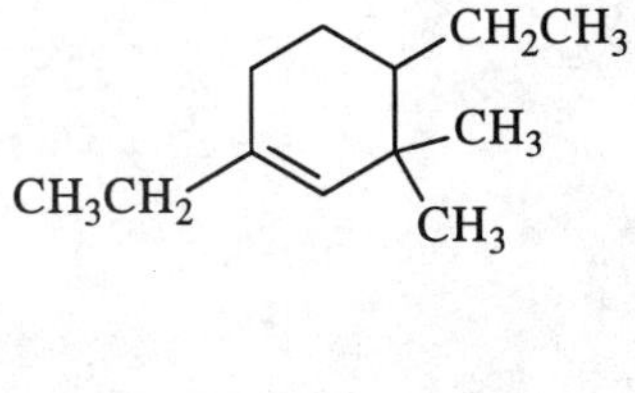

Alkanes, Cycloalkanes & Isomerism: *Degrees of Unsaturation*

For each of the following compounds, calculate the number of "degrees of unsaturation".

a.

b.

c.

d.

e.

f.

g.

h.

i.

Alkenes: *Carbocation Rearrangements*

For each of the following protonation reactions, draw the structure of the **initially formed carbocation** and the structure of the **most stable carbocation**, following rearrangement, if appropriate. Clearly show any rearrangements using "curved arrows".

a. [3-methyl-1-butene] $\xrightarrow{H^+}$ $\longrightarrow$

b. [1,1-dimethyl-2-cyclohexene] $\xrightarrow{H^+}$ $\longrightarrow$

c. [vinylcyclopentane] $\xrightarrow{H^+}$ $\longrightarrow$

d. $H_2C=CHC(CH_3)_3$ $\xrightarrow{H^+}$ $\longrightarrow$

e. [1-methylbicyclo[2.2.2]oct-2-ene] $\xrightarrow{H^+}$ $\longrightarrow$

Alkenes: *Addition of Halogen Acids*

For each of the reactions shown below, draw the structure of the major organic product. Remember that rearrangements may occur.

a. 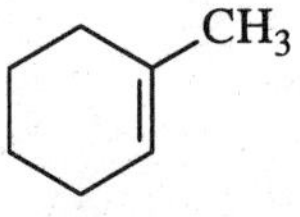 1-methylcyclohexene + HCl ⟶

b. 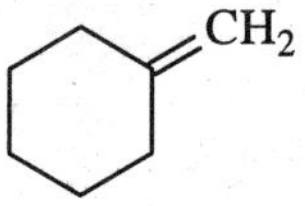 methylenecyclohexane + HBr ⟶

c. 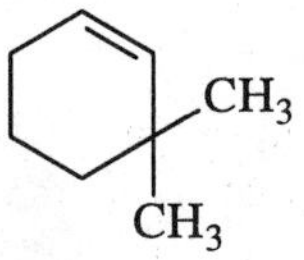 6,6-dimethylcyclohex-1-ene + HBr ⟶

d. 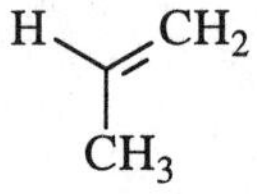 CH₂=C(H)(CH₃) (propene) + HCl ⟶

e. 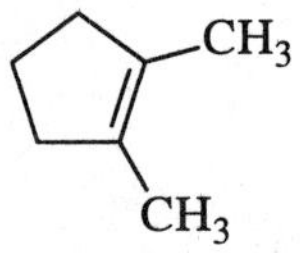 1,2-dimethylcyclopentene + HCl ⟶

f. 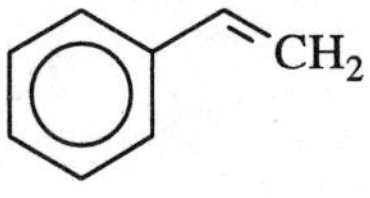 styrene (PhCH=CH₂) + HBr ⟶

g. 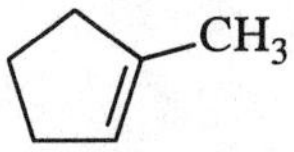 1-methylcyclopentene + HCl ⟶

h.  norbornene (bicyclo[2.2.1]hept-2-ene) + HBr ⟶

i. 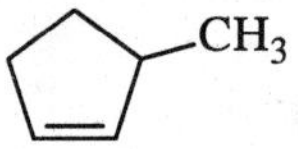 3-methylcyclopentene + HBr ⟶

j. 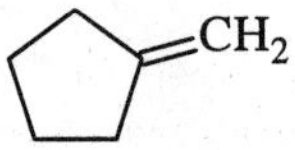 methylenecyclopentane + HCl ⟶

Alkenes: Addition and Oxidation Reactions

The exercises in this section focus on predicting the products of simple addition and oxidation reactions of alkenes, and on using these reactions as tools in simple organic synthesis problems. Distinct carbocation intermediates are only involved in two of the reactions which will be considered here, the addition of halogen acid (as covered in the previous section) and simple acid-catalyzed **hydration** of an alkene to give an alcohol. As in the addition of halogen acid, the intermediate carbocation can rearrange if there is a site in the molecule which would yield a more stable carbocation, and the overall regiochemistry can be described as "Markovnikov". The addition of HBr to an alkene in the presence of "peroxides" proceeds by an alternative mechanism in which bromide **radical** adds to the double bond to form the **most stable radical intermediate**, which will then abstract a hydrogen atom to form the alkyl halide. Since radical stability generally parallels carbocation stability, this radical will form on the same carbon which would form the most stable carbocation in an ionic addition and the bromine will be bonded to the *other* alkene carbon, making the net regiochemistry *anti*-Markovnikov.

Three of the reactions considered in this section involve other types of ionic intermediates, in which the carbons of the alkene are initially **bridged** by the electrophilic atom, and this cyclic intermediate then undergoes nucleophilic attack (ring-opening) to give the *trans* addition product; these are, the reactions with halogen, hypohalite and mercuric acetate. As an example of this type of reaction, consider the reaction with HOBr (HO⁻ as nucleophile), as shown below:

Since there is no distinct carbocation in this mechanism, rearrangements are not observed. You should note that the stereochemistry of the reaction shown is *trans* because the nucleophile must approach from the side *opposite* the bridged cation. Further, the regiochemistry shown is "Markovnikov" (the nucleophile is bonded to the center which would form the most stable carbocation). This is because the tertiary center in the bridged intermediate, and in the intermediate transition state, has more carbocation character than the secondary site, making attack at this center more favorable (it can accommodate more charge, hence it will carry more). For the addition of halogen (i.e., Br_2) the same intermediate is involved, with Br^- as the nucleophile. Mercuric acetate forms an intermediate mercurinium ion, which is attacked by water to give an organomercurial intermediate, which is reduced and removed by reaction with bisulfite anion, in a second step.

Hydroboration and reduction of an alkene (reaction with H_2) both involve nonionic reactions and result in *cis* addition to the double bond. Hydrogenation occurs on the metal surface and simply forms the hydrocarbon; hydroboration/oxidation involves a *concerted* addition of boron and hydride across the double bond, following *anti*-Markovnikov regiochemistry, as shown below.

The origin of the regiochemistry can be described as resulting from a *polar transition state* (but not an intermediate) so that the tertiary center, being the center which can best accommodate a positive charge, becomes partially cationic and reacts preferentially with the partially anionic "hydride" in BH_3. Hydroboration reactions are also very sensitive to steric bulk, and the regiochemistry can also be described as simply resulting from orientation of the boron away from the bulky tertiary center. The intermediate borane is oxidized in a second step to give the alcohol with retention of stereochemistry, so that the net addition of water is *anti*-Markovnikov and *cis*.

Chloroform in base generates **dichlorocarbene** which "inserts" into the double bond of alkenes to give dichlorocyclopropane derivatives; carbene itself can be generated from CH_2I_2 in the presence of Zn(Cu) and reacts with alkenes to form cyclopropane derivatives. Another reagent that adds *cis* across a double bond is alkaline permanganate. Both alkaline permanganate and OsO_4 react with alkenes to form *cis*-1,2-diols, or glycols. The reactions involve intermediate organometallic esters and both oxygens come from the metal, rationalizing the *cis* stereochemistry; with OsO_4, bisulfite must be used to cleave the osmate-ester intermediate.

Oxidation of alkenes can be utilized to prepare aldehydes, ketones and carboxylic acids. The mildest oxidant is ozone, O_3, and the reaction proceeds through the formation of an intermediate ozonide, which must be reduced with Zn dust in HCl to generate the carbonyl products. When you are considering the products of ozonolysis, you should mentally split the double bond, and convert each of the sp² carbons into a carbonyl. Thus, if one hydrogen and one alkyl group are attached to the alkene carbon, an aldehyde is formed, if two alkyl groups are attached, a ketone is formed.

Acidic permanganate is a much more powerful oxidant which will also split alkenes into carbonyl derivatives. A terminal alkene is converted by acidic permanganate into CO_2, an alkene carbon bearing one hydrogen is split, generating a carboxylic acid, and a disubstituted alkene carbon is converted into the ketone, as shown below.

Alkenes: *An Overview of Addition Reactions*

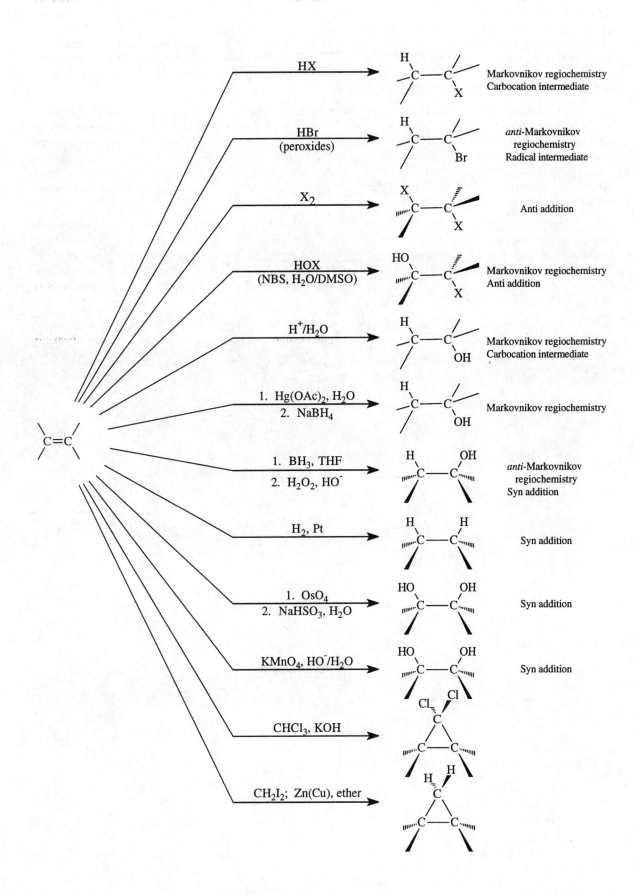

Alkenes: *An Overview of Oxidation Reactions*

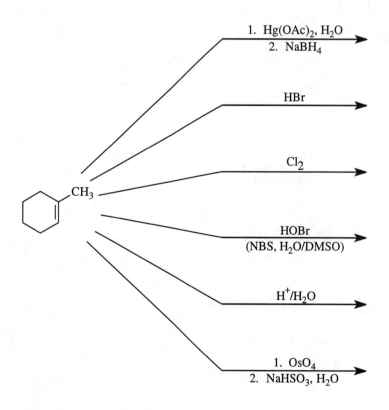

Alkenes: *Exercises in Regiochemistry*

For each of the reactions shown below, draw the structure of the major organic product. Clearly show all stereochemistry.

Alkenes: *Addition Reactions-I*

For each of the reactions shown below, draw the structure of the major organic product. Clearly show all stereochemistry.

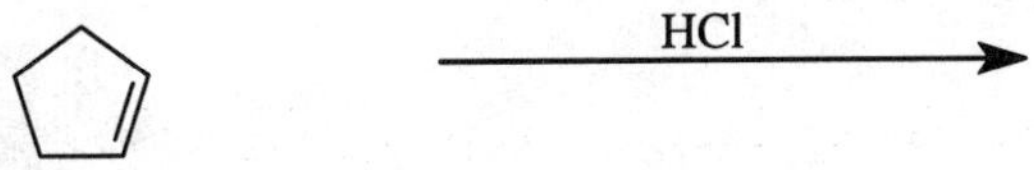

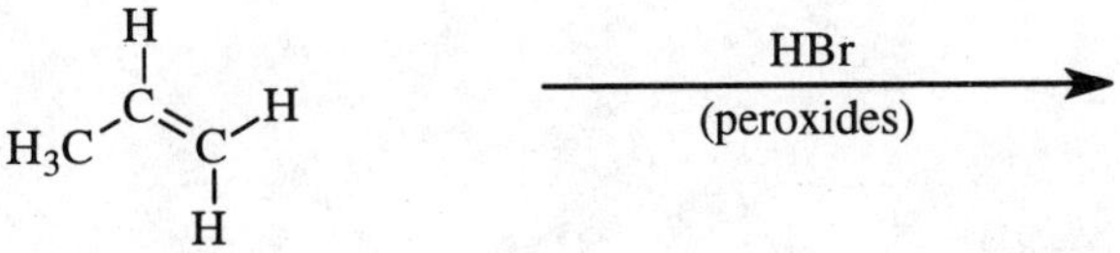

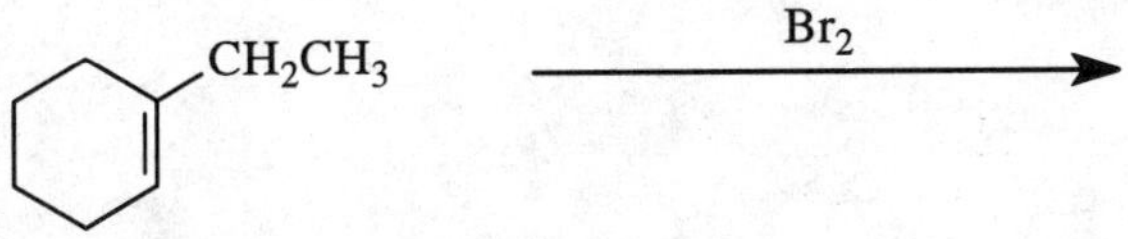

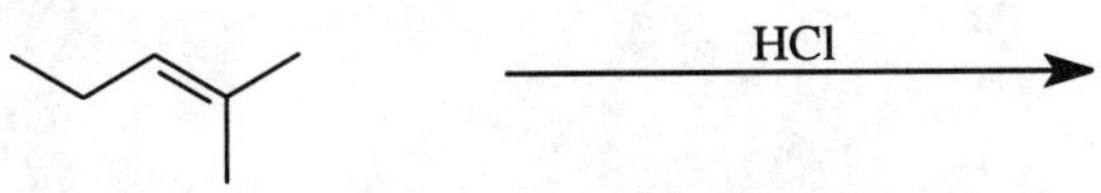

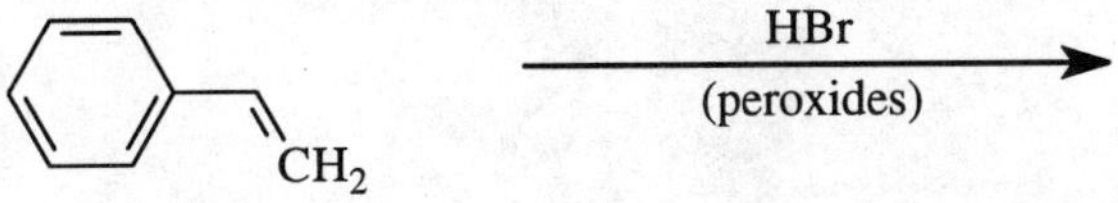

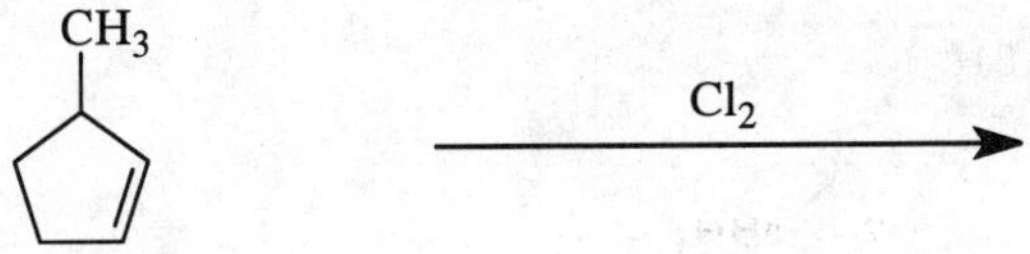

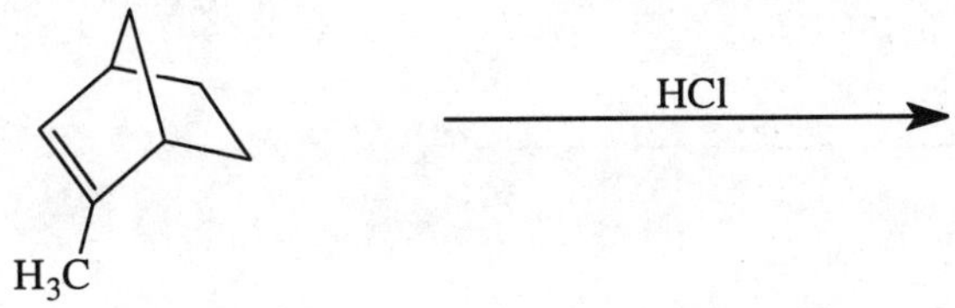

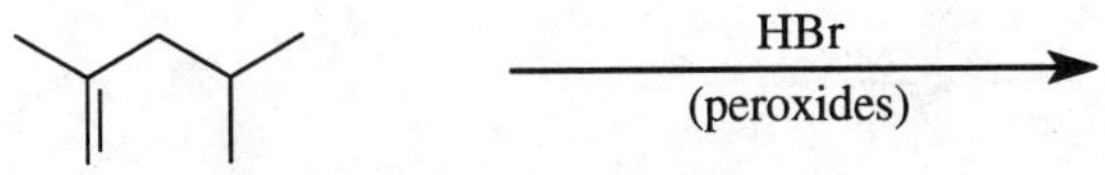

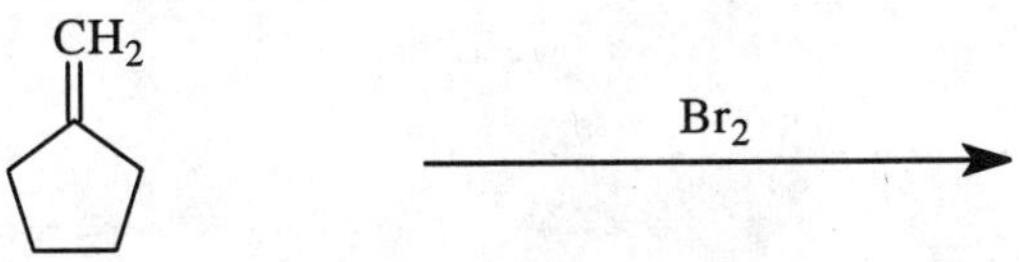

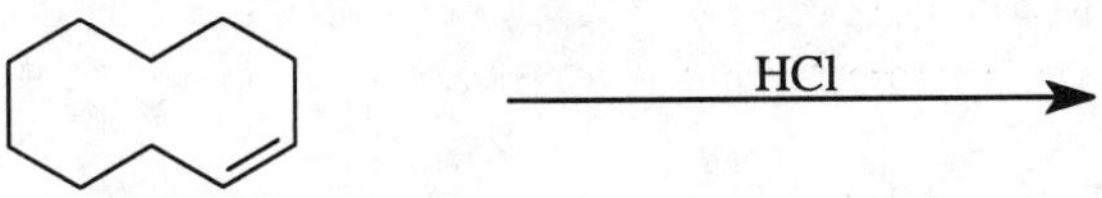

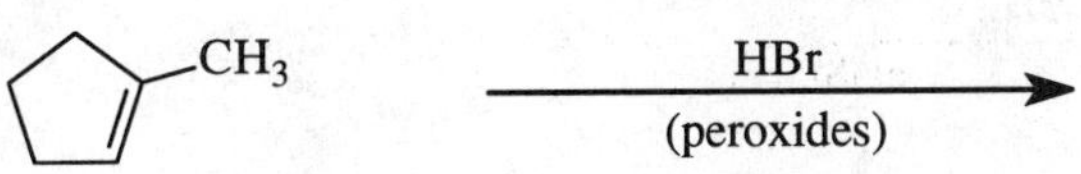

cyclopentene $\xrightarrow{\text{HCl}}$

cis-CH₃CH=CHH (2-butene type, with H₃C and H cis) $\xrightarrow[\text{(peroxides)}]{\text{HBr}}$

1-ethylcyclohexene $\xrightarrow{\text{Br}_2}$

2-methyl-2-pentene $\xrightarrow{\text{HCl}}$

styrene (PhCH=CH₂) $\xrightarrow[\text{(peroxides)}]{\text{HBr}}$

3-methylcyclopentene $\xrightarrow{\text{Cl}_2}$

2-methylnorbornene $\xrightarrow{\text{HCl}}$

2,4-dimethyl-1-pentene $\xrightarrow[\text{(peroxides)}]{\text{HBr}}$

methylenecyclopentane $\xrightarrow{\text{Br}_2}$

cis-cyclodecene $\xrightarrow{\text{HCl}}$

1-methylcyclopentene $\xrightarrow[\text{(peroxides)}]{\text{HBr}}$

Alkenes: Addition Reactions-II

For each of the reactions shown below, draw the structure of the major organic product. Clearly show all stereochemistry.

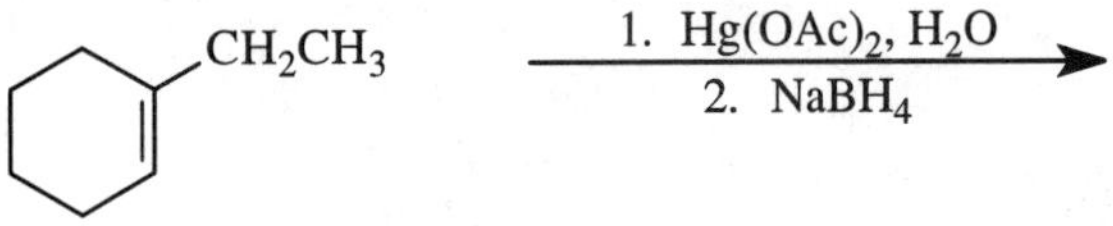

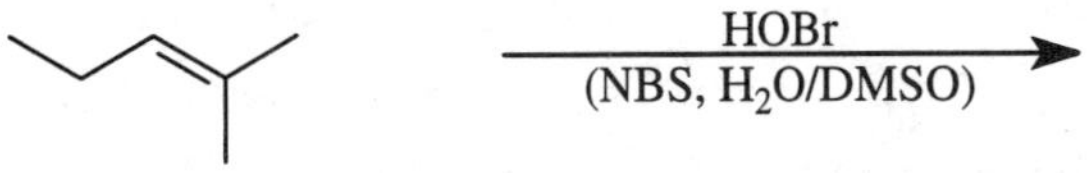

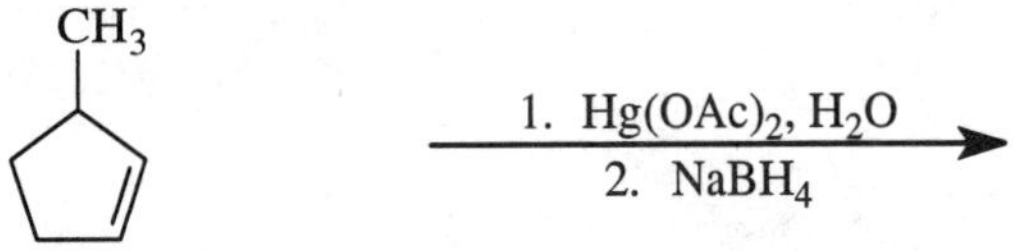

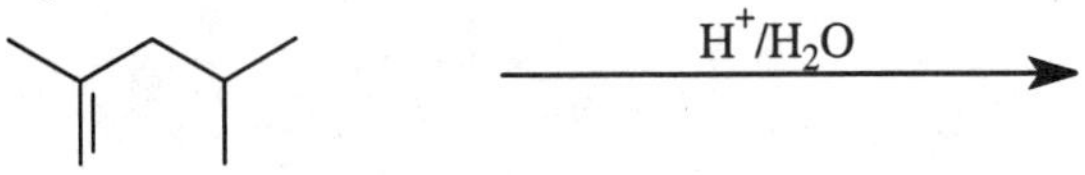

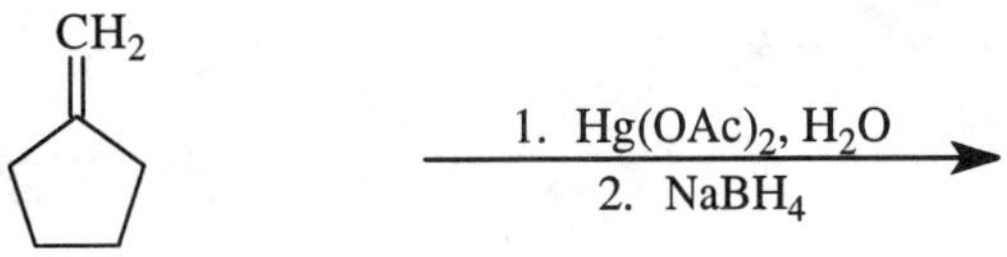

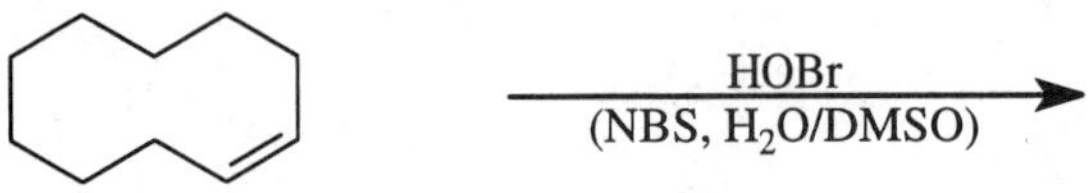

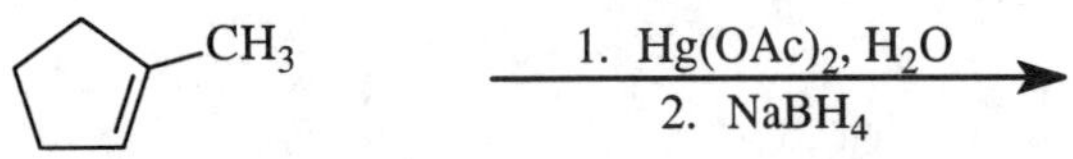

cyclopentene → HOBr (NBS, H$_2$O/DMSO) →

(Z)-2-butene (H$_3$C and H cis) → H$^+$/H$_2$O →

1-ethylcyclohexene → 1. Hg(OAc)$_2$, H$_2$O; 2. NaBH$_4$ →

2-methyl-2-pentene → HOBr (NBS, H$_2$O/DMSO) →

styrene → H$^+$/H$_2$O →

3-methylcyclopentene → 1. Hg(OAc)$_2$, H$_2$O; 2. NaBH$_4$ →

2-methylnorbornene → HOBr (NBS, H$_2$O/DMSO) →

2,4-dimethyl-1-pentene → H$^+$/H$_2$O →

methylenecyclopentane → 1. Hg(OAc)$_2$, H$_2$O; 2. NaBH$_4$ →

(Z)-cyclodecene → HOBr (NBS, H$_2$O/DMSO) →

1-methylcyclopentene → 1. Hg(OAc)$_2$, H$_2$O; 2. NaBH$_4$ →

Alkenes: *Addition Reactions-III*

For each of the reactions shown below, draw the structure of the major organic product. Clearly show all stereochemistry.

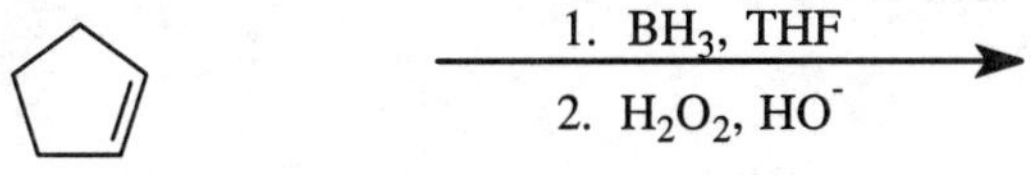

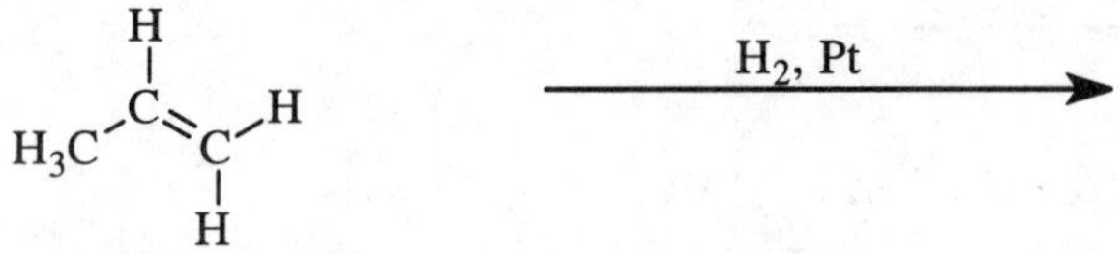

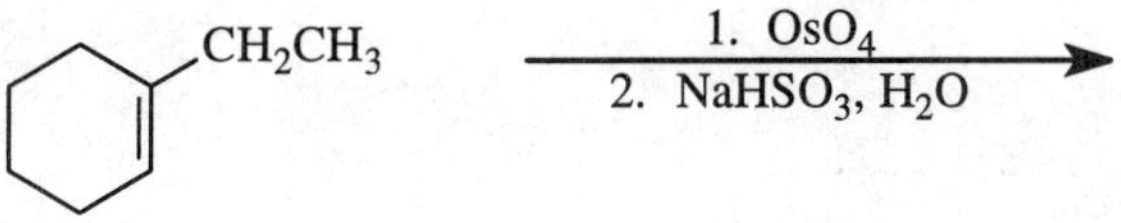

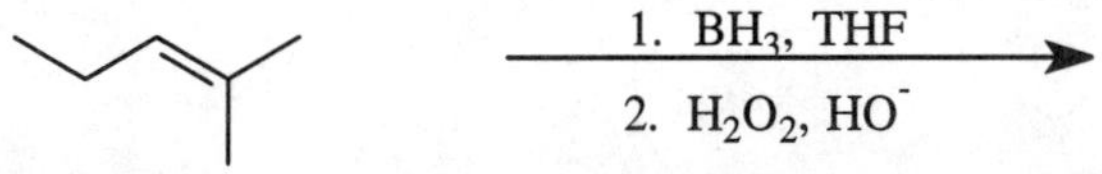

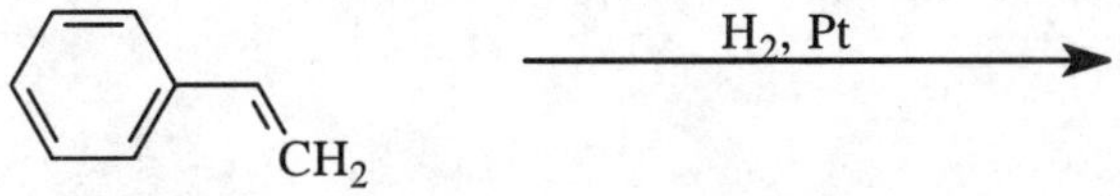

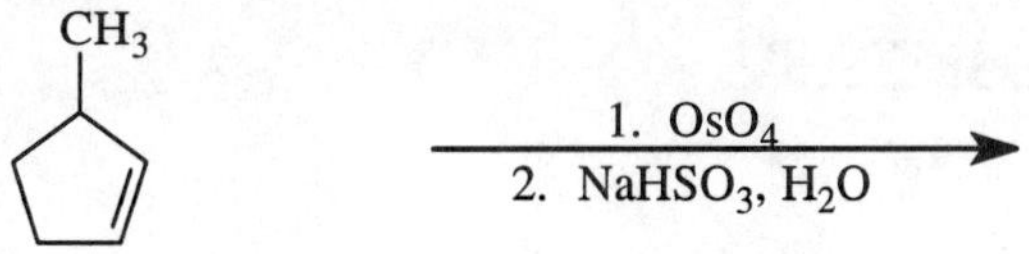

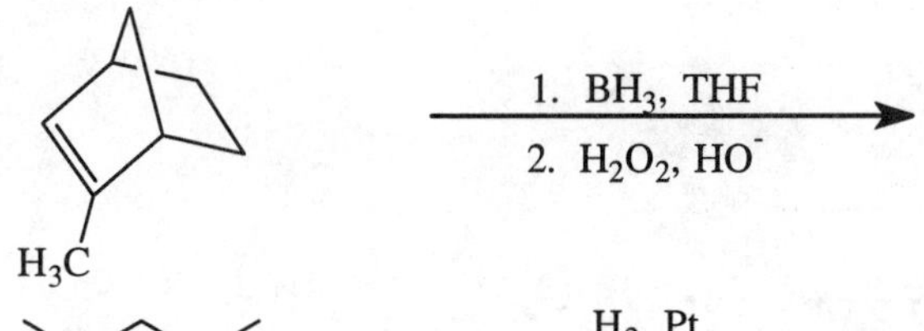

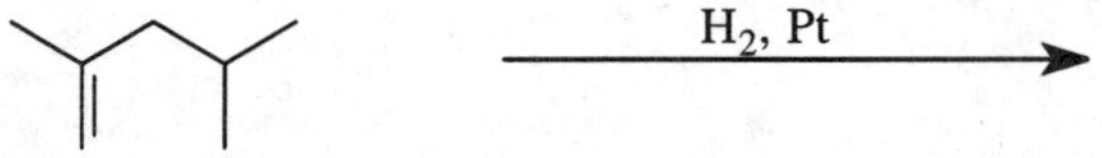

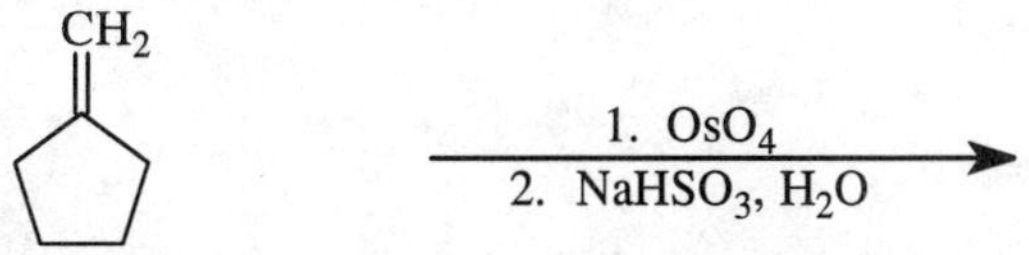

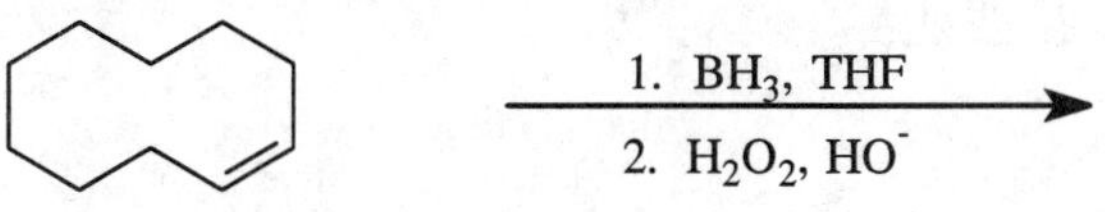

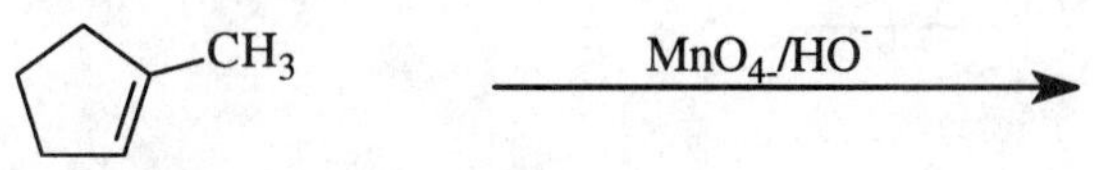

cyclopentene $\xrightarrow{\text{1. BH}_3\text{, THF} \quad \text{2. H}_2\text{O}_2\text{, HO}^-}$

(E)-2-butene (H$_3$C–CH=CH–H, trans) $\xrightarrow{\text{H}_2\text{, Pt}}$

1-ethylcyclohexene $\xrightarrow{\text{1. OsO}_4 \quad \text{2. NaHSO}_3\text{, H}_2\text{O}}$

2-methyl-2-pentene $\xrightarrow{\text{1. BH}_3\text{, THF} \quad \text{2. H}_2\text{O}_2\text{, HO}^-}$

styrene (PhCH=CH$_2$) $\xrightarrow{\text{H}_2\text{, Pt}}$

3-methylcyclopentene $\xrightarrow{\text{1. OsO}_4 \quad \text{2. NaHSO}_3\text{, H}_2\text{O}}$

2-methylnorbornene $\xrightarrow{\text{1. BH}_3\text{, THF} \quad \text{2. H}_2\text{O}_2\text{, HO}^-}$

2,4-dimethyl-1-pentene $\xrightarrow{\text{H}_2\text{, Pt}}$

methylenecyclopentane $\xrightarrow{\text{1. OsO}_4 \quad \text{2. NaHSO}_3\text{, H}_2\text{O}}$

(Z)-cyclodecene $\xrightarrow{\text{1. BH}_3\text{, THF} \quad \text{2. H}_2\text{O}_2\text{, HO}^-}$

1-methylcyclopentene $\xrightarrow{\text{MnO}_4^-/\text{HO}^-}$

Alkene Reactions-IV: *Addition & Oxidation Reactions*

For each of the reactions shown below, draw the structure of the major organic product. Clearly show all stereochemistry.

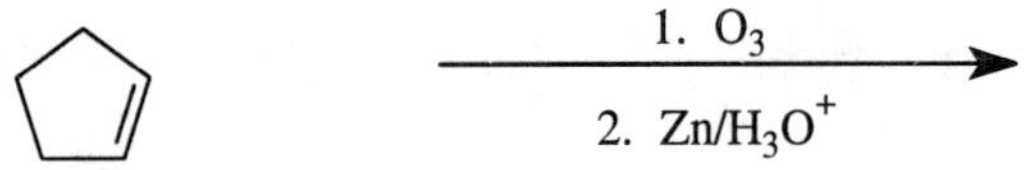

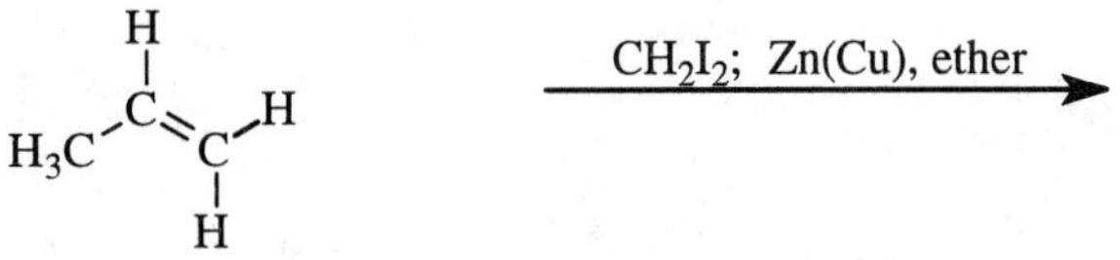

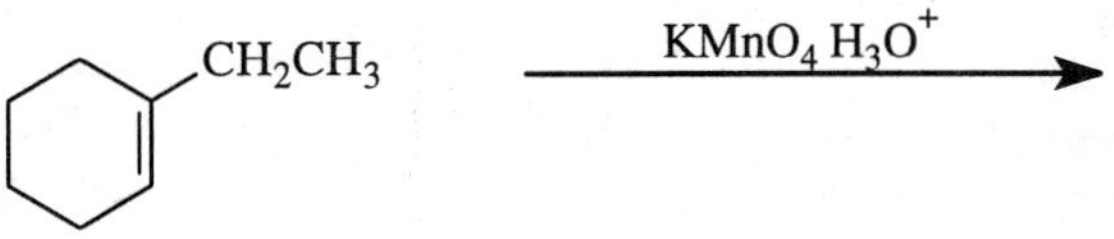

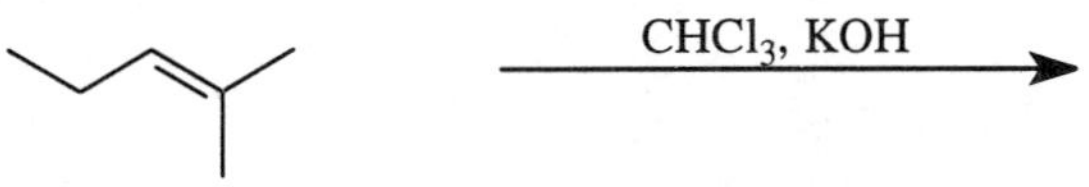

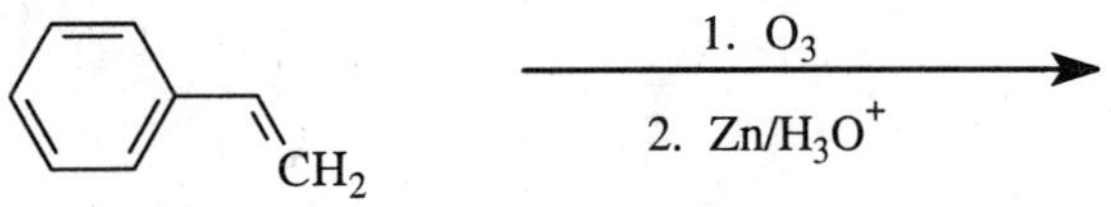

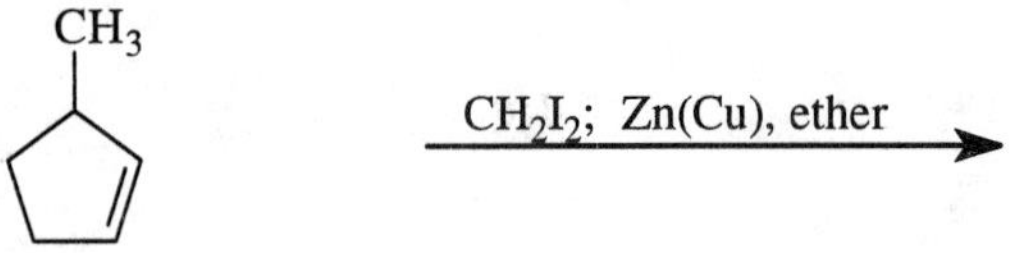

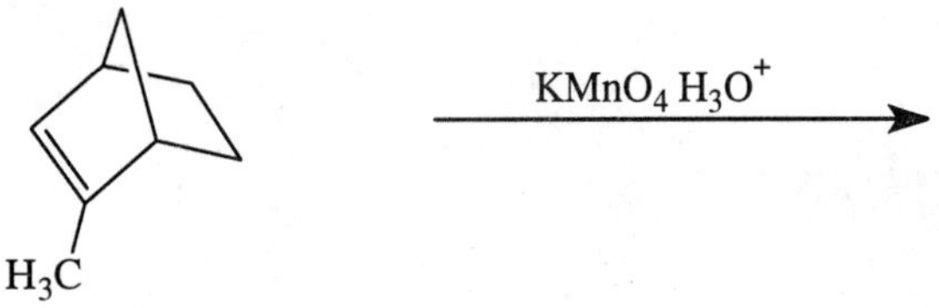

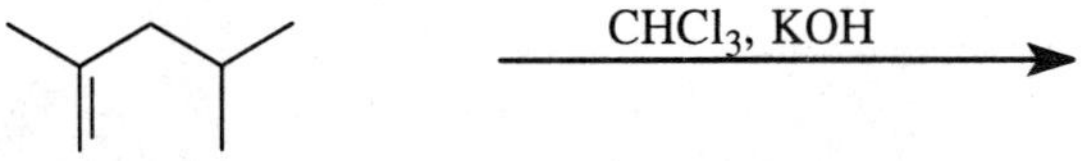

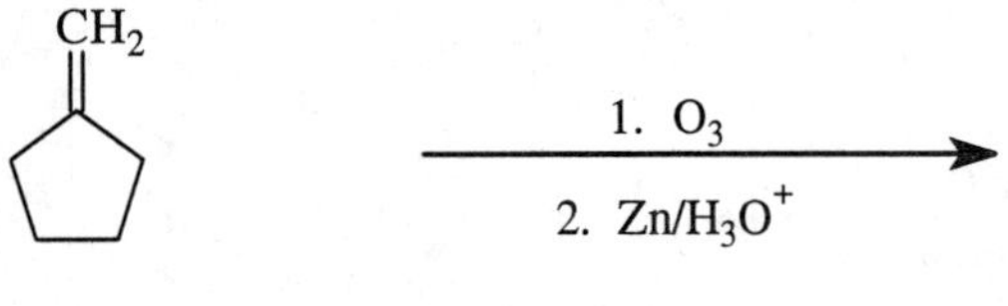

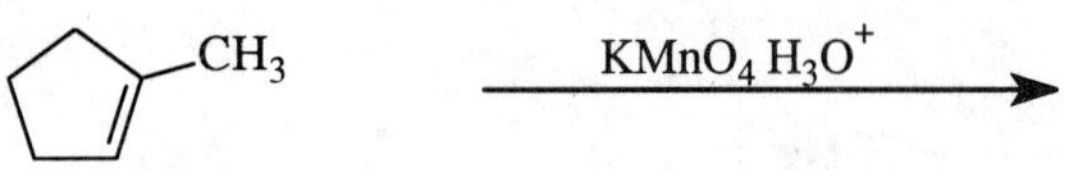

cyclopentene → 1. O$_3$ 2. Zn/H$_3$O$^+$

cis-2-butene (H$_3$C and H on one carbon, H and H on other — drawn as cis-2-butene with H's cis) → CH$_2$I$_2$; Zn(Cu), ether

1-ethylcyclohexene (cyclohexene with CH$_2$CH$_3$ substituent) → KMnO$_4$ H$_3$O$^+$

2-methyl-2-pentene → CHCl$_3$, KOH

styrene (PhCH=CH$_2$) → 1. O$_3$ 2. Zn/H$_3$O$^+$

3-methylcyclopentene → CH$_2$I$_2$; Zn(Cu), ether

3-methylnorbornene (norbornene with CH$_3$ on alkene carbon) → KMnO$_4$ H$_3$O$^+$

2,4-dimethyl-1-pentene (=CH$_2$ with isopropyl and methyl) → CHCl$_3$, KOH

methylenecyclopentane → 1. O$_3$ 2. Zn/H$_3$O$^+$

cis-cyclodecene → CH$_2$I$_2$; Zn(Cu), ether

1-methylcyclopentene → KMnO$_4$ H$_3$O$^+$

Alkenes: *Synthesis*

Suggest a simple synthesis for each of the molecules shown below, beginning with any one of the alkenes utilized in the problems shown on the previous pages. Clearly show all required reagents and reaction conditions.

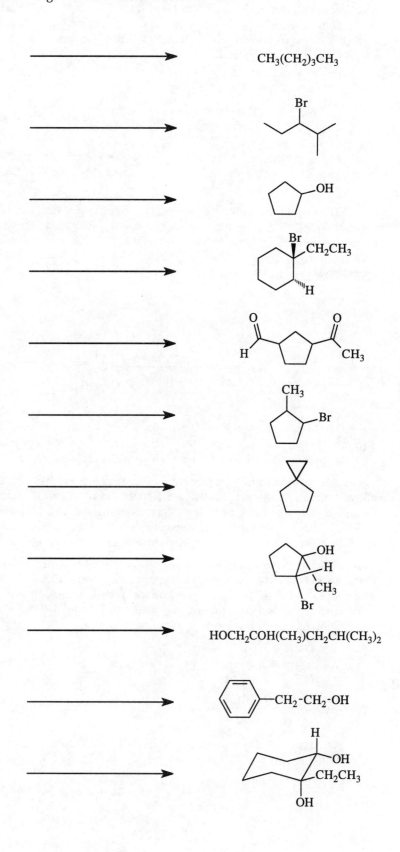

Alkynes: Addition and Oxidation Reactions

The exercises in this section focus on predicting the products of simple addition and oxidation reactions of alkynes, and on using these reactions as tools in simple organic synthesis problems. Reactions of alkynes do not involve distinct carbocation intermediates and rearrangements are not a problem. In the addition of halogen acids, the overall regiochemistry is "Markovnikov", suggesting a transition state in which some charge has developed on the alkyne carbons, giving rise to a selectivity and Markovnikov regiochemistry. As shown below, alkynes will add **two moles** of HX to form a geminal dihalide, with the alkenyl halide as an intermediate.

$$CH_3-\!\!\equiv \quad \xrightarrow{2\ HCl} \quad \underset{Cl}{\overset{CH_3}{>\!\!=\!\!<}} \quad \longrightarrow \quad CH_3\!-\!\underset{Cl}{\overset{Cl}{C}}\!-\!CH_3$$

Halogens (X_2) add **twice** to alkynes to form tetrahalide addition products. As with alkenes, the mechanism involves the formation of an intermediate bridged halonium ion and the intermediate in the reaction is a (*trans*) dihaloalkene.

The hydration of alkynes requires Hg^{2+} as a (Lewis acid) catalyst and results in the addition of **one mole** of water to form an intermediate **enol**. In this case, a second mole of water does not add to the intermediate double bond because the enol rapidly undergoes *tautomerism* and is converted into the corresponding carbonyl compound. The regiochemistry of the addition is Markovnikov, suggesting a polar transition state, but rearrangements do not occur.

$$H_3C-\!\!\equiv \quad \xrightarrow[Hg^{++}]{H_2SO_4} \quad \text{(enol intermediate)} \quad \longrightarrow \quad H_3C\overset{O}{-\!\!\overset{\|}{C}\!\!-}CH_3$$

Hydroboration/oxidation of an alkyne also results in the formation of an intermediate enol, which also rapidly tautomerizes to give the corresponding carbonyl compound. As with alkenes, the addition is *anti*-Markovnikov; thus with terminal alkynes, the carbonyl compound formed is an **aldehyde**.

Alkynes can be fully reduced to the hydrocarbon using H_2/Pt, and can be partially reduced to the *cis* alkene using a "poisoned catalyst", or a special catalyst called "Lindlar catalyst". *Trans* alkenes can be produced from alkynes by partial reduction using a dissolving metal reaction. Typically Na or Li, **dissolving in** liquid NH_3 is used; the stereochemistry results from the fact that the reaction involves the stepwise addition of electrons to the alkyne, allowing the more stable (E)-alkenyl radical to be formed preferentially.

The fact that terminal alkynes are weak carbon acids means that strong bases, such as sodium amide (from Na **previously dissolved** in liquid NH_3) to give the alkyne anion. This anion can then be utilized as a nucleophile, displacing a halogen from a methyl- or primary alkyl halide to give a new substituted alkyne as product. The reaction is limited to primary and methyl halides since the strong base tends to promote elimination reactions in secondary and tertiary halides.

$$H_3C-\!\!\equiv\!\!-H \quad \xrightarrow{NaNH_2} \quad H_3C-\!\!\equiv\!\!:^{\ominus} \quad \xrightarrow{Ph-CH_2Br} \quad H_3C-\!\!\equiv\!\!-CH_2Ph$$

Oxidation of alkynes with either ozone or acidic MnO_4^- will produce two moles of carboxylic acid, splitting the triple bond. Terminal alkynes will yield carbonic acid (CO_2) in this reaction.

Alkynes: *An Overview of Reactions*

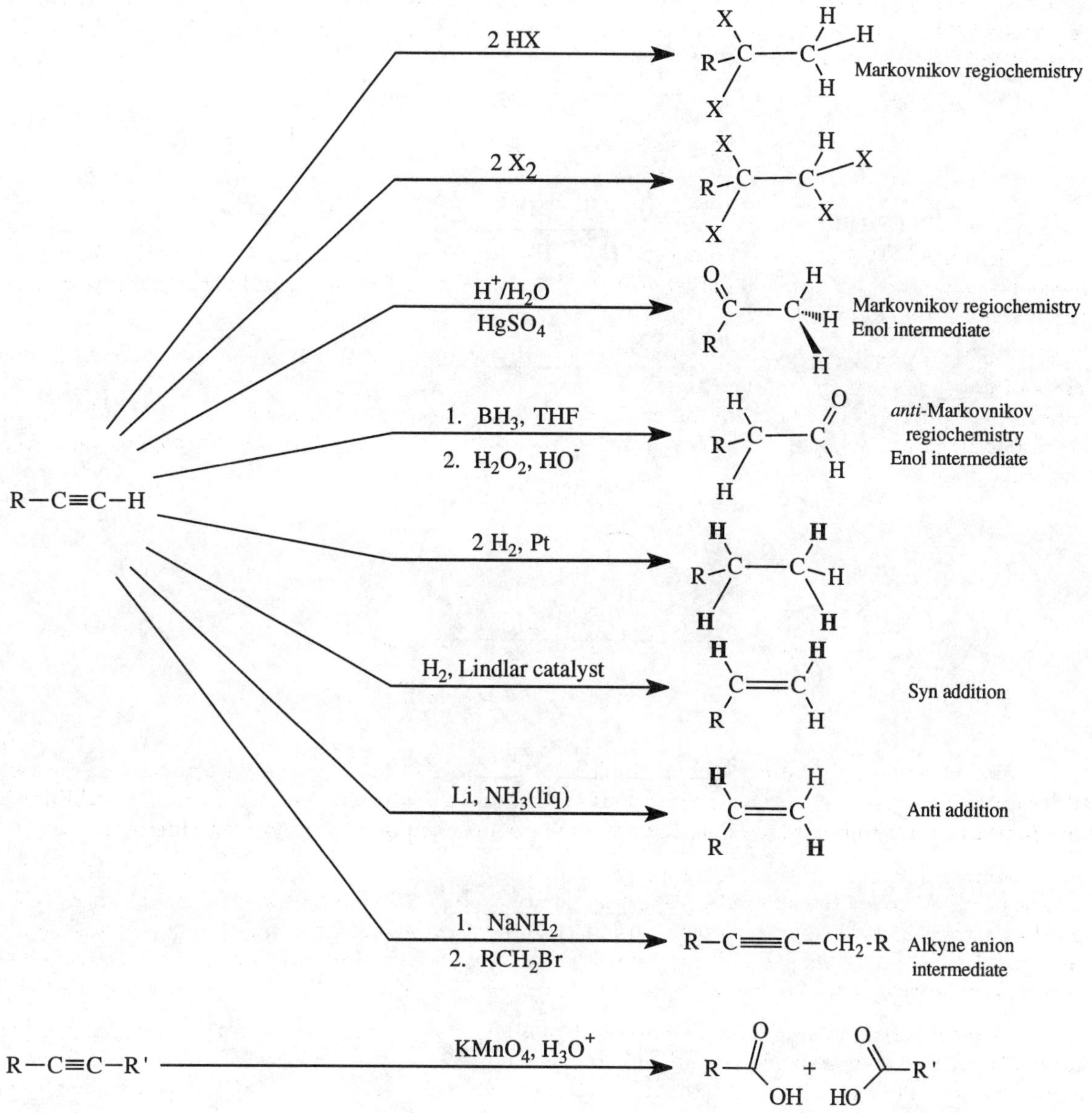

$R-C\equiv C-H$

- $\xrightarrow{2\ HX}$ $R-CX_2-CH_3$ — Markovnikov regiochemistry
- $\xrightarrow{2\ X_2}$ $R-CX_2-CHX_2$
- $\xrightarrow{H^+/H_2O,\ HgSO_4}$ $R-C(=O)-CH_3$ — Markovnikov regiochemistry, Enol intermediate
- $\xrightarrow{1.\ BH_3,\ THF;\ 2.\ H_2O_2,\ HO^-}$ $R-CH_2-CHO$ — anti-Markovnikov regiochemistry, Enol intermediate
- $\xrightarrow{2\ H_2,\ Pt}$ $R-CH_2-CH_3$
- $\xrightarrow{H_2,\ Lindlar\ catalyst}$ cis-$RCH=CH_2$ — Syn addition
- $\xrightarrow{Li,\ NH_3(liq)}$ trans-$RCH=CH_2$ — Anti addition
- $\xrightarrow{1.\ NaNH_2;\ 2.\ RCH_2Br}$ $R-C\equiv C-CH_2-R$ — Alkyne anion intermediate

$R-C\equiv C-R' \xrightarrow{KMnO_4,\ H_3O^+} R-COOH + HOOC-R'$

Enol interconversion:

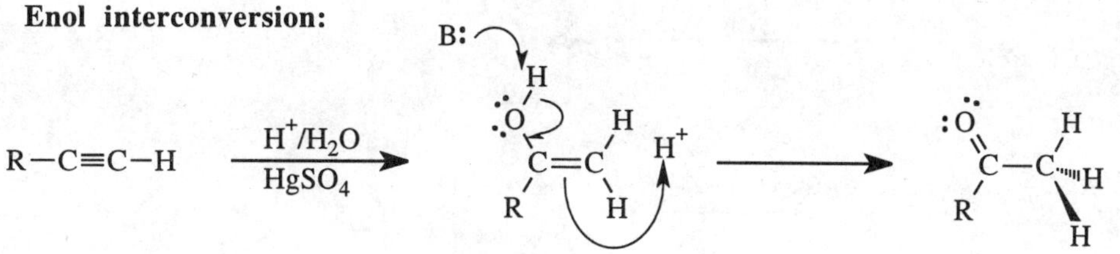

$R-C\equiv C-H \xrightarrow{H^+/H_2O,\ HgSO_4}$ [enol intermediate with B: abstracting O–H proton and H⁺ adding to carbon] $\longrightarrow$ $R-C(=\ddot{O})-CH_2H$ (ketone)

Alkynes: *Reactions*

For each of the reactions shown below, draw the structure of the major organic product. Clearly show all stereochemistry.

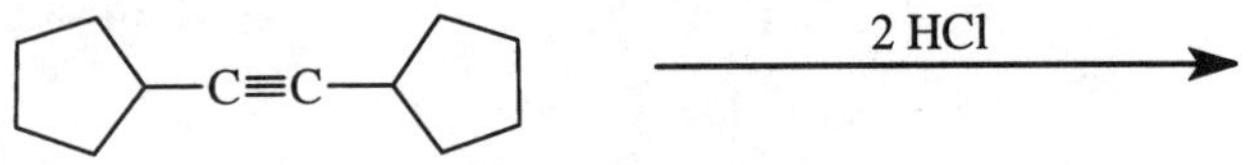

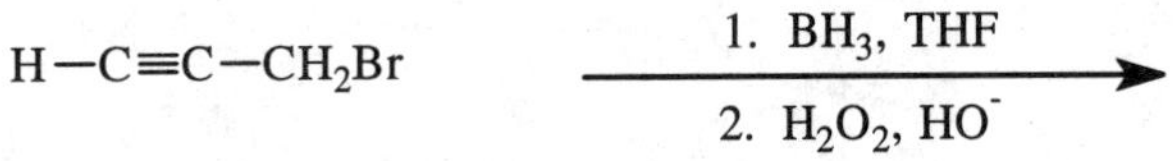

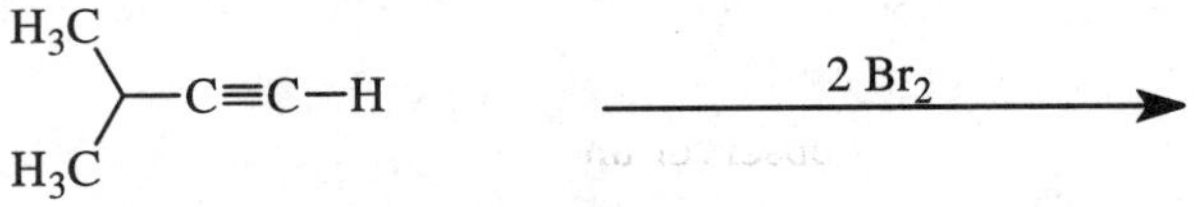

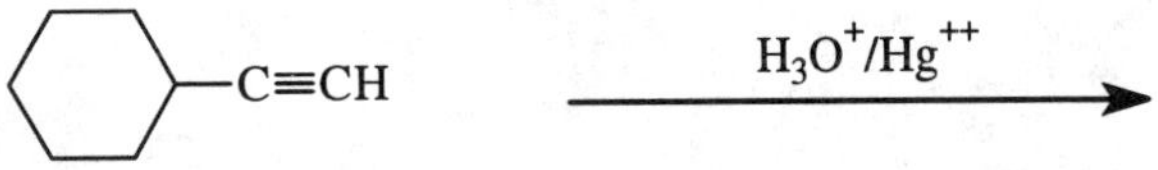

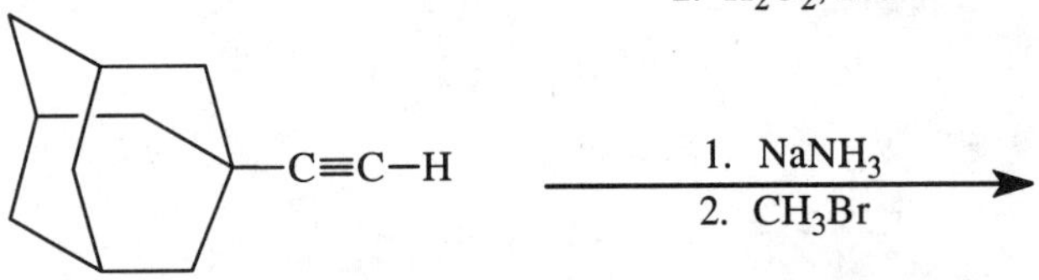

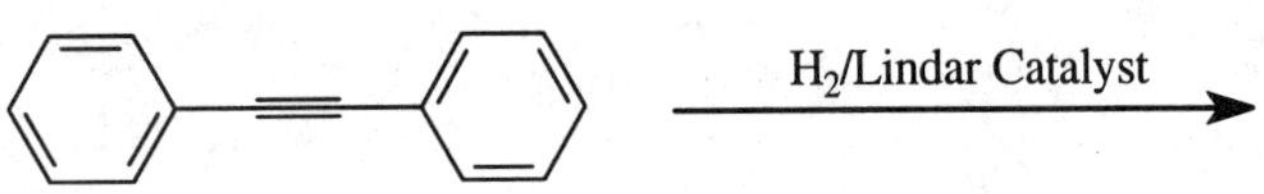

cyclopentyl−C≡C−cyclopentyl → 2 HCl

H−C≡C−CH$_2$Br → 1. BH$_3$, THF 2. H$_2$O$_2$, HO$^-$

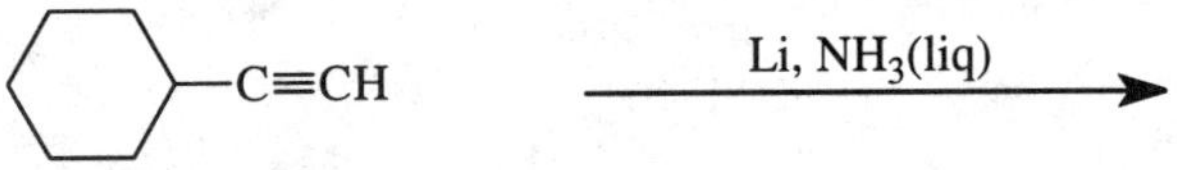

cyclohexyl−C≡CH → Li, NH$_3$(liq)

(H$_3$C)$_2$CH−C≡C−H → 2 Br$_2$

cyclohexyl−C≡CH → H$_3$O$^+$/Hg^{++}

HC≡C−CH$_2$−CH(CH$_3$)$_2$ → 1. BH$_3$, THF 2. H$_2$O$_2$, HO$^-$

adamantyl−C≡C−H → 1. NaNH$_2$ 2. CH$_3$Br

H$_3$C−C≡C−CH$_3$ → Li, NH$_3$(liq)

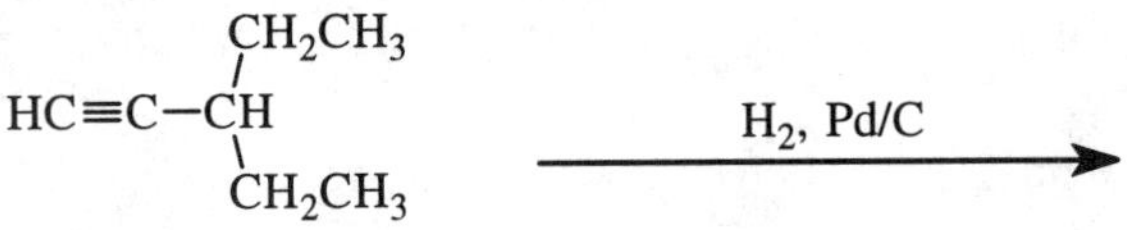

HC≡C−CH(CH$_2$CH$_3$)$_2$ → H$_2$, Pd/C

Ph−C≡C−Ph → H$_2$/Lindlar Catalyst

Stereochemistry

The exercises in this section focus on the skills necessary to identify, categorize and manipulate stereoisomers. As a consequence of tetrahedral geometry, carbons which are attached to **four different substituents** posses internal asymmetry. Because of this asymmetry, these molecules are **not superimposible on their mirror images** and carbons bearing four different substituents are termed "chiral" (from the Greek for "hand") or are more properly denoted as "stereocenters". Compounds which are related to each other as non-superimposible mirror images are called **enantiomers**. These stereoisomers have identical physical properties, except for their interactions with other chiral molecules (and surfaces) and for their interaction with plane polarized light. A solution of a pure enantiomer will rotate plane polarized light by a fixed amount (termed the "specific rotation", [α]) and the other enantiomer will rotate the plane of the light by an equal amount, in the opposite direction. This property is often useful in characterizing enantiomers, and in testing the enantiomeric purity of an unknown mixture (since equal quantities of each enantiomer would yield no net rotation, an observed rotation, a, can be compared with the specific rotation, [α], to get "enantiomeric purity").

The absolute orientation of a stereocenter in space can be determined by x-ray diffraction methods, and that configuration can be identified using **R** and **S** nomenclature. This system of nomenclature utilized the Cahn-Ingold-Prelog sequence rules to assign priorities to the substituents surrounding a stereocenter, based on atomic numbers. The lowest priority group is then oriented **away** from the viewer and the resulting order of the groups (by decreasing priority) is determined to be either clockwise (**R**) or counterclockwise (**S**). Exercises in this area utilize the Chemscape Chime™ plugin from MDL Industries to allow direct visualization of the required manipulations.

A useful tool in working with chiral carbons is the Fisher Projection, in which a chiral carbon is oriented with the two "side" groups coming towards the observer and the "top" and "bottom" groups pointing away. The molecule is then "flattened" and the resulting structure (with no central carbon shown) is simply drawn as a "cross".

conversion to a Fisher Projection...

Fisher projections are a simple way to represent stereochemistry, and are also useful in assigning absolute configuration. If the lowest priority group in a Fisher projection is oriented towards **the top or bottom of the structure**, an arrow connecting the remaining three groups (in order of decreasing priority) will define the center as **R** or **S** (clockwise or counterclockwise).

If the lowest priority group is oriented towards one of the "sides", the structure must be "rearranged" before a determination can be made (**a Fisher projection cannot be rotated 90° since this rotation produces an enantiomeric structure**; prove it to yourself!). The process of "rearranging" groups around a Fisher projection is sometimes called the "exchange method". In practice, you simply redraw the structure with any two groups exchanged; this operation generates an enantiomeric structure. A second exchange generates the enantiomer of the enantiomer, or simply, regenerates the original absolute stereochemistry. To move the lowest priority group from the side to the top or bottom, simply perform any two exchanges, moving the group as desired. This method is also useful in comparing two Fisher projections to determine if they represent the same compound or an isomeric structure.

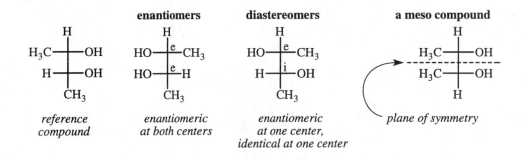

Compounds which contain more than one chiral center can exist as stereoisomers in which all chiral center are enantiomeric (and the molecules would therefore be enantiomers), or as stereoisomers in which or more chiral centers had identical absolute configuration, while one or more centers were enantiomeric. Compounds such as these are not enantiomers, but are termed **diastereomers**. Diastereomers, unlike enantiomers, have totally different physical and chemical properties. A special class of compounds which contain more than one chiral center are **meso compounds**. Meso compounds contain two or more chiral centers (always an even number), but are not themselves chiral. This is because they also posses an internal plane of symmetry, such that one half of the molecule is the mirror image of the other half. In identifying meso compounds, rotate the molecules (or do an even number of exchanges on a Fisher projection) to visualize this internal plane of symmetry.

Stereochemistry: *Recognizing Symmetry*

For each of the molecules show below, determine the **total number of internal planes of symmetry**, and sketch their locations.

Stereochemistry: *Identifying Chiral Centers*

Indicate with an astrisk (*) each of the chiral centers in the following molecules:

a, b, c, d, e, f, g, h, i, j, k, l, m, n, o, p, q, r.

Stereochemistry: *Conversion to Fisher Projections*

Convert each of the molecules show on the left into the corresponding Fisher Projection, completing the partial structure shown on the right.

a.

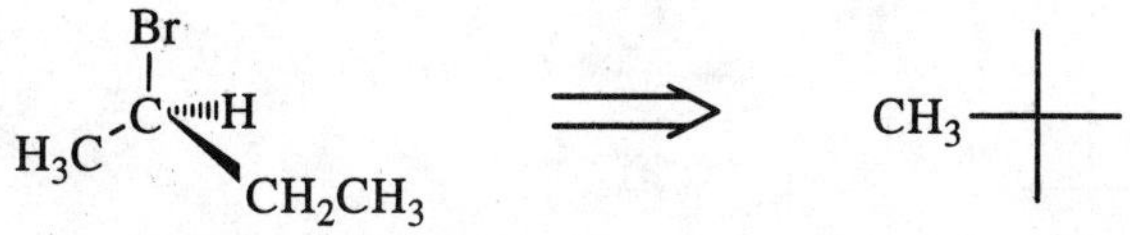

b.

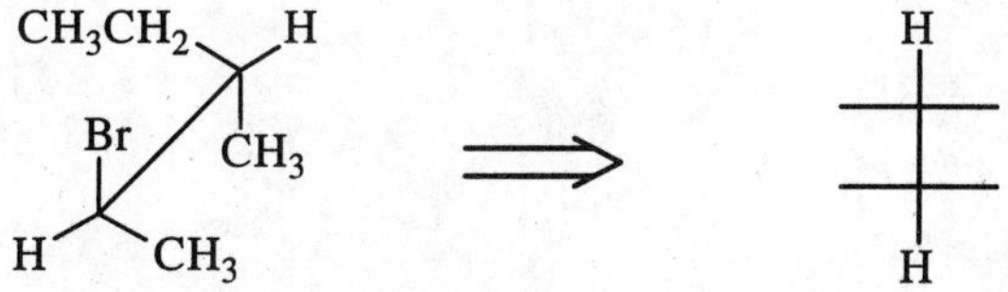

(bromine on top carbon)

c.

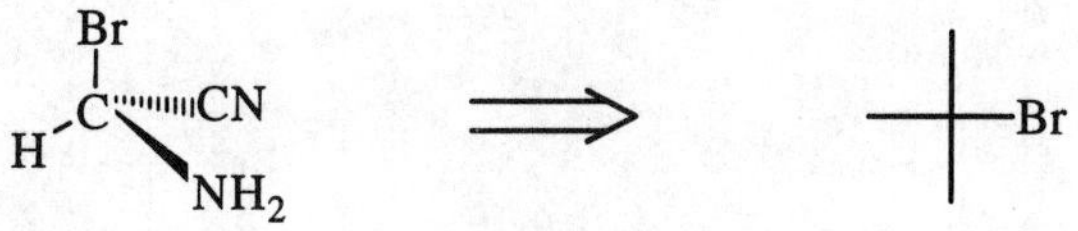

d.

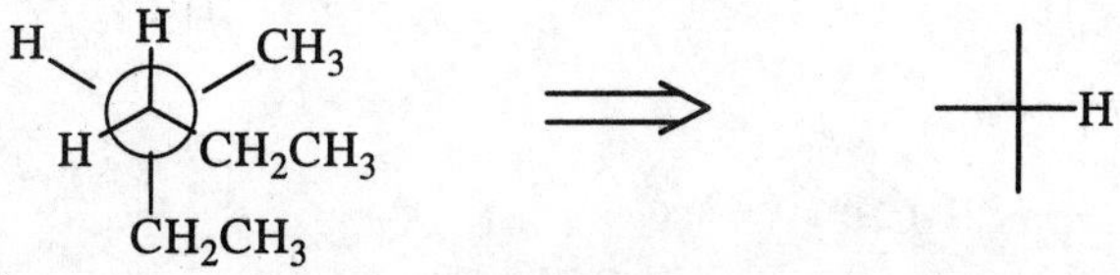

e.

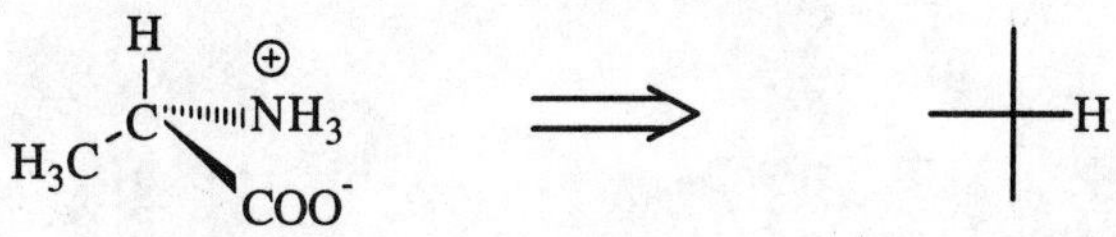

f.

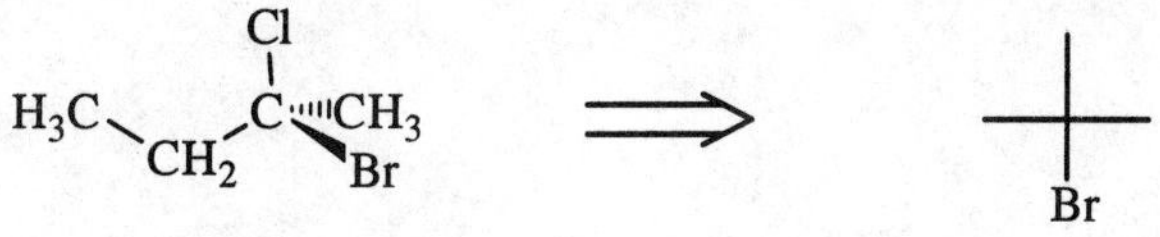

g.

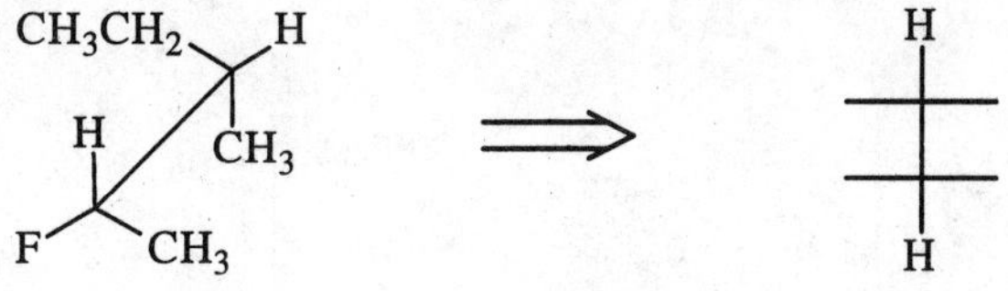

(fluorine on top carbon)

h.

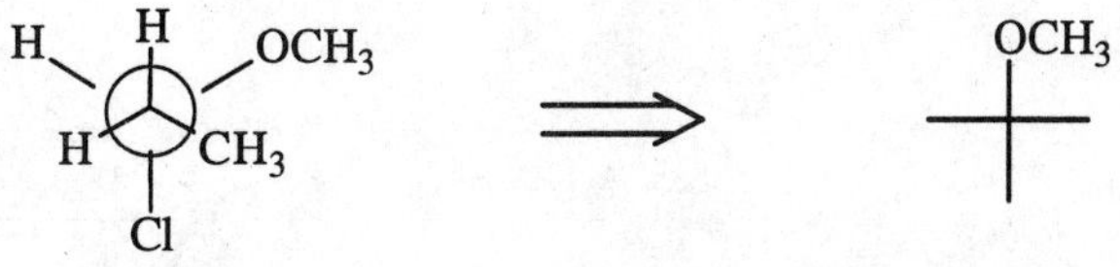

i.

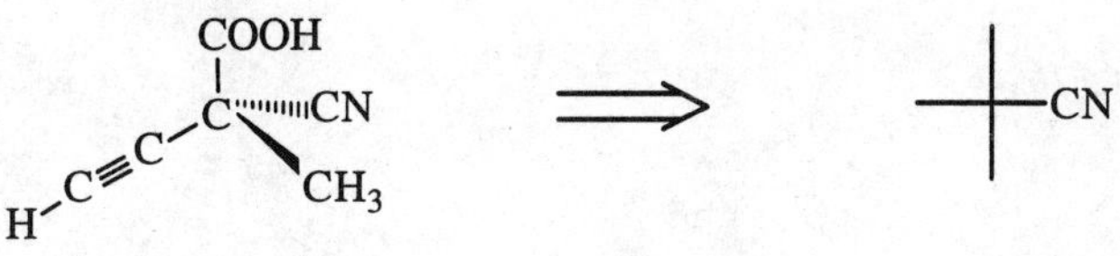

j.

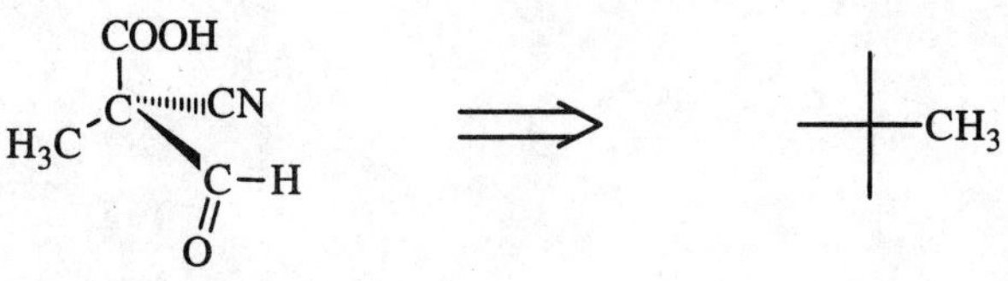

Stereochemistry: *Identifying Isomerism*

For each of the following sets of structures, indicate if the compounds are:
a) identical b) constitutional isomers c) enantiomers d) diastereomers
e) identical and meso f) different compounds and not isomeric

a. ___ and

b. ___ and

c. ___ and

d. ___ and

e. ___ and

f. ___ and

g. ___ and

h. ___ and

i. ___ and

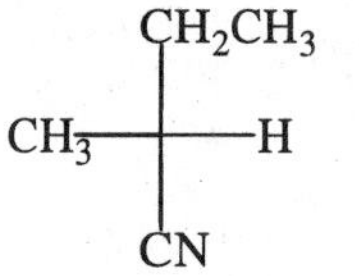

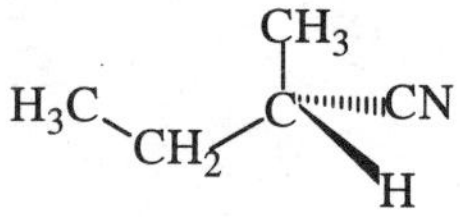

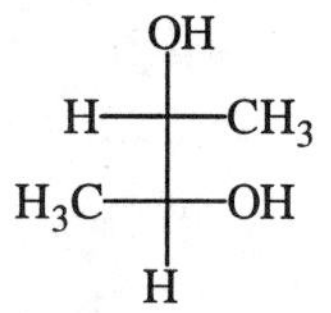

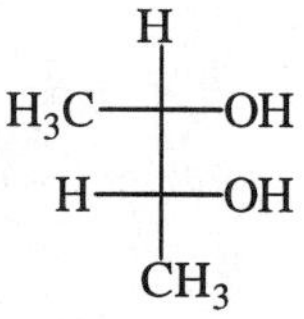

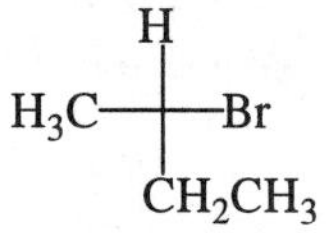

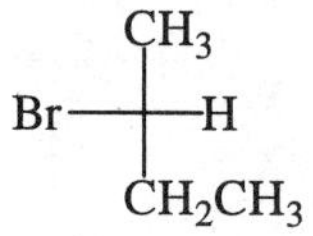

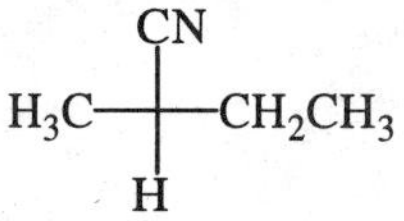

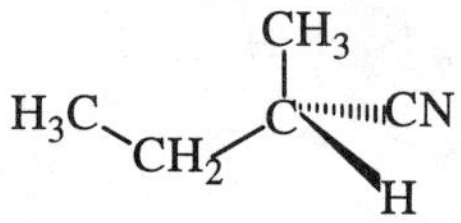

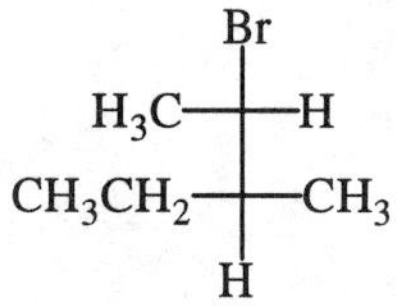

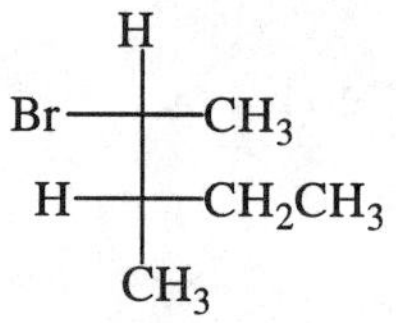

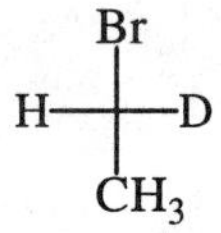

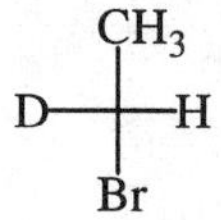

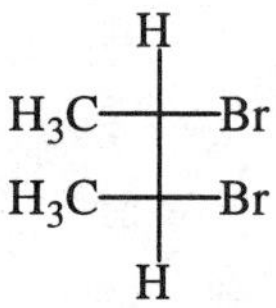

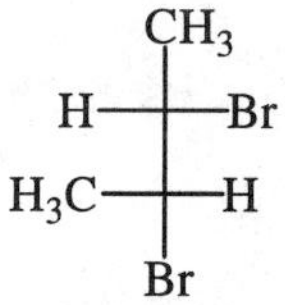

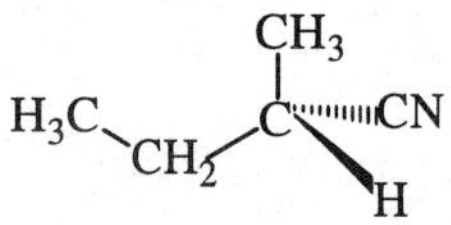

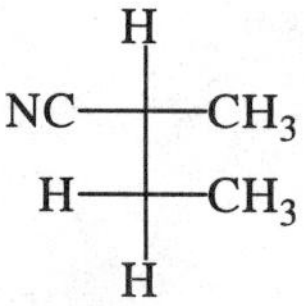

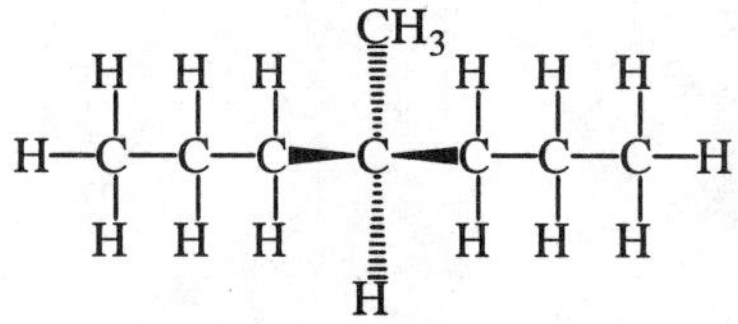

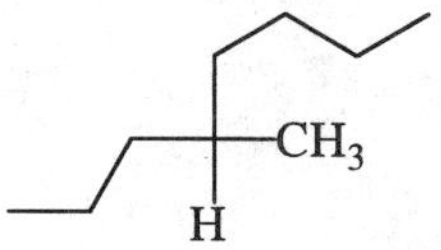

Stereochemistry: *Identifying Isomerism, con't*

For each of the following sets of structures, indicate if the compounds are:
  a) identical   b) constitutional isomers   c) enantiomers   d) diastereomers
  e) identical and meso   f) different compounds and not isomeric

j.

k.

l.

m.

n.

o.

Stereochemistry: *Assigning Absolute Configuration*

For each of the molecules show below, identify any chiral centers and assign each center as **R** or **S**. For molecules with symmetry, indicate if the molecule is **Meso**. Note: these problems are designed to utilize the Chime™ plugin to render the molecule in three dimensions; please see your CD or the WWW site.

Stereochemistry: *Assigning Absolute Configuration, continued*

For each of the molecules show below, identify any chiral centers and assign each center as **R** or **S**. For molecules with symmetry, indicate if the molecule is **Meso**.

Stereochemistry: *Assigning Absolute Configuration, continued*

For each of the molecules show below, identify any chiral centers and assign each center as **R** or **S**. For molecules with symmetry, indicate if the molecule is **Meso**.

Stereochemistry: *Optical Activity*

1. The (+) enantiomer of a compound has an observed optical rotation of 1.72° when measured in a one dm tube at a concentration of 0.3 g/15 mL. Calculate the specific rotation for this molecule.

2. The (-) enantiomer of a compound has a specific rotation of $[\alpha] = 123°$. What rotation, α, will be observed for a sample of this pure enantiomer at a concentration of 0.25 g/10 mL, when measured in the polarimeter in a two dm sample tube?

3. The (-) enantiomer of a compound has a specific rotation of $[\alpha] = 86°$. A sample of this pure enantiomer, when measured in the polarimeter in a two dm sample tube, gives an observed rotation of $\alpha = -4.3°$. What is the concentration of this sample in g/mL?

4. The specific rotation for a pure enantiomer is known to be +139° g^{-1} mL^{-1} dm^{-1}. A sample containing both enantiomers is found to have an observed rotation of +0.87° in a one dm tube at a concentration of 0.025 g/mL. What is the optical purity of the sample?

5. The specific rotation for a pure enantiomer is known to be +158° g^{-1} mL^{-1} dm^{-1}. What will be the specific rotation of mixtures containing:

   a. 25% of the (+) enantiomer and 75% of the (-) enantiomer,
   b. 50% of the (+) enantiomer and 50% of the (-) enantiomer, and,
   c. 75% of the (+) enantiomer and 25% of the (-) enantiomer?

6. The solvolysis of 2-bromooctane proceeds by an "S_N1-like mechanism, but when a single enantiomer is used as substrate, the product mixture is found to have a significant amount of optical activity, indicating that the mechanism does not involve full racemization. If the specific rotation of (S)(+)-2-octanol is +89° and the reaction mixture after solvolysis has an apparent specific rotation of $[\alpha]_{apparent}$ = +59°, what percentage of the reaction is proceeding through a mechanism involving stereochemical inversion?

7. The specific rotation of pure (-)cholesterol is -39° g^{-1} mL^{-1} dm^{-1}. What is the apparent specific rotation of a sample containing 90% of the (-)enantiomer and 10% of the (+)enantiomer?

8. The specific rotation of pure (+)2-octanol is +10° g^{-1} mL^{-1} dm^{-1}. A sample containing both enantiomers is found to have an apparent specific rotation of +4.0° g^{-1} mL^{-1} dm^{-1}. What is the optical purity of the sample?

Stereochemistry: *Reactions Generating Stereocenters*

For each of the reactions shown below, predict the **major organic products** and predict the **stereochemical relationship between the products of the reaction** (i.e., enantiomers, diastereomers, or identical).

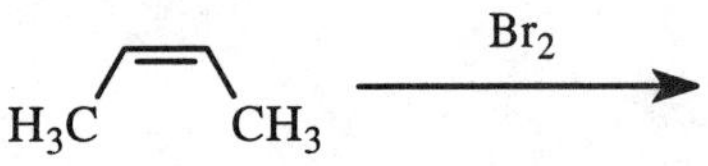

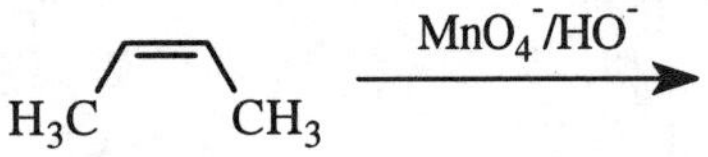

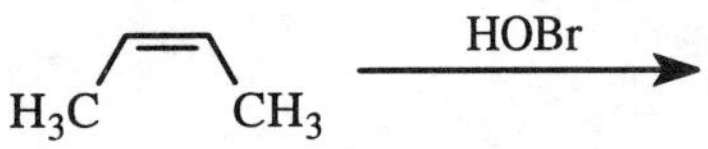

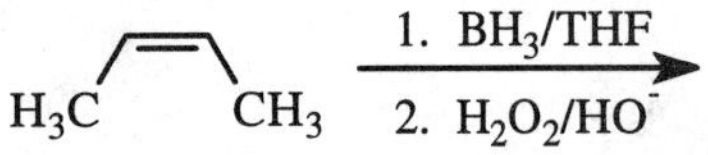

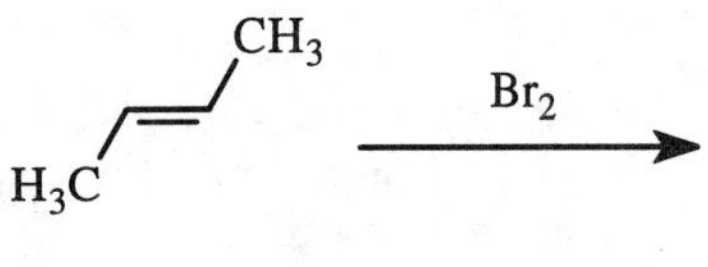

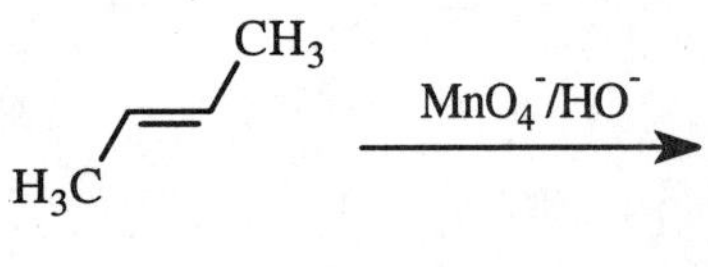

H₃C–CH=CH–CH₃ (cis) $\xrightarrow{Br_2}$

H₃C–CH=CH–CH₃ (cis) $\xrightarrow{MnO_4^-/HO^-}$

H₃C–CH=CH–CH₃ (cis) $\xrightarrow{HOBr}$

H₃C–CH=CH–CH₃ (cis) $\xrightarrow{\text{1. BH}_3/\text{THF} \quad \text{2. H}_2O_2/HO^-}$

H₃C–CH=CH–CH₃ (trans) $\xrightarrow{Br_2}$

H₃C–CH=CH–CH₃ (trans) $\xrightarrow{MnO_4^-/HO^-}$

Stereochemistry: *Reactions Generating Stereocenters, continued*

For each of the reactions shown below, predict the **major organic products** and predict the **stereochemical relationship between the products of the reaction** (i.e., enantiomers, diastereomers, or identical).

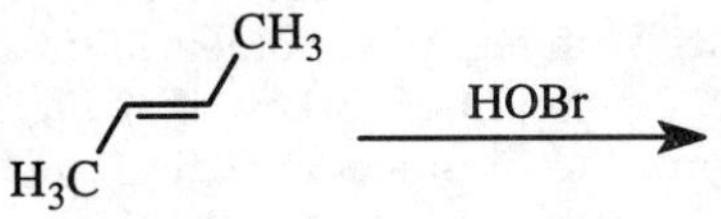

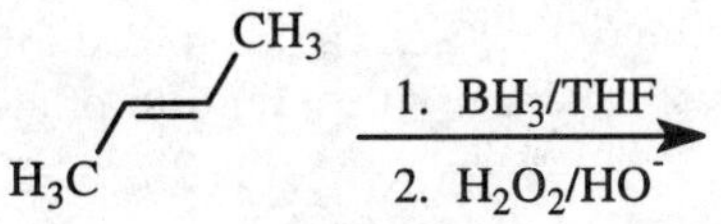

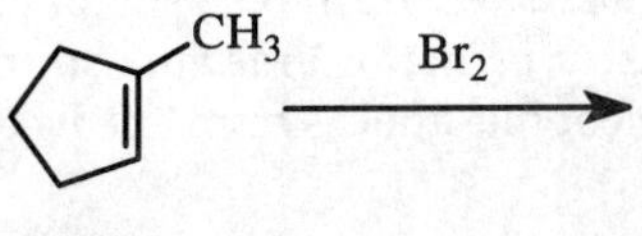

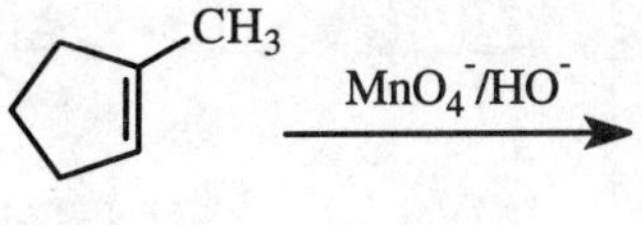

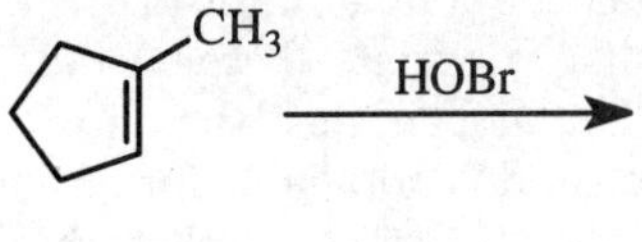

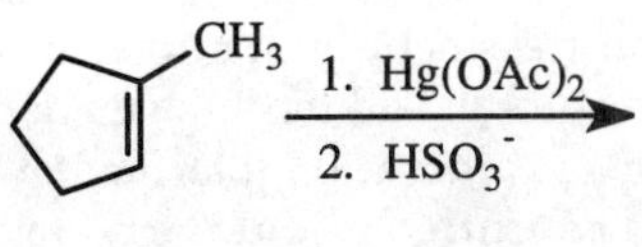

$H_3C-CH=CH-CH_3$ (cis) → HOBr

$H_3C-CH=CH-CH_3$ (cis) → 1. BH_3/THF   2. H_2O_2/HO⁻

1-methylcyclopentene → Br_2

1-methylcyclopentene → MnO_4^-/HO⁻

1-methylcyclopentene → HOBr

1-methylcyclopentene → 1. $Hg(OAc)_2$ 2. HSO_3^-

Alkyl Halides: Preparation; Substitution & Elimination Reactions

Within this section, simple nomenclature of alkyl halides is briefly revisited and methods for their preparation are reviewed. The main utility of alkyl halides in organic synthesis resides in their reactions with nucleophiles, yielding **substitution** products, and in their reactions with base to generate alkenes by **elimination** reactions.

As we have seen in earlier sections, alkyl halides are named as simple substituents on the parent hydrocarbon, and are among the lowest priority substituents, with regard to nomenclature. Alkyl halides can be prepared from alkenes by simple ionic addition reactions, yielding Markovnikov regiochemistry, by a radical mechanism in which bromine radical adds to form the most stable radical, yielding net *anti*-Markovnikov regiochemistry, by the addition of halogen and hypohalite, and by the addition of dichlorocarbene to yield dichlorocyclopropane derivatives. Alcohols can be converted into alkyl halides by treatment with concentrated HCl (since this involves a carbocation intermediate, it only works well for tertiary alcohols), and by reaction with PBr_3 and with $SOCl_2$. Both of these latter reactions involve inorganic ester formation, followed by displacement with halide anion to give the alkyl halide with **inversion** of stereochemistry. In a non-polar solvent, however, the reaction with $SOCl_2$ can be forced into an alternate mechanism where the intermediate sulfite ester decomposes directly with **retention** of stereochemistry (the SN_i mechanism), allowing stereochemical control during synthesis.

One of the most important reactions of alkyl halides is their tendency to undergo substitution reactions with nucleophiles to release halide anion and form alkylated nucleophiles. This reaction can proceed by two mechanisms; a concerted displacement reaction in which the nucleophile attacks at the same time that the bond to the leaving halide is being broken (the S_N2 mechanism), and a stepwise mechanism in which the halide leaves in the slow step to form a carbocation intermediate, which reacts with halide anion in a second, rapid step (the S_N1 mechanism). The stereochemistry of the S_N2 mechanism is to generate a stereochemical **inversion** at the reacting carbon, since the nucleophile attacks from one face, and the halide departs from the opposite side.

The S_N1 mechanism involves a carbocation (which is planar) and yields products arising from attack at either face. In working simple substitution reaction problems, the simple rule is to break the bond to the halide, attach the nucleophile and invert the stereochemistry, if appropriate. Because the S_N2 mechanisms requires the nucleophile to collide with the carbon bearing the halogen, the reaction proceeds best with primary and methyl halides, will *work* with secondary centers, and is not viable for tertiary centers. The S_N1 mechanism involves a carbocation, hence those alkyl halides which form stable carbocations are most likely to undergo this mechanism (i.e., tertiary centers and centers which tend to form stabilized carbocations).

Alkyl halides can also undergo **elimination** of HX on treatment with strong base. Again, two mechanisms are observed, a concerted mechanism in which the proton and the halogen depart at the same time (the E2 mechanism) and a stepwise mechanism in which a carbocation intermediate is formed. In practice, substitution and elimination reactions often proceed concurrently, but there are external factors which can be utilized to favor one over the other. In general, **a bulky base**, typically a tertiary alkoxide (*tert*-butoxide in *tert*-butanol), will be a poor nucleophile because of steric factors and will favor **elimination** over substitution. A non-polar solvent, or a polar, non-protic solvent (DMSO or DMF) will favor bimolecular process (S_N2 or E2), while a polar, protic solvent (water or alcohol) will favor the formation of ions in a stepwise, unimolecular process (S_N1 or E1). More details on factors affecting this selectivity are given on the following pages.

Alkyl Halides: *Trends in S_N2 and E2 Reactivity*

Substrate Effects

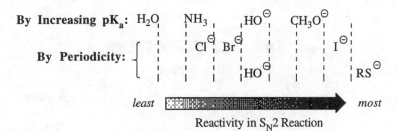

Reactivity in S_N2 Reaction

Nucleophile Effects

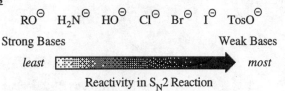

Reactivity in S_N2 Reaction

Leaving Group Effects

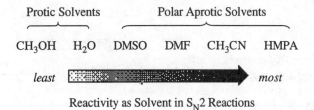

Reactivity in S_N2 Reaction

Solvent Effects

Protic Solvents   Polar Aprotic Solvents

CH$_3$OH   H$_2$O   DMSO   DMF   CH$_3$CN   HMPA

least ➝ most

Reactivity as Solvent in S_N2 Reactions

S_N2 and E2 Selectivity

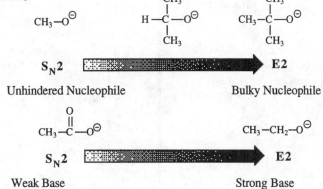

Alkyl Halides: *Trends in S_N1 and E1 Reactivity*

Substrate Effects

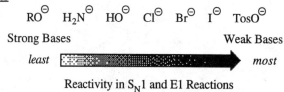

least ➔ most

Reactivity in S_N1 Reaction

Reactivity in S_N1 and E1 processes parallels **Carbocation Stability**

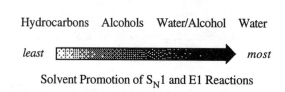

least ➔ most

Carbocation Stability

Nucleophile Effects

S_N1 and E1 reactions are independent of the nature of the nucleophile since the nucleophile is not involved in the rate-limiting step

Leaving Group Effects

$RO^\ominus$ $H_2N^\ominus$ $HO^\ominus$ $Cl^\ominus$ $Br^\ominus$ $I^\ominus$ $TosO^\ominus$

Strong Bases   Weak Bases

least ➔ most

Reactivity in S_N1 and E1 Reactions

Solvent Effects

Hydrocarbons   Alcohols   Water/Alcohol   Water

least ➔ most

Solvent Promotion of S_N1 and E1 Reactions

Free-Energy Profiles

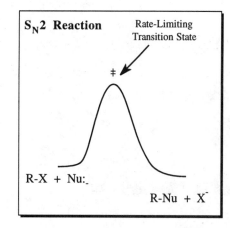

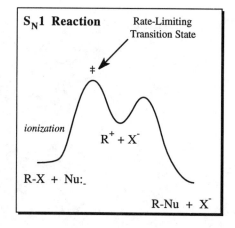

Preparation of Alkyl Halides: *An Overview*

From Alcohols:

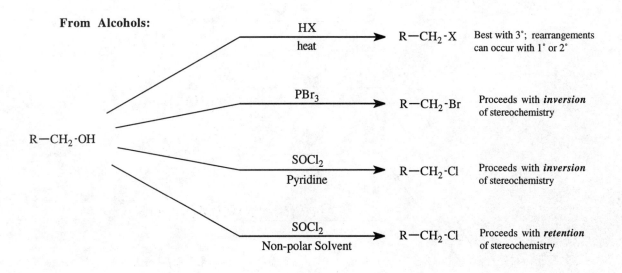

From Alkenes

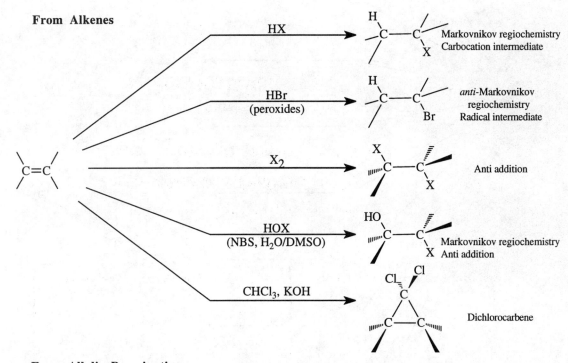

From Allylic Bromination

Alkyl Halides: *An Overview of S_N2 Reactions*

$R-CH_2-Br$ reacts with:

- $HO^{\ominus}$ → $R-CH_2-OH$ — Alcohols
- $RO^{\ominus}$ → $R-CH_2-OR$ — Ethers
- $X^{\ominus}$ → $R-CH_2-X$ — Alkyl Halides
- $RCOO^{\ominus}$ → $R-CH_2-O-C(=O)-R$ — Esters
- $HS^{\ominus}$ → $R-CH_2-SH$ — Thiols
- $RS^{\ominus}$ → $R-CH_2-SR$ — Sulfides
- $NC^{\ominus}$ → $R-CH_2-CN$ — Nitriles
- $RC\equiv C^{\ominus}$ → $R-CH_2-C\equiv C-R$ — Alkynes
- $N_3^{\ominus}$ → $R-CH_2-N_3$ — Alkyl Azides
- NH_3 → $R-CH_2-\overset{\oplus}{N}H_3$ — Primary Amines
- $(CH_3)_2NH$ → $R-CH_2-\overset{\oplus}{N}H(CH_3)_2$ — Tertiary Amines
- $(CH_3)_3N$ → $R-CH_2-\overset{\oplus}{N}(CH_3)_3$ — Quarternary Ammonium Salts

Alkyl Halides: *Nomenclature*

Provide the proper IUPAC name for each of the molecules shown below:

a.

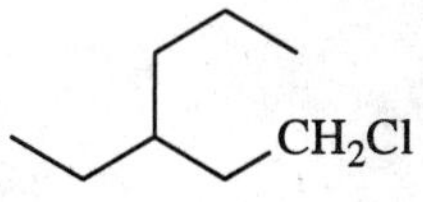

b.

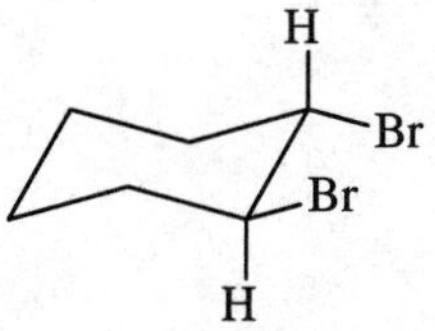

c.

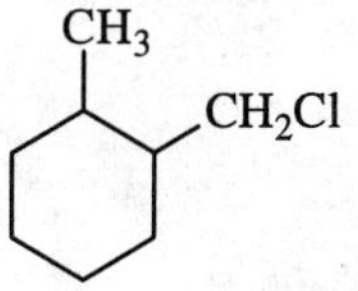

d.

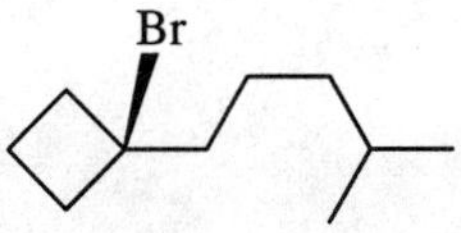

e.

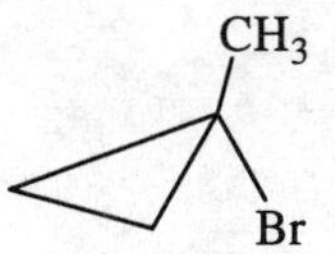

f.

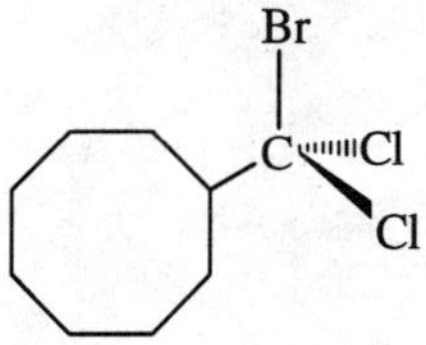

g.

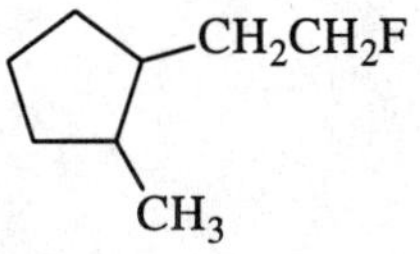

h.

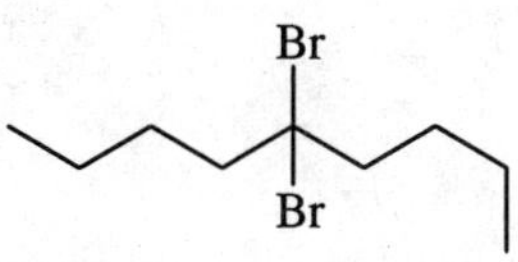

i.

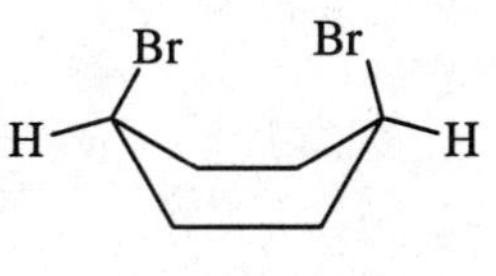

Alkyl Halides: *Substitution Reactions*

For each of the substitution reactions shown below, draw the structure of the major organic product. Clearly show all stereochemistry. You should assume stochiometric concentrations have been utilized, unless otherwise indicated.

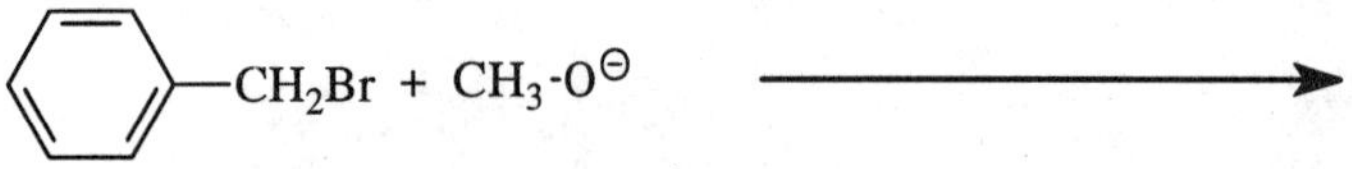

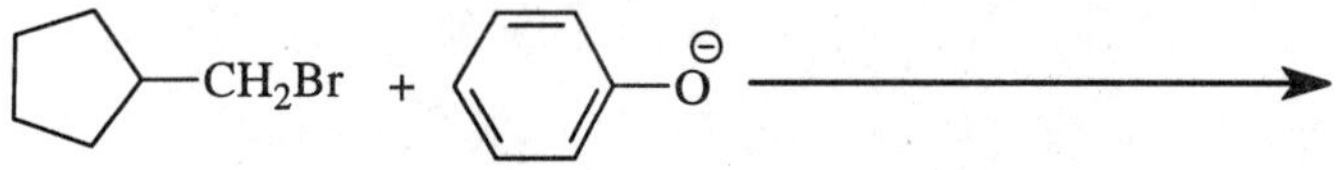

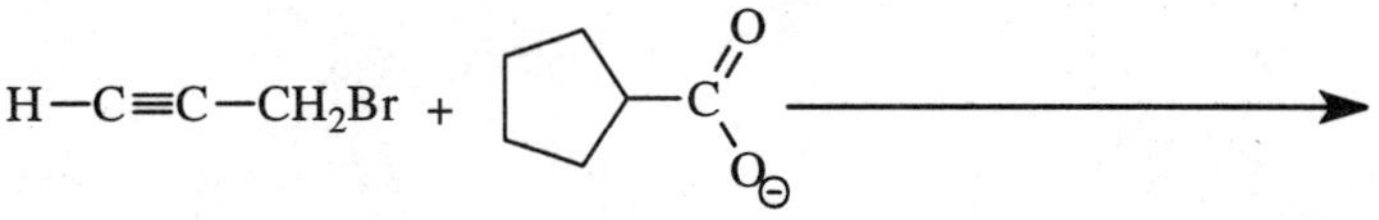

Ph—CH$_2$Br + CH$_3$-O$^{\ominus}$ ⟶

(cyclopentyl)—CH$_2$Br + PhO$^{\ominus}$ ⟶

H—C≡C—CH$_2$Br + (cyclopentyl)-C(=O)-O$^{\ominus}$ ⟶

Excess H$_3$C—CH$_2$-Br + NH$_3$ ⟶

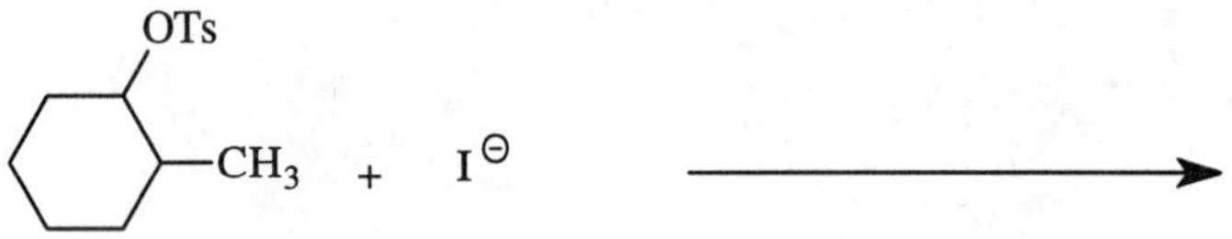

(cyclohexyl with OTs and CH$_3$ on adjacent carbons) + I$^{\ominus}$ ⟶

(cyclohexyl)—Cl + HS$^{\ominus}$ ⟶

CH$_3$CH$_2$CH(Br)CH$_3$ + (pyrrolidine) NH ⟶

Ph—Br + CH$_3$-O$^{\ominus}$ ⟶

(H$_3$C)$_2$C(OTs)(CH$_3$) + Ph—NH$_2$ ⟶

Br—CH(CH$_2$CH$_3$)(CH$_2$CH$_3$) + CH$_3$—NH$_2$ ⟶

(adamantyl)—O$^{\ominus}$ + CH$_3$Br ⟶

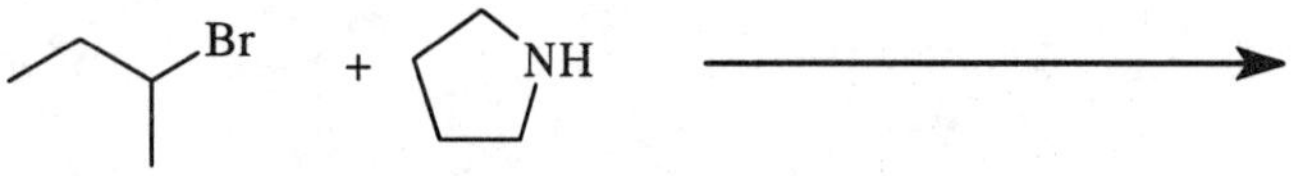

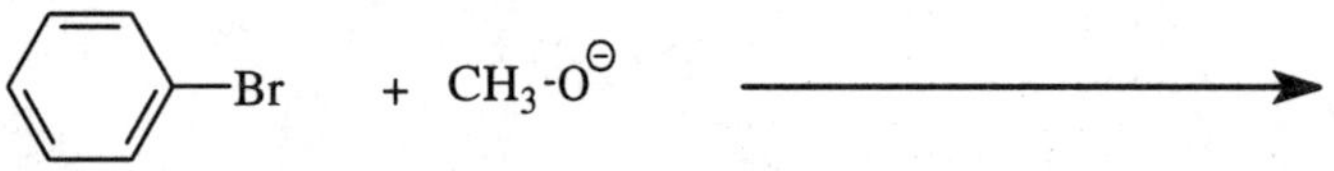

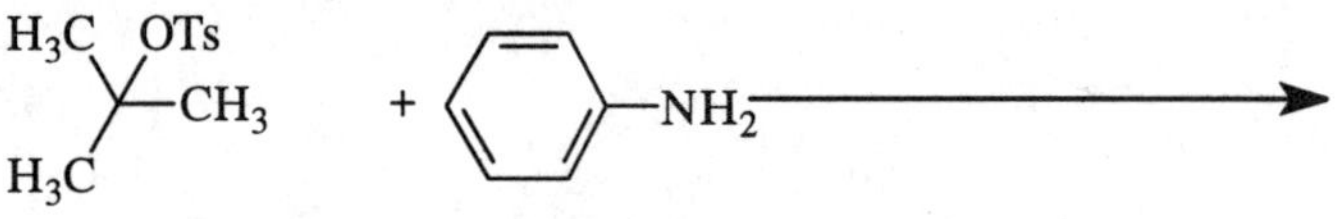

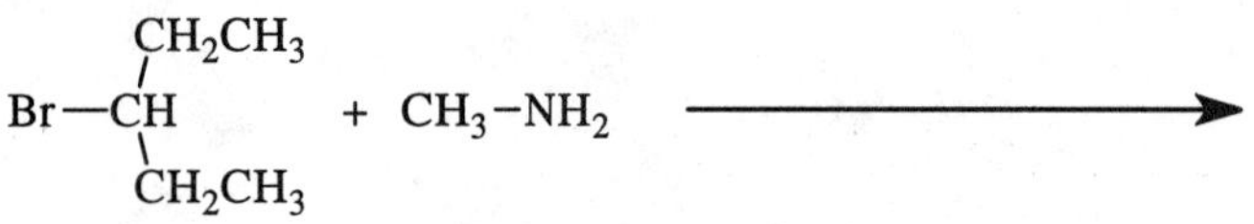

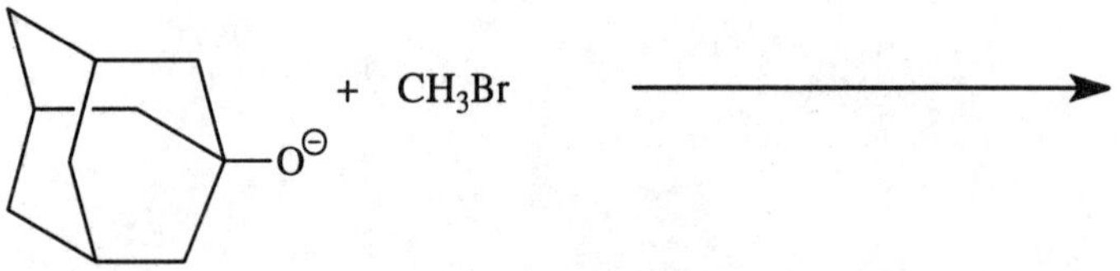

Alkyl Halides: *Elimination Reactions*

For each of the elimination reactions shown below, draw the structure of the major organic product. Clearly show all stereochemistry.

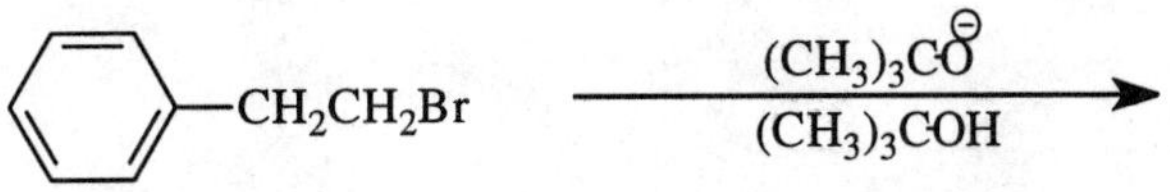

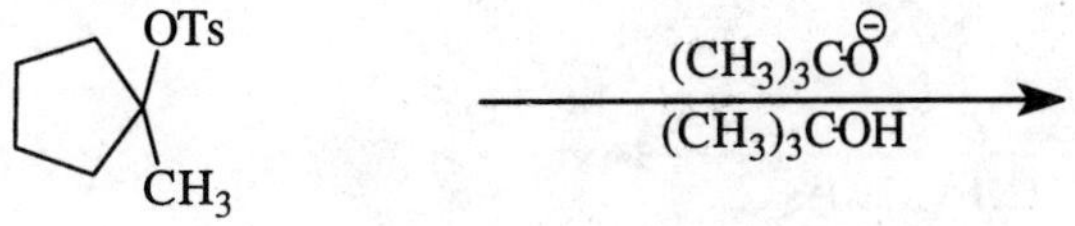

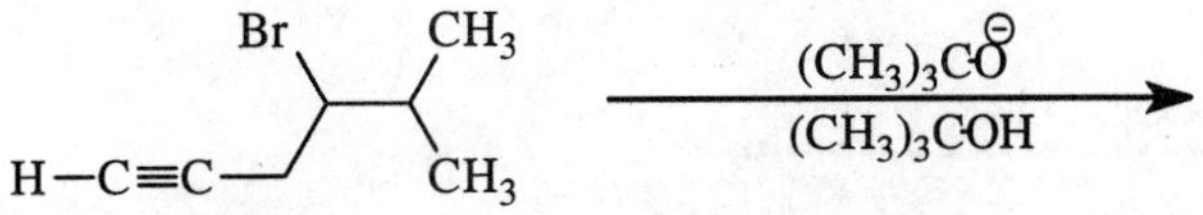

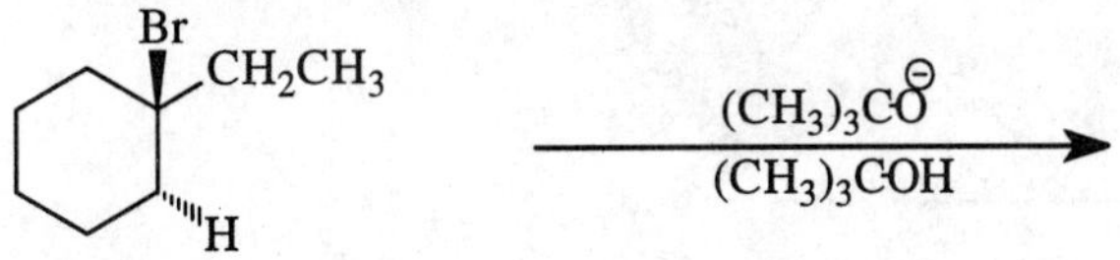

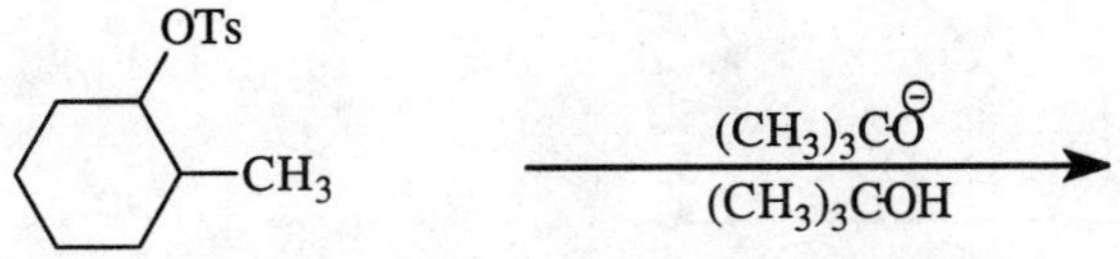

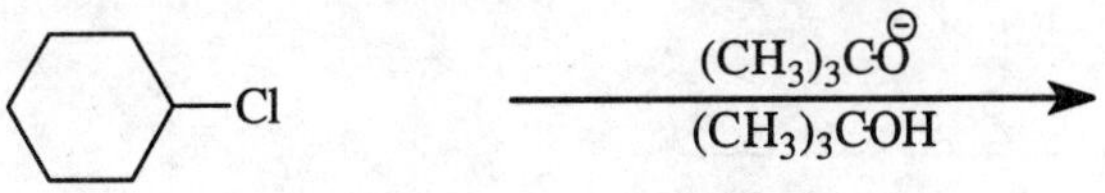

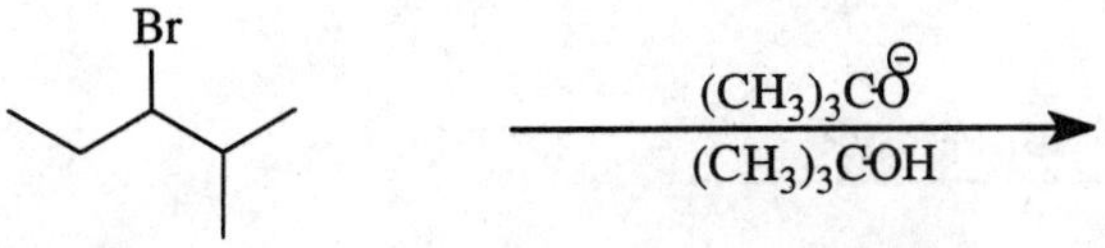

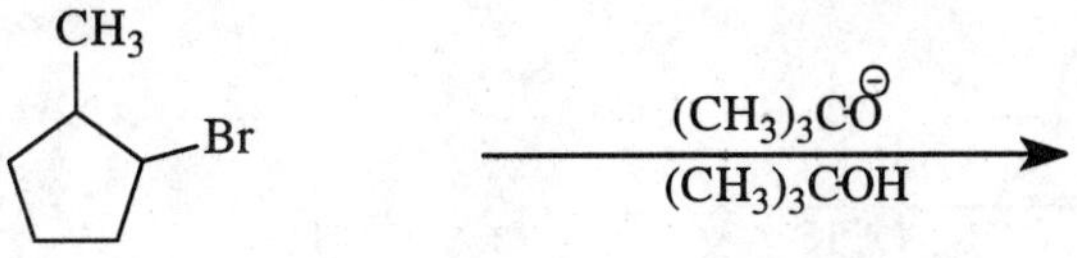

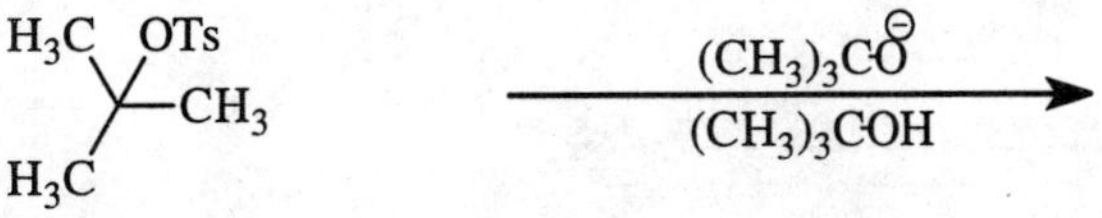

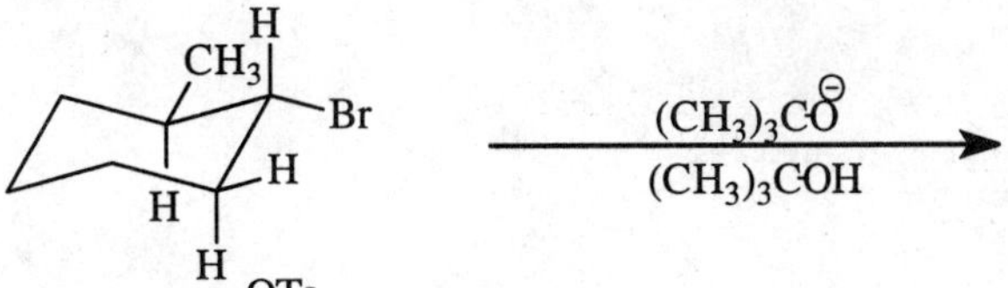

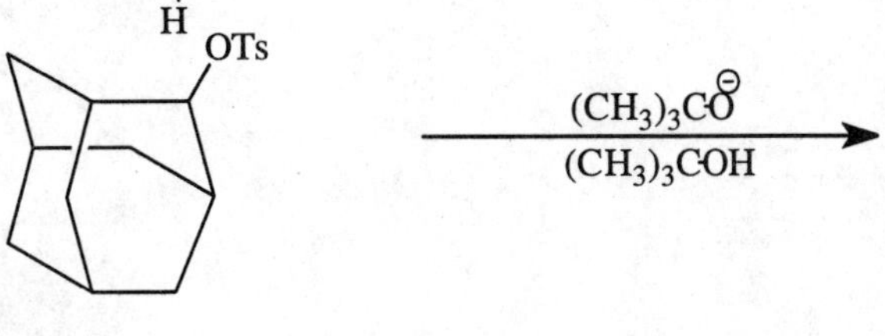

Ph–CH$_2$CH$_2$Br $\xrightarrow{\underset{(CH_3)_3COH}{(CH_3)_3CO^{\ominus}}}$

1-methylcyclopentyl tosylate $\xrightarrow{\underset{(CH_3)_3COH}{(CH_3)_3CO^{\ominus}}}$

HC≡C–CH(Br)–CH(CH$_3$)$_2$ $\xrightarrow{\underset{(CH_3)_3COH}{(CH_3)_3CO^{\ominus}}}$

1-bromo-1-ethyl-2-H cyclohexane (trans) $\xrightarrow{\underset{(CH_3)_3COH}{(CH_3)_3CO^{\ominus}}}$

2-methylcyclohexyl tosylate $\xrightarrow{\underset{(CH_3)_3COH}{(CH_3)_3CO^{\ominus}}}$

cyclohexyl chloride $\xrightarrow{\underset{(CH_3)_3COH}{(CH_3)_3CO^{\ominus}}}$

3-bromo-2-methylpentane $\xrightarrow{\underset{(CH_3)_3COH}{(CH_3)_3CO^{\ominus}}}$

1-methyl-2-bromocyclopentane $\xrightarrow{\underset{(CH_3)_3COH}{(CH_3)_3CO^{\ominus}}}$

(CH$_3$)$_3$C–OTs $\xrightarrow{\underset{(CH_3)_3COH}{(CH_3)_3CO^{\ominus}}}$

trans-1-methyl-2-bromocyclohexane (chair) $\xrightarrow{\underset{(CH_3)_3COH}{(CH_3)_3CO^{\ominus}}}$

2-adamantyl tosylate $\xrightarrow{\underset{(CH_3)_3COH}{(CH_3)_3CO^{\ominus}}}$

Alkyl Halides: *More Reactions...*

For each of the reactions shown below, draw the structure of the major organic product. Clearly show all stereochemistry. You should assume stochiometric concentrations have been utilized, unless otherwise indicated.

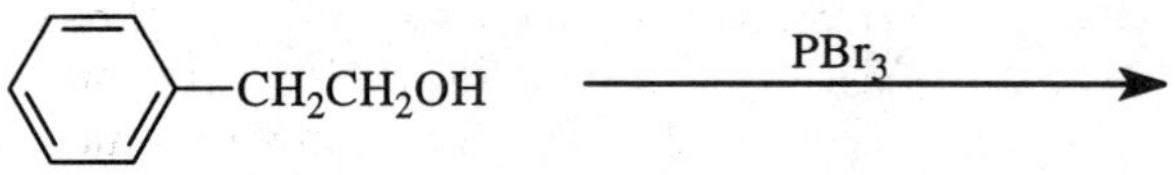

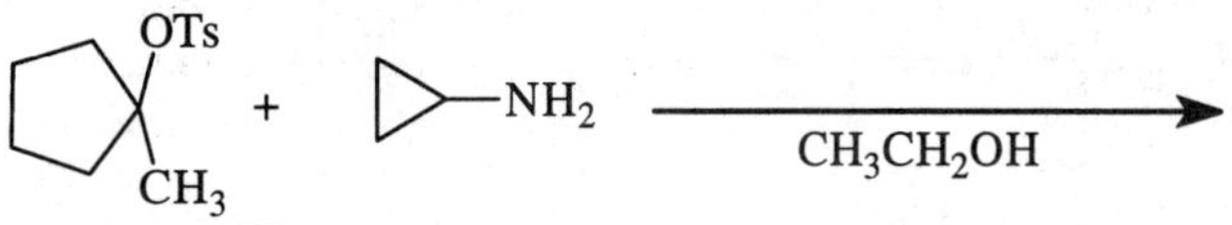

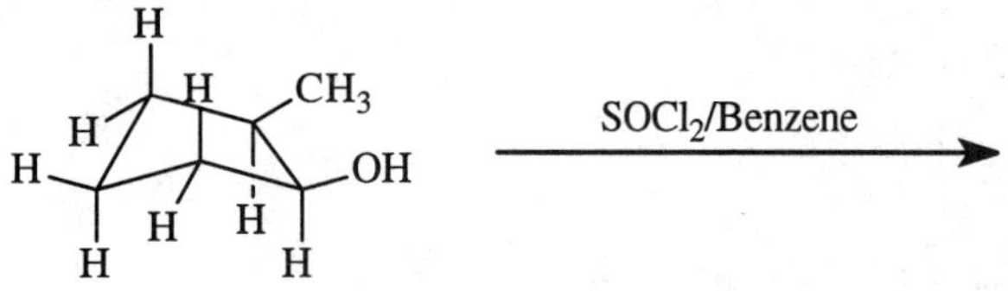

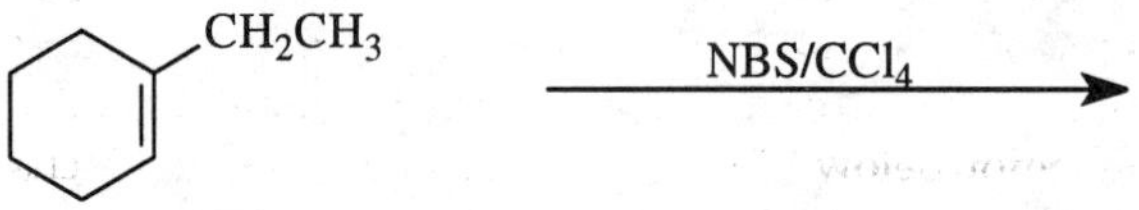

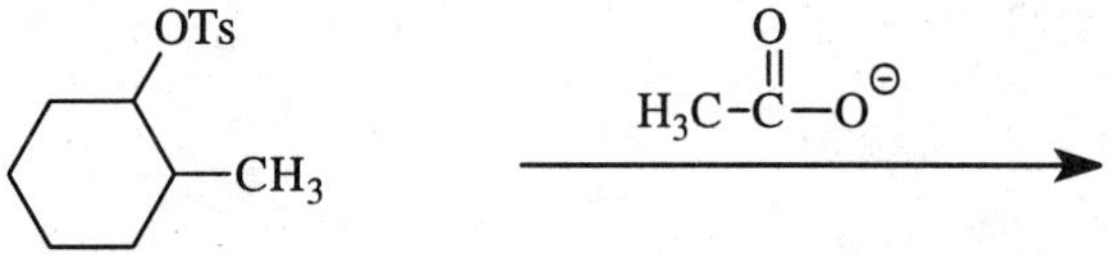

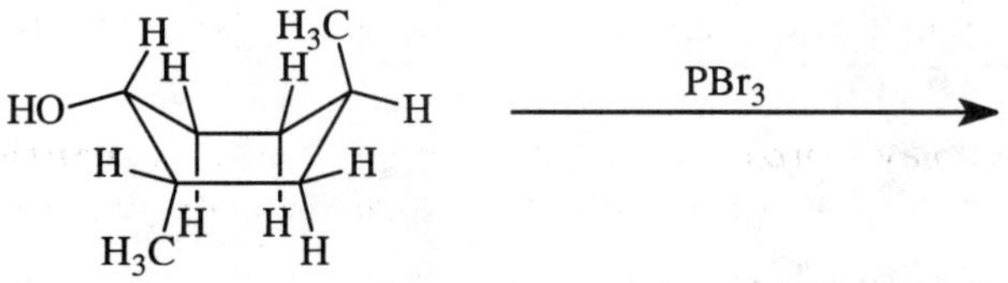

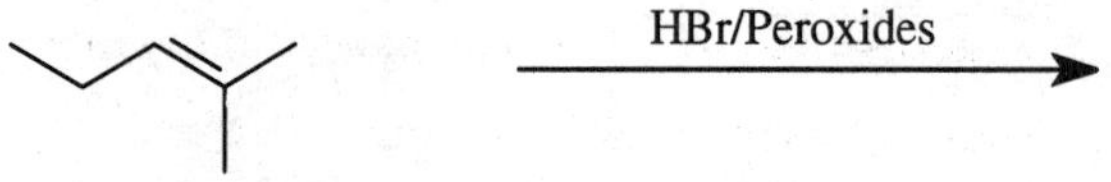

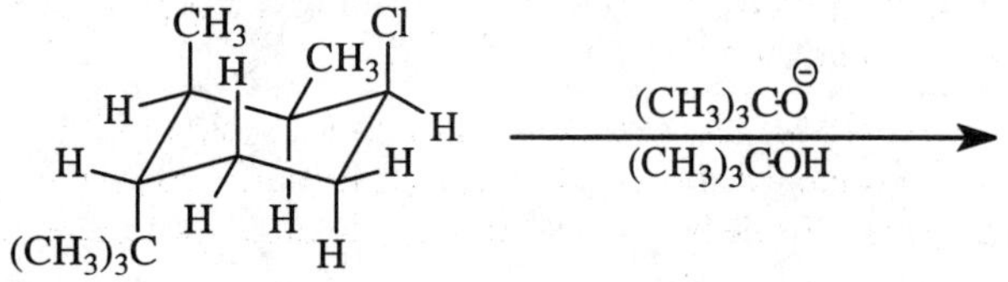

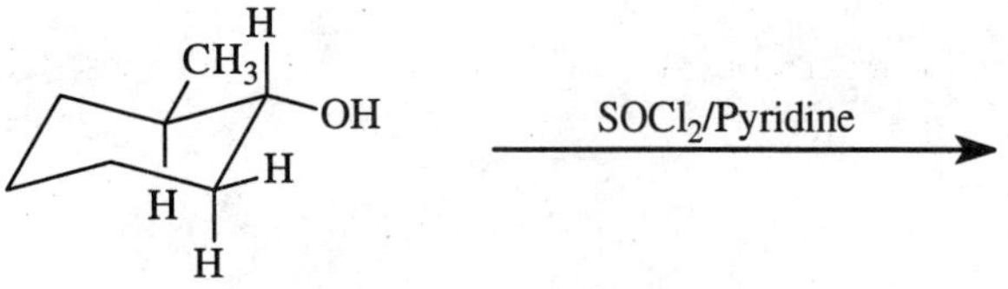

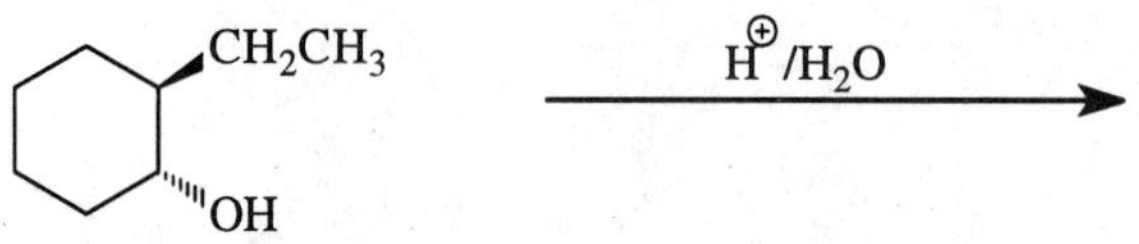

Ph—CH$_2$CH$_2$OH $\xrightarrow{\text{PBr}_3}$

1-methyl-1-(OTs)cyclopentane + cyclopropyl-NH$_2$ $\xrightarrow{\text{CH}_3\text{CH}_2\text{OH}}$

(cyclohexane with CH$_3$ and OH, stereochemistry shown) $\xrightarrow{\text{SOCl}_2/\text{Benzene}}$

1-ethylcyclohexene $\xrightarrow{\text{NBS/CCl}_4}$

1-methyl-2-(OTs)cyclohexane $\xrightarrow{\text{H}_3\text{C-C(=O)-O}^{\ominus}}$

(dimethylcyclohexanol, stereochemistry shown) $\xrightarrow{\text{PBr}_3}$

(CH$_3$)$_2$C=CHCH$_2$CH$_3$ (2-methyl-2-butene) $\xrightarrow{\text{HBr/Peroxides}}$

(substituted cyclohexane with Cl, CH$_3$, C(CH$_3$)$_3$) $\xrightarrow[\text{(CH}_3\text{)}_3\text{COH}]{\text{(CH}_3\text{)}_3\text{CO}^{\ominus}}$

(CH$_3$)$_2$CH—CH=CH$_2$ $\xrightarrow{\text{NBS/CCl}_4}$

(1-methyl-2-hydroxycyclohexane, stereochemistry shown) $\xrightarrow{\text{SOCl}_2/\text{Pyridine}}$

(trans-1-ethyl-2-hydroxycyclohexane) $\xrightarrow{\text{H}^{\oplus}/\text{H}_2\text{O}}$

Spectroscopy

Infrared Spectroscopy: In general, the region of the infrared spectrum which is of greatest interest to organic chemists is the wavelength range 2.5 to ≈ 15 micrometers (μ). In practice, units proportional to *frequency*, (wave number in units of cm^{-1}) rather than wavelength, are commonly used and the region 2.5 to ≈ 15 μ corresponds to approximately 4000 to 600 cm^{-1}.

Absorption of radiation in this region by a typical organic molecule results in the excitation of vibrational, rotational and bending modes, while the molecule itself remains in its electronic ground state. Molecular asymmetry is a requirement for excitation by infrared radiation and fully symmetric molecules do not display absorbances in this region unless asymmetric stretching or bending transitions are possible. For the purpose of routine organic structure determination, using a battery of spectroscopic methods, the most important absorptions in the infrared region are the simple stretching vibrations. The stretching vibrations of typical organic molecules tend to fall within distinct regions of the infrared spectrum, as shown below:

- 3700 - 2500 cm^{-1}:   X-H stretching (X = C, N, O, S)
- 2300 - 2000 cm^{-1}:   C≡X stretching (X = C or N)
- 1900 - 1500 cm^{-1}:   C=X stretching (X = C, N, O)
- 1300 - 800 cm^{-1}:   C-X stretching (X = C, N, O)

Since most organic molecules have single bonds, the region below 1500 cm^{-1} can become quite complex and is often referred to as the '**fingerprint region**': that is, if you are dealing with an unknown molecule which has the same 'fingerprint' in this region, that is considered evidence that the two molecules may be identical. Because of the complexity of the region below 1500 cm^{-1}, in this review, we will focus on functional group stretching bands in the higher frequency region. You should note that for many of these bands, the IR spectrum may give equivocal structural information; quite often the *absence* of a band is as informative as the *presence* of a particular band. Some of the common IR bands which will be covered in this tutorial include:

• **Alcohols and amines** display strong broad O-H and N-H stretching bands in the region 3400-3100 cm^{-1}. The bands are broadened due to hydrogen bonding and a sharp 'non-bonded' peak can often be seen at around 3400 cm^{-1}.

• **Alkene and alkyne C-H bonds** display sharp stretching absorptions in the region 3100-3000 cm^{-1}. The bands are of medium intensity and are often obscured by other absorbances in the region (i.e., OH).

• **Alkane C-H bonds** display sharp stretching absorptions in the region 3000-2900 cm^{-1}. The bands are of medium-strong intensity but are sometimes obscured by other absorbances in the region.

• **Triple bond** stretching absorptions occur in the region 2400-2200 cm^{-1}. Absorptions from nitriles are generally of medium intensity and are clearly defined. Alkynes absorb weakly in this region unless they are highly asymmetric; symmetrical alkynes do not show absorption bands.

• **Carbonyl** stretching bands occur in the region 1800-1700 cm^{-1}. The bands are generally very strong and broad. Carbonyl compounds which are more reactive in nucleophilic addition reactions (acyl halides, esters) are generally at higher wave number than simple ketones and aldehydes, and amides are the lowest, absorbing in the region 1700-1650 cm^{-1}.

• **Carbon-carbon double bond** stretching occurs in the region around 1650-1600 cm^{-1}. The bands are generally sharp and of medium intensity. Aromatic compounds will typically display a series of sharp bands in this region.

• **Carbon-oxygen single bonds** display stretching bands in the region 1200-1100 cm^{-1}. The bands are generally strong and broad. You should note that many other functional groups have bands in this region which appear similar.

Spectroscopy: *Infrared Spectroscopy Problem Set*

Correlation Table:

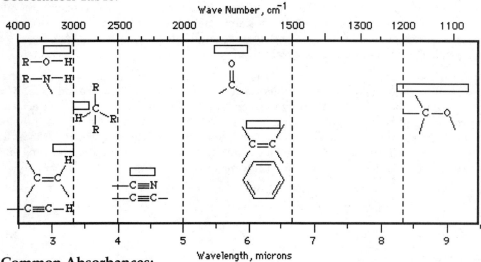

Common Absorbances:

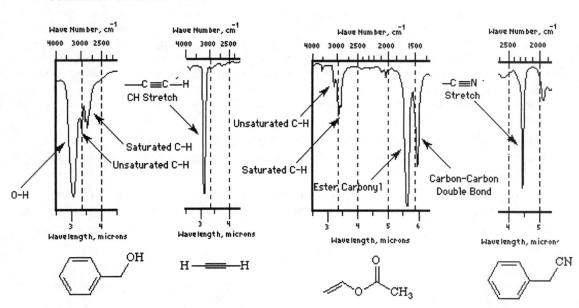

Suggest a structure which is consistent with each of the IR spectra shown below.

IR #1: $C_5H_{10}O$

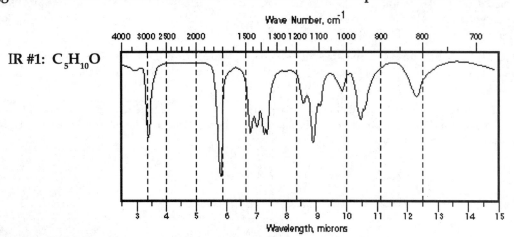

Spectroscopy: *Infrared Spectroscopy Problem Set, continued*

Suggest a structure which is consistent with each of the IR spectra shown below.

IR #2: C_8H_8O

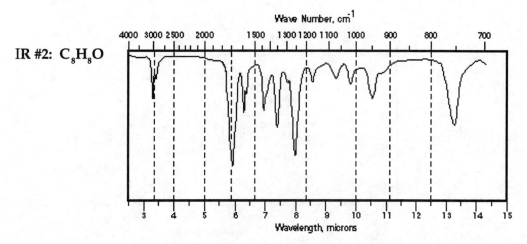

IR #3: C_7H_8O

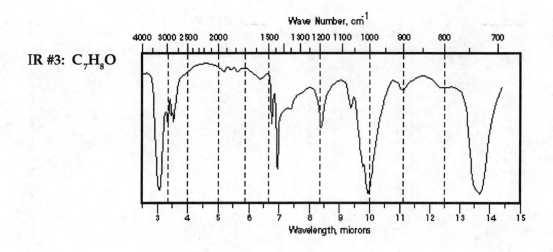

IR #4: C_8H_7N

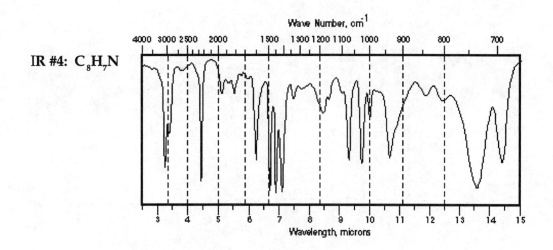

Spectroscopy: *Infrared Spectroscopy Problem Set, continued*

Suggest a structure which is consistent with each of the IR spectra shown below.

IR #5: C_7H_6O

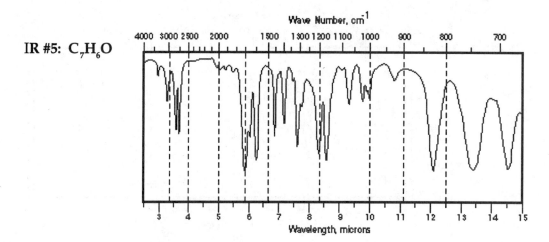

IR #6: C_3H_7NO

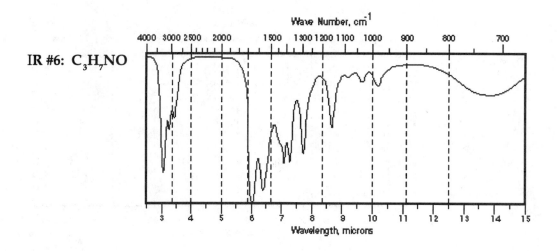

IR #7: $C_4H_8O_2$

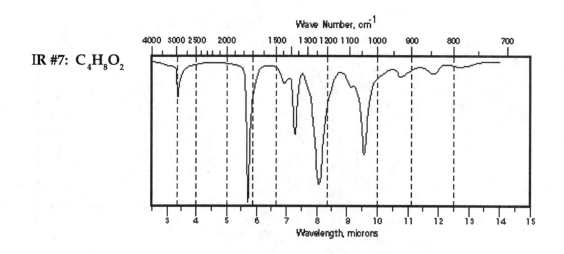

Spectroscopy: *Infrared Spectroscopy Problem Set, continued*

Suggest a structure which is consistent with each of the IR spectra shown below.

IR #8: C_7H_5OCl

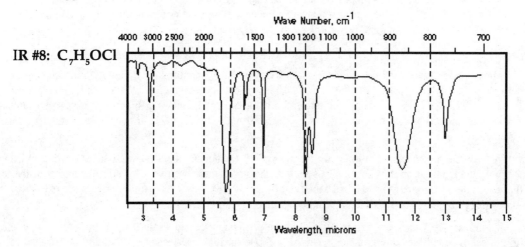

IR #9: C_6H_6S

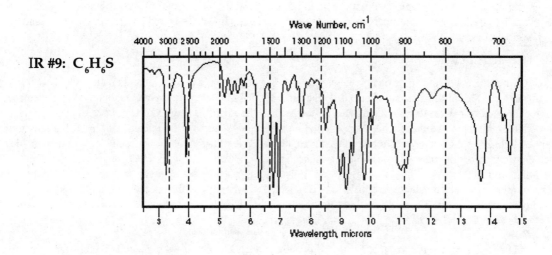

IR #10: C_4H_6

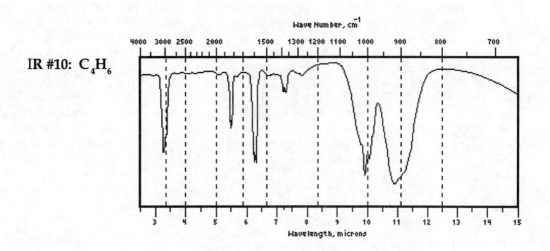

Nuclear Magnetic Resonance

Background: Nuclei of isotopes which possess an odd number of protons, an odd number of neutrons, or both, exhibit mechanical spin phenomena which are associated with angular momentum. This angular momentum is characterized by a nuclear spin quantum number, I such that,

$$I = \tfrac{1}{2}n, \text{ where } n \text{ is an integer } 0,1,2,3...\text{etc.}$$

Those nuclei for which $I = 0$ do not possess spin angular momentum and do not exhibit magnetic resonance phenomena. The nuclei of ^{12}C and ^{16}O fall into this category. Nuclei for which $I = \tfrac{1}{2}$ include ^1H, ^{19}F, ^{13}C, ^{31}P and ^{15}N, while ^2H and ^{14}N have $I = 1$. Since atomic nuclei are associated with charge, a spinning nucleus generates a small electric current and has a finite magnetic field associated with it. The magnetic dipole, μ, of the nucleus varies with each element.

When a spinning nucleus is placed in a magnetic field, the nuclear magnet experiences a torque which tends to align it with the external field. For a nucleus with a spin of $\tfrac{1}{2}$, there are two allowed orientations of the nucleus; parallel to the field (low energy) and against the field (high energy). Since the parallel orientation is lower in energy, this state is slightly more populated than the anti-parallel, high energy state.

If the oriented nuclei are now irradiated with electromagnetic radiation of the proper frequency, the lower energy state will absorb a quantum of energy and spin-flip to the high energy state. When this spin transition occurs, the nuclei are said to be in resonance with the applied radiation, hence the name nuclear magnetic resonance. The amount of electromagnetic radiation necessary for resonance depends on both the strength of the external magnetic field and on the characteristics of the nucleus being examined. The nucleus of the proton, placed in 14,100 gauss field, undergoes resonance when irradiated with radiation in the 60 MHz range (microwave radiation); higher magnetic fields, such as those common in superconducting magnets, require higher energy radiation and give a correspondingly higher resolution.

The Chemical Shift

Since the magnetic dipole of a given nucleus (μ) is a constant, you might predict that all nuclei of a given type would undergo the spin-flip transition at exactly the same applied frequency in a given magnetic field. Fortunately, in a typical organic molecule this is not the case. This is because the electrons in the molecule have small magnetic fields associated with them and these tend to oppose the applied field, screening the nuclei from the full strength of the applied field. The greater the electron density, the greater this "shielding" will be, hence nuclei which are in electron rich environments will undergo transition at a higher applied field than nuclei in electron poor environments. The resulting shift in the NMR signal for a given nuclei is referred to as the **chemical shift**, and, in general, protons or carbons adjacent to electronegative atoms will be deshielded and moved to a higher chemical shift (undergo transition at a lower applied field). The scale utilized for measuring chemical shifts is defined by the equation shown below:

$$\text{Chemical Shift } (\delta) = (\text{shift observed}/\text{oscillator frequency}) \times 10^6 \text{ ppm}$$

The factor of 10^6 is introduced into the equation to give a simple whole number scale for convenience. Experimentally, for both ^1H and ^{13}C NMR, the δ scale is anchored at zero by the NMR absorption of the molecule tetramethyl silane ($(CH_3)_4Si$) in which the carbons and protons are more highly shielded than those observed in most common organic molecules. For ^1H NMR, the δ scale generally extends from 0-12 ppm; the δ scale for ^{13}C nuclei, however, is much larger and covers the range 0-220 ppm. On the δ scale for ^1H NMR, simple hydrocarbon protons tend to absorb in the region δ 0.5-1.5, protons on a carbon adjacent to a carbonyl are shifted to $\delta \approx 2$, electronegative atoms (oxygen or halogens) move α-protons to $\delta \approx 3-4$, alkene protons are shifted to $\delta \approx 5-6$, aromatic protons to δ 7-8, aldehydic protons to $\delta \approx 10$, and the most highly shifted protons are generally those of carboxylic acids, with values of $\delta \approx 12$. For ^{13}C NMR, simple methyl carbons tend to absorb in the region δ 15-30, simple methylene carbons are shifted to $\delta \approx 20-65$, electronegative atoms (oxygen or halogens) move attached carbons to $\delta \approx 40-80$, alkyne carbons are shifted to $\delta \approx 70-90$, alkene carbons to $\delta \approx 100-150$, aromatic carbons to $\delta \approx 120-170$, and the most highly shifted carbons are generally those of carbonyls, with values of $\delta \approx 180-220$. One further feature of the proton NMR is the fact that the intensity of the absorbance of a given class of nuclei (with a certain chemical shift) is proportional to the number of protons giving rise to the signal; that is, the area under a given peak (the **integration**) is directly proportional to the number of that type of proton in the molecule. Integrations are typically given as simplest whole-number ratios, hence, acetic acid, CH_3COOH, will have two peaks in the proton NMR, one at $\delta = 2$, area = 3, and a second at $\delta = 12$, area = 1. Methyl acetate, CH_3COOCH_3, will also have two peaks in the proton NMR, one at $\delta = 2$, area = 1, and a second at $\delta = 4$, area = 1 (the relative areas or both peaks are the same, but each one represents three hydrogens).

Spin-Coupling, or "Splitting"

Proton NMR: For a molecule such as diethyl ether, $CH_3CH_2OCH_2CH_3$, two types of protons would be predicted to appear in the NMR spectrum; a "simple" CH_3 in the area of $\delta \approx 1$, and a CH_2 shifted down to about $\delta \approx 4$ by the electronegative oxygen. The NMR spectrum of diethyl ether, however, displays *seven peaks*. This multiplicity is due to the phenomena known as **spin coupling** and arises because of interaction of the proton magnetic field with bonding electrons. In essence, each proton can have one of two possible spin orientations in the applied field, so that the magnetic field sensed by adjacent protons can have one of two possible values. The result is that ***n* protons will split adjacent protons into (*n* + 1) peaks**. Thus, in the spectrum for diethyl ether, the CH_3 group is split by the two protons on the adjacent CH_2 group into three peaks (a *triplet*) and the absorbance for the CH_2 is split by the three protons on the methyl group into $(n + 1) = 4$ peaks (a *quartet*). The two ethyl groups in diethyl ether show as a single absorbance since the molecule has a plane of symmetry, that is, both ethyl groups are chemically and magnetically equivalent.

^{13}C NMR: Spin coupling, while possible between adjacent ^{13}C nuclei, is not typically observed since the natural abundance of ^{13}C is very low (1.1%), making it unlikely that two ^{13}C nuclei will reside next to each other in a given molecule. Attached protons, however, with a spin of $1/2$, will couple with the ^{13}C nucleus to generate spin-coupling. As described above, the signal from the carbon will be split into $(n + 1)$ peaks, where n is the number of attached protons. Typically, however, the spectrometer is set up in a mode referred to as proton noise-decoupling in which the sample is irradiated at a second frequency which promotes all of the protons in the molecule to high spin states, disallowing the spin-coupling process. All of the split ^{13}C peaks are thereby reduced to sharp singlets. While this operating mode provides less information than the non-decoupled mode, it is commonly used because it results in a significant signal enhancement due to a phenomena known as the **Nuclear Overhauser Effect**. While the details of this process are beyond the scope of this simple tutorial, the important fact to note is that the NOE results in the transfer of energy to attached ^{13}C nuclei, resulting in a significant enhancement in the NMR signal. Unfortunately, the NOE enhancement is not uniform, hence **the integration of a ^{13}C NMR spectrum is typically meaningless**. In the examples given in this tutorial, we will provide coupling information (singlet, doublet, etc.) along with the chemical shift, but without intensity data.

Spectroscopy: *^{1}H NMR Spectroscopy Problem Set*

Correlation Chart

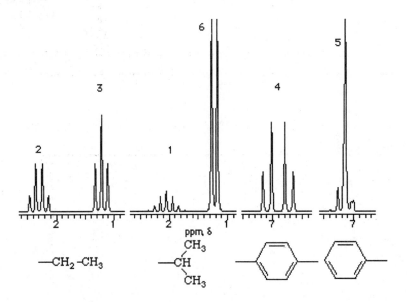

Common Splitting Patterns

Spectroscopy: ¹H NMR Spectroscopy Problem Set

Suggest a structure which is consistent with each of the NMR spectra shown below.

NMR #1: $C_3H_6O_2$

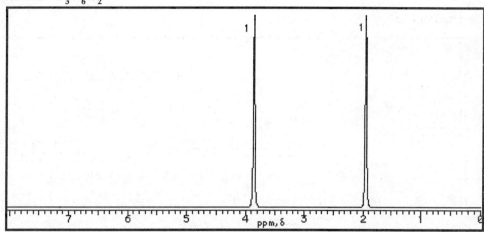

NMR #2: $C_5H_{10}O$

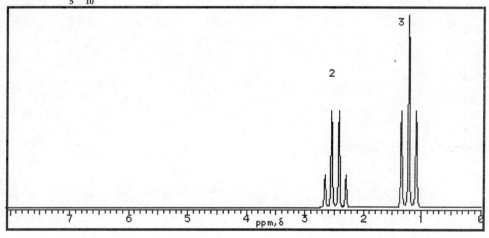

NMR #3: C_2H_4O

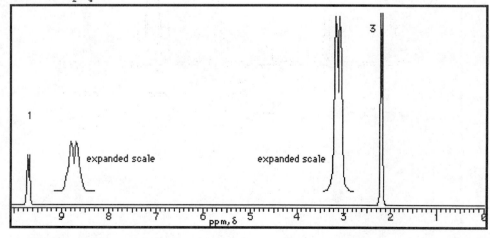

Spectroscopy: ¹H NMR Spectroscopy Problem Set, continued

Suggest a structure which is consistent with each of the NMR spectra shown below.

NMR #4: $C_4H_8O_2$

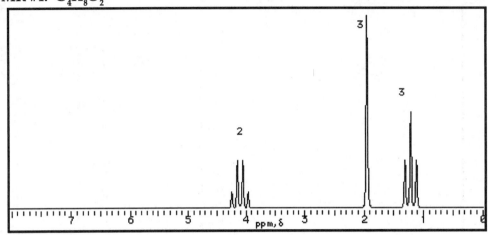

NMR #5: C_7H_8

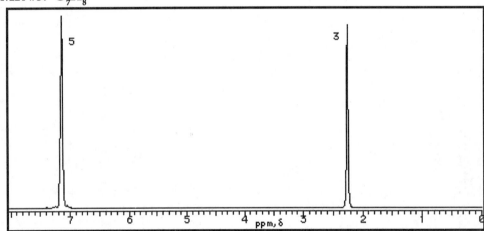

NMR #6: C_8H_8O

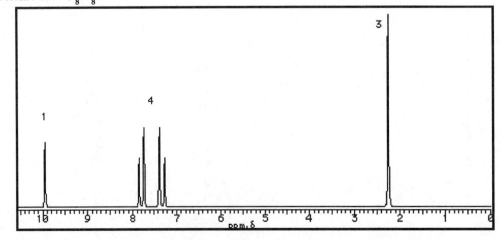

Spectroscopy: ¹H NMR Spectroscopy Problem Set, continued

Suggest a structure which is consistent with each of the NMR spectra shown below.

NMR #7: C_3H_6O

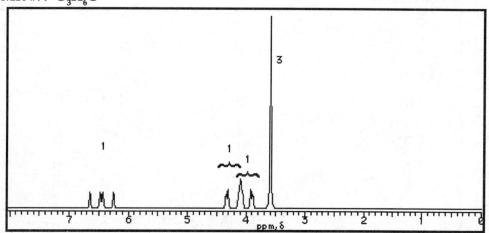

NMR #8: $C_4H_{10}O_2$

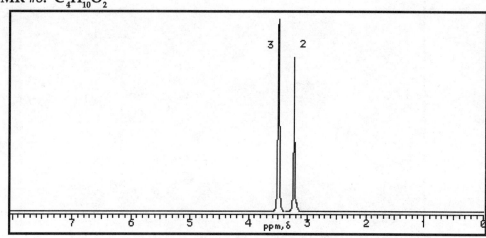

NMR #9: $C_{10}H_{14}O$

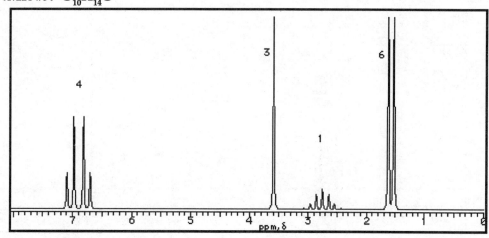

Spectroscopy: *¹H NMR Spectroscopy Problem Set, continued*

Suggest a structure which is consistent with each of the NMR spectra shown below.

NMR #10: C_8H_9Br

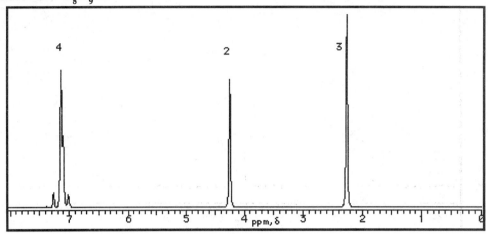

Spectroscopy: *¹³C NMR Spectroscopy Problem Set*

Correlation Table

¹³C NMR #1: $C_4H_{10}O_2$

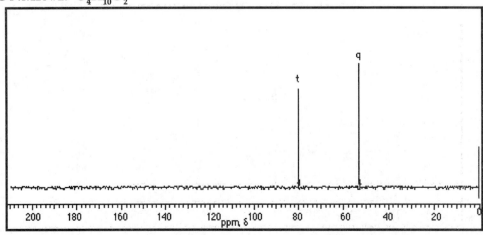

Spectroscopy: ¹³C NMR Spectroscopy Problem Set

Suggest a structure which is consistent with each of the NMR spectra shown below.

¹³C NMR #2: $C_5H_7O_2N$

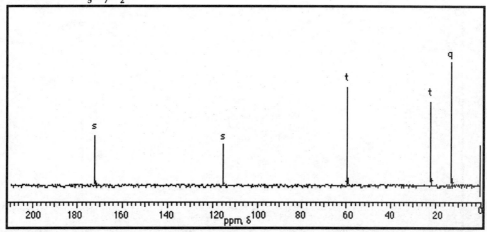

¹³C NMR #3: $C_6H_{10}O$

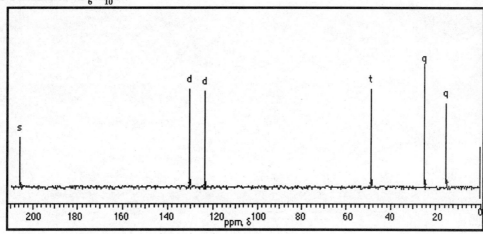

¹³C NMR #3: C_8H_8O

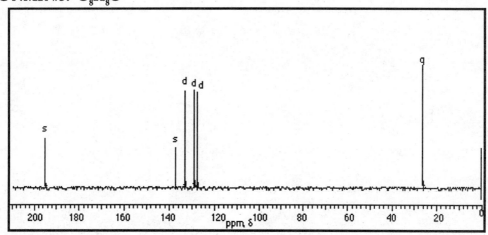

Spectroscopy: ¹³C NMR Spectroscopy Problem Set

Suggest a structure which is consistent with each of the NMR spectra shown below.

¹³C NMR #5: C_6H_8O

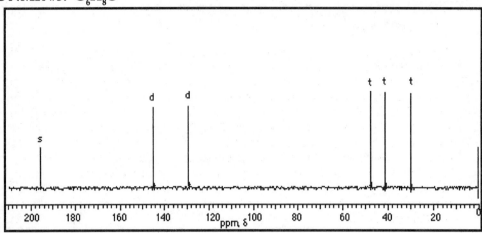

Notes:

Mass Spectrometry - Background

In mass spectrometry, a substance is bombarded with an electron beam having sufficient energy to fragment the molecule. The positive fragments which are produced (cations and radical cations) are accelerated in a vacuum through a magnetic field and are sorted on the basis of mass-to-charge ratio (**m/e**). Since the bulk of the ions produced in the mass spectrometer carry a unit positive charge, the value m/e is equivalent to the molecular weight of the fragment. The analysis of mass spectroscopy information involves the re-assembling of fragments, working backwards to generate the original molecule.

In a mass spectrometer, a very low concentration of sample molecules are allowed to leak into the ionization chamber (which is under a very high vacuum) where they are bombarded by a high-energy electron beam. The molecules fragment and the positive ions produced are accelerated through a charged array into an analyzing tube. The path of the charged molecules in this tube is bent by an applied magnetic field. Ions having low mass (low momentum) will be deflected most by this field and will collide with the walls of the analyzer. Likewise, high momentum ions will not be deflected enough and will also collide with the analyzer wall. Ions having the proper mass-to-charge ratio, however, will follow the path of the analyzer, exit through the slit and collide with the **collector**. This generates an electric current, which is then amplified and detected. By varying the strength of the magnetic field, the mass-to-charge ratio which is analyzed can be continuously varied. The output of the mass spectrometer shows a plot of relative intensity *vs* the mass-to-charge ratio (m/e). The most intense peak in the spectrum is termed the **base peak** and all others are reported relative to it's intensity. The peaks themselves are typically very sharp, and are often simply represented as vertical lines.

The process of fragmentation follows simple and predictable chemical pathways and the ions which are formed will reflect **the most stable cations and radical cations which that molecule can form**. The highest molecular weight peak observed in a spectrum will typically represent the parent molecule, minus an electron, and is termed the **molecular ion** (m^+). Generally, small peaks are also observed above the calculated molecular weight due to the natural isotopic abundance of ^{13}C, ^2H, etc. Many molecules with especially labile protons do not display molecular ions; an example of this is alcohols, where the highest molecular weight peak occurs at m/e one less than the molecular ion (m-1). Fragments can be identified by their mass-to-charge ratio, but it is often more informative to identify them by the mass which has been lost. That is, loss of a methyl group will generate a peak at m-15; loss of an ethyl, m-29, etc.

The mass spectrum of toluene (methyl benzene) is a classic example. The spectrum displays a strong molecular ion at m/e = 92, small m+1 and m+2 peaks, a base peak at m/e = 91 and an assortment of minor peaks m/e = 65 and below. The molecular ion, again, represents loss of an electron and the peaks above the molecular ion are due to isotopic abundance. The base peak in toluene is due to loss of a hydrogen atom to form the relatively stable **benzyl cation**. This is thought to undergo rearrangement to form the very stable **tropylium cation**, and this strong peak at m/e = 91 is a hallmark of compounds containing a benzyl unit. The minor peak at m/e = 65 represents loss of neutral acetylene from the tropylium ion and the minor peaks below this arise from more complex fragmentation.

Fragmentations of Common Functional Groups

Alkanes: Simple alkanes tend to undergo fragmentation by the initial loss of a methyl group to form a (m-15) species. This carbocation can then undergo stepwise cleavage down the alkyl chain, expelling neutral two-carbon units (ethene). Branched hydrocarbons form more stable secondary and tertiary carbocations, and these peaks will tend to dominate the mass spectrum.

Aromatic Hydrocarbons: The fragmentation of the aromatic nucleus is somewhat complex, generating a series of peaks having m/e = 77, 65, 63, etc. While these peaks are difficult to describe in simple terms, they do form a pattern (the "aromatic cluster") that becomes recognizable with experience. If the molecule contains a benzyl unit, the major cleavage will be to generate the benzyl carbocation, which rearranges to form the tropylium ion. Expulsion of acetylene (ethyne) from this generates a characteristic m/e = 65 peak.

Aldehydes and Ketones: The predominate cleavage in aldehydes and ketones is loss of one of the side-chains to generate the substituted oxonium ion. This is an extremely favorable cleavage and this ion often represents the base peak in the spectrum. The methyl derivative (CH_3CO^+) is commonly referred to as the "acylium ion".

Esters, Acids and Amides: As with aldehydes and ketones, the major cleavage observed for these compounds involves expulsion of the group attached to the acyl center, to form the substituted oxonium ion. For carboxylic acids and unsubstituted amides, characteristic peaks at m/e = 45 and 44 are also often observed.

Alcohols: In addition to losing a proton and hydroxyl radical, alcohols tend to lose one of the α-alkyl groups (or hydrogens) to form the oxonium ions shown below. For primary alcohols, this generates a peak at m/e = 31; secondary alcohols generate peaks with m/e = 45, 59, 73, etc., according to substitution.

Ethers: Following the trend of alcohols, ethers will fragment, often by loss of an alkyl radical, to form a substituted oxonium ion.

Halides: Organic halides fragment with simple expulsion of the halogen. The molecular ions of chlorine and bromine-containing compounds will show multiple peaks due to the fact that each of these exists as two isotopes in relatively high abundance. Thus for chlorine, the $^{35}Cl/^{37}Cl$ ratio is roughly 3:1 and for bromine, the $^{79}Br/^{81}Br$ ratio is about 1:1. The molecular ion of a chlorine-containing compound will have two peaks, separated by two mass units, in the ratio 3:1, and a bromine-containing compound will have two peaks, again separated by two mass units, having approximately equal intensities.

Primary Fragmentations Associated with Some Common Functional Groups

Functional Group	Fragmentation
Amine	$[R_1(CH_2)_nN(R_2)-CH_2R_3]^{+\cdot} \longrightarrow R_1(CH_2)_n\overset{\oplus}{N}(R_2)=CH_2 + R_3^{\cdot} \longrightarrow \overset{\oplus}{H}N(R_2)=CH_2 +$ alkene
Aldehydes (R_2 = H) and Ketones	$[R_1\text{-CO-}R_2]^{+\cdot} \longrightarrow R\text{-}C\equiv\overset{\oplus}{O} + R_2^{\cdot} \longrightarrow R_1^{\oplus} + CO$
Halides	$[R\text{-}X]^{+\cdot} \longrightarrow R^{\oplus} + X^{\cdot}$
Acyl Compounds	$[R_1\text{-CO-}X]^{+\cdot} \longrightarrow R\text{-}C\equiv\overset{\oplus}{O} + X^{\cdot}$
Alcohols and Thiols	$[R_1CH_2\text{-}XH(R_2)]^{+\cdot} \longrightarrow R_1CH_2=\overset{\oplus}{X}H + R_2^{\cdot}$

Fragments Commonly Lost from Molecular Ions

Mass	Group	Mass	Group
15	CH_3	32	CH_3OH
16	NH_2	44	C_3H_5
17	OH	42	CH_2CO
18	H_2O	42	C_3H_6
19	F	43	C_3H_7
20	HF	43	CH_3CO
26	C_2H_2	44	CO_2
29	CHO	44	C_3H_8
29	CH_2CH_3	45	CO_2H
30	CH_2O	45	OCH_2CH_3
31	OCH_3	46	CH_3CH_2OH

Masses and Possible Structures of Common Fragment Ions

m/z	Associated Structures
29	CHO^+, $CH_3CH_2^+$
30	$H_2C=NH_2^+$
31	$H_2C=OH^+$
41	$H_2C=CH-CH_2^+$
42	$H_2C=CH-CH_3^{+\cdot}$
43	$CH_3C\equiv O^+$, $CH_3{}^+CHCH_3$
44	$CH_3CHNH_2^+$, $CO_2^{+\cdot}$, $O=C=NH_2^+$, $C_3H_8^{+\cdot}$, $H_2C=CHOH^{+\cdot}$
45	$H_2C=\overset{+}{O}CH_3$, $CH_3CH=OH^+$, $CO_2H^{+\cdot}$
54	$H_2C=CH-CH=CH_2^{+\cdot}$
57	$C_4H_9^+$, $CH_3CH_2C\equiv O^+$
58	$CH_3CH_2\overset{+}{N}H=CH_2$, $H_2C=C(OH)CH_3^{+\cdot}$, $CH_3CH_2CH=NH_2^+$
59	$CH_3CH_2CH=OH^+$, $H_2C=C(OH)NH_2^{+\cdot}$, $H_2C=\overset{+}{O}-CH_2CH_3$, $CH_3O-C\equiv O^+$
60	$H_2C=C(OH)OH^{+\cdot}$
65	$C_5H_5^+$, H–(cyclopentadienyl cation)
66	(cyclopentadiene cation)$^{+\cdot}$
67	(cyclopentenyl cation)$^+$
71	$CH_3CH_2CH_2C\equiv O^+$ (and isomers)
72	$H_2C=C(OH)CH_2CH_3^{+\cdot}$
73	$^+O\equiv C-OCH_2CH_3$
74	$H_2C=C(OH)OCH_3^{+\cdot}$
76	$C_6H_4^{+\cdot}$
77	(phenyl cation)$^+$
78	$C_6H_6^{+\cdot}$
79 & 81	Br^+
80 & 82	$HBr^{+\cdot}$
85	$C_4H_9C\equiv O^+$
88	$H_2C=C(OH)OCH_2CH_3^{+\cdot}$
91	(tropylium cation)$^+$ (and isomers)
105	$^+O\equiv C-$(phenyl), CH_3-(tropylium)$^+$ (and isomers)

Spectroscopy: *Mass Spectroscopy Problem Set*

Suggest a structure which is consistent with each of the Mass spectra shown below.

MS#1: $C_5H_{12}O$

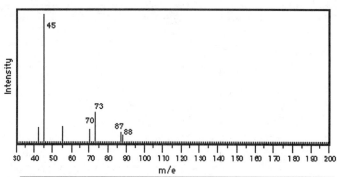

MS #2: C_7H_7Br

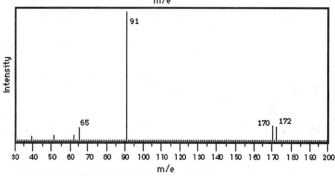

MS #3: $C_9H_{10}O$

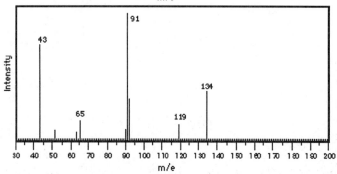

MS #4: $C_{11}H_{12}O_3$

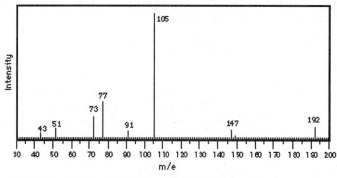

MS #5: $C_5H_8O_2$

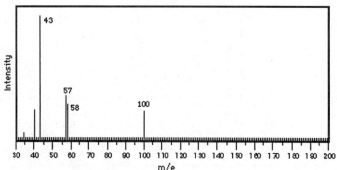

Spectroscopy: *Integrated Spectroscopy Problem Sets*

Problem #1: Suggest a structure which is consistent with all of the spectral data given below.

MS data: C_3H_8O

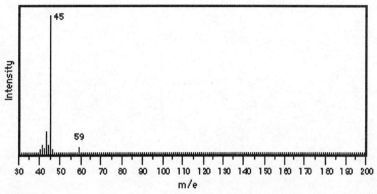

^{13}C NMR data: q-26.3; d-64.9

^{1}H NMR data:

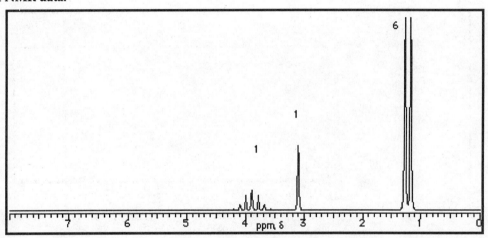

IR data:

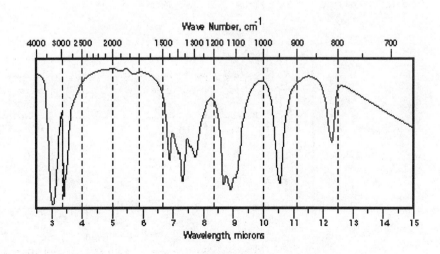

Spectroscopy: *Integrated Spectroscopy Problem Sets*

Problem #2: Suggest a structure which is consistent with all of the spectral data given below.

MS data: C_6H_{12}

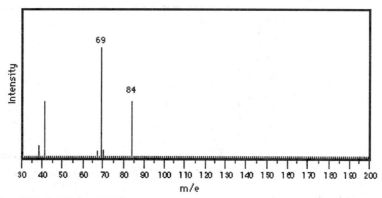

13C NMR data: q-30.6; s-45.2; d-149.3; t-108.5

1H NMR data:

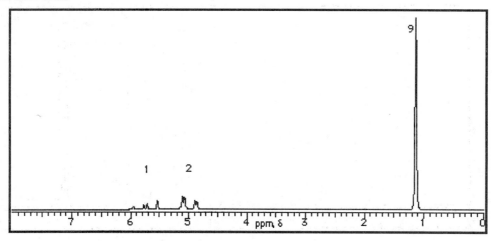

IR data:

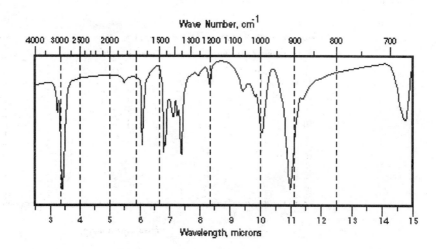

Spectroscopy: *Integrated Spectroscopy Problem Sets*

Problem #3: Suggest a structure which is consistent with all of the spectral data given below.

MS data: $C_5H_{10}O$

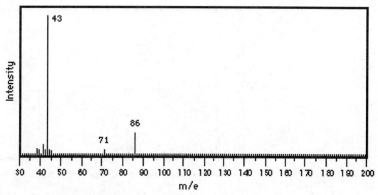

^{13}C NMR data: q-22.0; q-16.7; d-45.2; s-210.2

^{1}H NMR data:

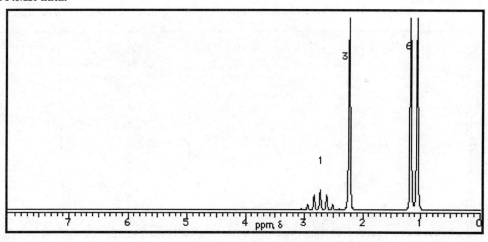

IR data:

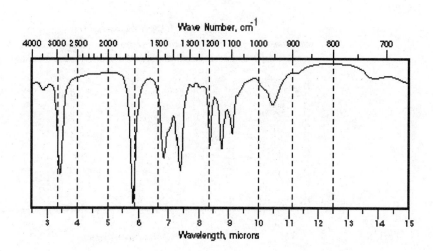

Spectroscopy: *Integrated Spectroscopy Problem Sets*

Problem #4: Suggest a structure which is consistent with all of the spectral data given below.

MS data: $C_6H_{10}O_3$

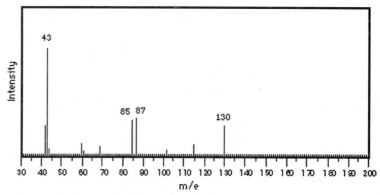

¹³C NMR data: q-13.6; q-24.2; t-59.2; t-46.6; s-172.0; s-207.1

¹H NMR data:

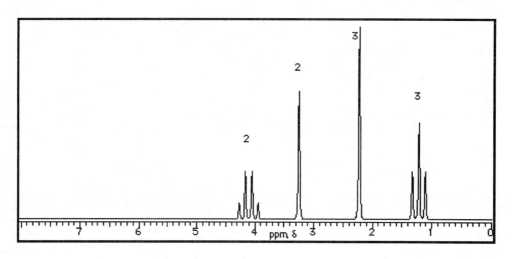

IR data:

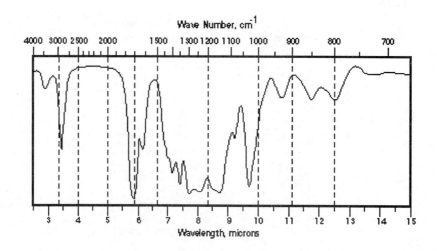

Spectroscopy: *Integrated Spectroscopy Problem Sets*

Problem #5: Suggest a structure which is consistent with all of the spectral data given below.

MS data: $C_8H_{10}O_2$

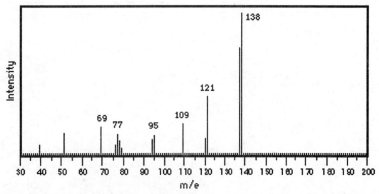

^{13}C NMR data: q-56.0; t-71.0; d-114.3; d-128.3; s-160.9; s-133.2

^{1}H NMR data:

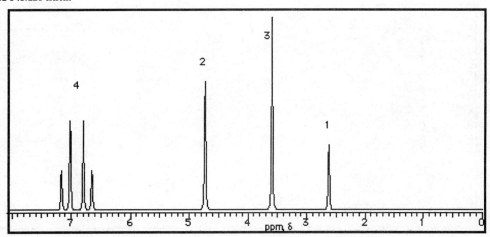

IR data:

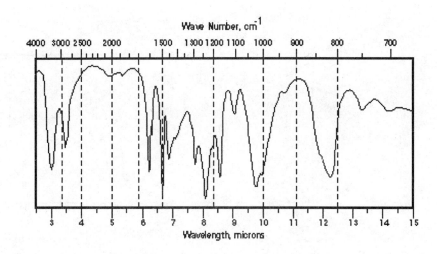

Spectroscopy: *Integrated Spectroscopy Problem Sets*

Problem #6: Suggest a structure which is consistent with all of the spectral data given below.

MS data: $C_{14}H_{18}O_4$

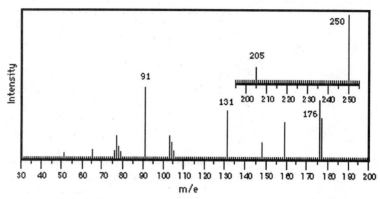

^{13}C NMR data: q-13.6; t-35.7; t-59.5; d-57.8; s-141.1; d-128.7; d-127.6; d-125.3; s-174.5

^{1}H NMR data:

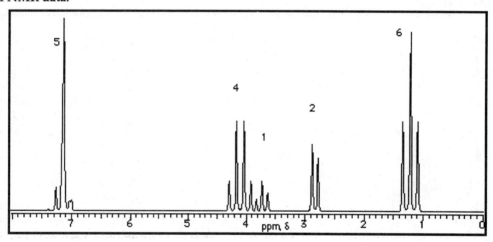

IR data:

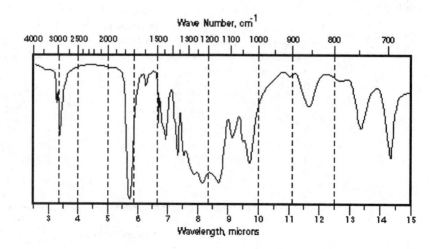

Spectroscopy: *Integrated Spectroscopy Problem Sets*

Problem #7: Suggest a structure which is consistent with all of the spectral data given below.

MS data: C_8H_7N

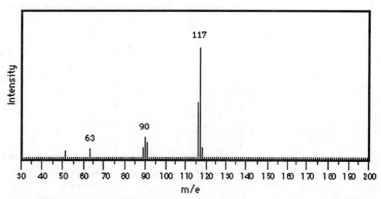

13C NMR data: q-22.0; s-116.5; d-131.9; d-129.9; s-142.0

1H NMR data:

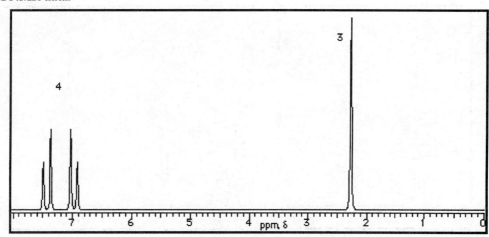

IR data:

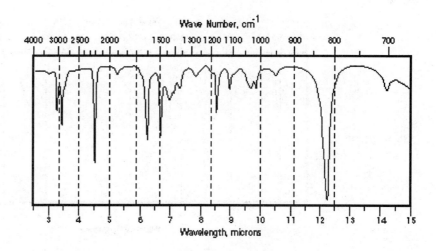

Spectroscopy: *Integrated Spectroscopy Problem Sets*

Problem #8: Suggest a structure which is consistent with all of the spectral data given below.

MS data: $C_8H_8O_2$

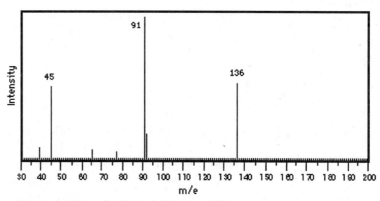

13C NMR data: t-43.8; s-178.0; d-129.9; d-128.9; d-127.3; s-135.0

1H NMR data:

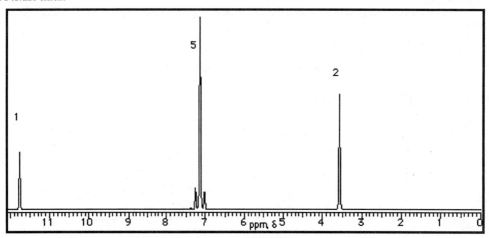

IR data:

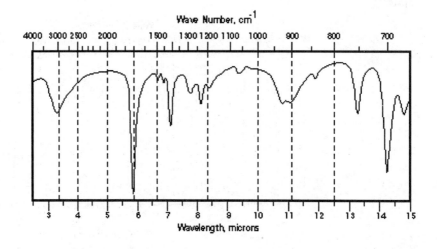

Spectroscopy: *Integrated Spectroscopy Problem Sets*

Problem #9: Suggest a structure which is consistent with all of the spectral data given below.

MS data: C_9H_{10}

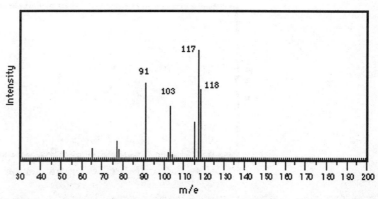

^{13}C NMR data: t-44.9; s-137.3; t-115.1; s-140.2; d-127.9; d-128.4; d-125.7

^{1}H NMR data:

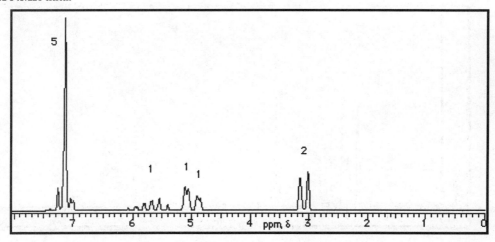

IR data:

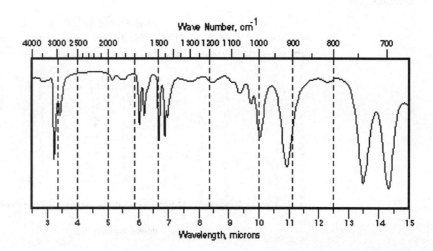

Spectroscopy: *Integrated Spectroscopy Problem Sets*

Problem #10: Suggest a structure which is consistent with all of the spectral data given below.

MS data: C_7H_5OBr

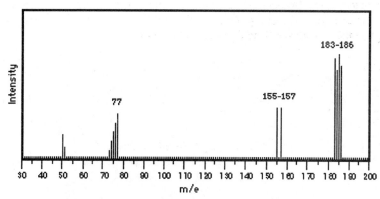

13C NMR data: s-137.4; s-103.0; d-129.7; d-137.8; d-133.2; d-190.0

1H NMR data:

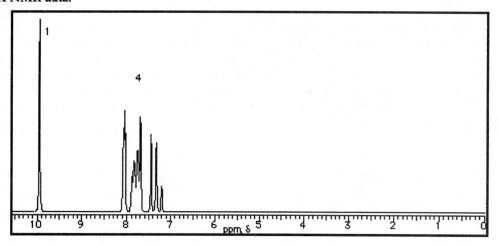

IR data:

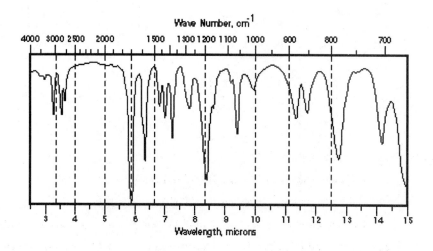

Conjugated Dienes:

Ionic Addition Reactions

When compounds containing conjugated double bonds undergo typical ionic alkene addition reactions (addition of HBr or Br_2, for example), the products which are obtained are not those which would be expected for addition to the individual double bonds in the molecule. For the addition of HBr to 1,3-butadiene, two products are obtained, 3-bromo-1-butene and 1-bromo-2-butene. These products can be seen to arise from a "standard" 1,2-addition to one of the terminal double bond (Markovnikov-style), and from a 1,4-addition of HBr to the two terminal carbons, with relocation of the double bond onto the central two carbons.

The formation of these products can be readily understood by examining the mechanism of the addition reaction. Protonation of the conjugated diene on either terminal carbon will generate a carbocation on the adjacent secondary carbon. This carbocation, however, can be stabilized by resonance with the adjacent double bond to give a delocalized carbocation (an allylic carbocation) in which there is positive character on both a secondary center and on the terminal, primary carbon. Since both of these centers share positive character in the resonance hybrid, both are subject to nucleophilic attack by bromide anion; attack on the secondary carbon gives the 1,2-addition product, and attack on the terminal carbon gives the 1,4-addition product.

Some further observations on this reaction:

- The 1,2- addition product forms rapidly at low temperatures;
- the 1,4-addition product is predominant at higher temperatures;
- even at low temperatures, 1,4-addition products will predominate if given enough time;
- the addition of HBr to butadiene is reversible and isolated 1,2-addition product will convert to the 1,4-product at higher temperatures or at longer times.

These data can be explained using the reaction coordinates shown below. The pathway to form the 1,2-product must have a lower activation energy, because it forms more rapidly than the 1,4-product. The 1,4-product, however, must be more stable than the 1,2-product because it accumulates at equilibrium (note that the reaction appears freely reversible, since isolated 1,2-product reverts to 1,4-, given enough time).

Effect of Temperature and Time on Product Distribution

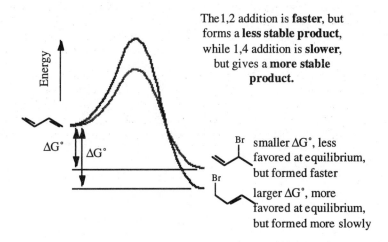

The 1,2-addition product is referred to as the kinetic product since it is formed faster. The 1,4-product is the thermodynamic product since it is thermodynamically more stable. A similar product distribution is observed for Br_2 addition, through a similar mechanism.

Cycloaddition Reactions

Conjugated dienes react with alkenes to yield cyclohexene derivatives. The reaction is termed a 4+2 cycloaddition and is generally referred to as the **Diels-Alder Reaction**. The reactants in the cycloaddition are referred to, generically, as a **diene** and a **dienophile**. The reaction usually requires heat and pressure to give good yields and is promoted by electron withdrawing groups on the dienophile and electron donating groups on the diene.

a diene   a dieneophile

The mechanism of the reaction is generally described as concerted involving an electrocyclic transition state in which the two new sigma bonds form simultaneously; this is usually represented by showing the electron movement with "curved arrows", as shown above. Since both bonds form at the same time, it is necessary for the diene to be in the proper conformation prior to the reaction, that is, the *s-cis* conformation is required, and dienes which cannot adopt this conformation will not react.

Examination of the animation, shown on the CD, for the reaction of ethene with butadiene clearly shows that the initial product of the reaction is the **boat cyclohexene**. Converting this structure to a "chair" by rotating one end down (and rotating the molecule slightly along the Z-axis) gives an idealized cyclohexene "chair". In fact, the geometry of the double bond distorts the molecule, but the idealized chair is useful to establish stereochemical relationship between substituents on the diene and dienophile, as they appear in the cyclohexene product.

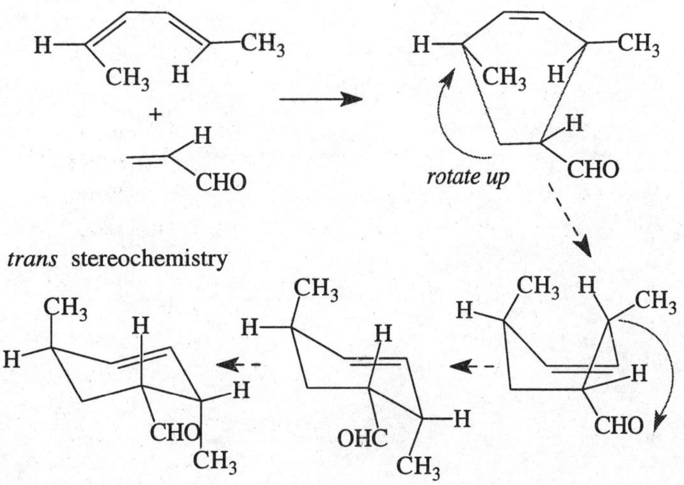

As an example of this, the product of the reaction of *cis-trans*-2,3-hexadiene with propenal is the dimethyl cyclohexene carbaldehyde shown above. The methyl groups of the diene are 1,4-relative to each other on the cyclohexene ring and they are *trans*- (both are axial). A general rule can be established, as shown below, that the stereochemistry of the dienophile is **retained** and that a *trans-trans*-diene yields *cis*-substituents, while a *cis-trans*-diene yields *trans*-substituents (a *cis-cis*-diene will not react since it cannot achieve the *s-cis* conformation).

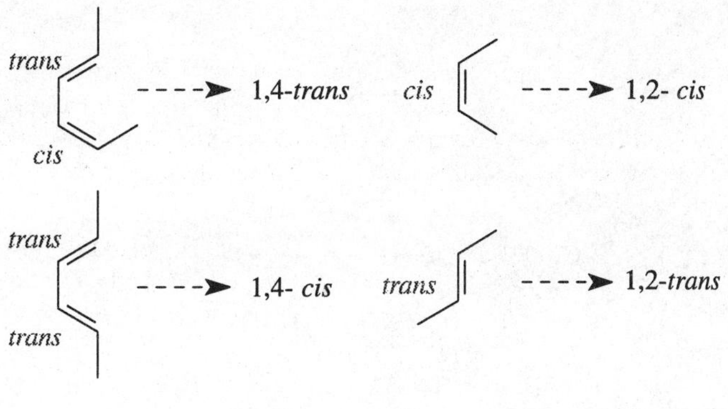

Conjugated Dienes: *Cycloaddition Reactions*

Give the major organic product for each of the following (4+2) cycloaddition reactions.
Where appropriate, you should draw the products in their most stable conformation.

Conjugated Dienes: *Diels-Alder Reactions - Synthesis*

Using a 4 + 2 cycloaddition reaction, suggest a synthesis for each of the molecules shown below

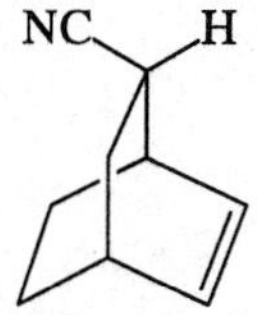

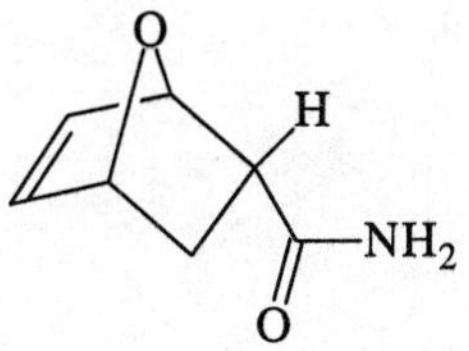

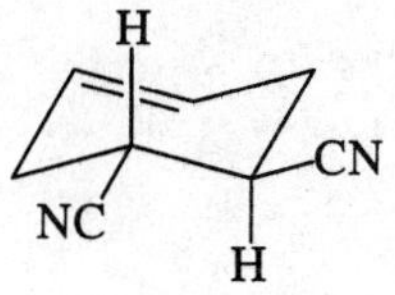

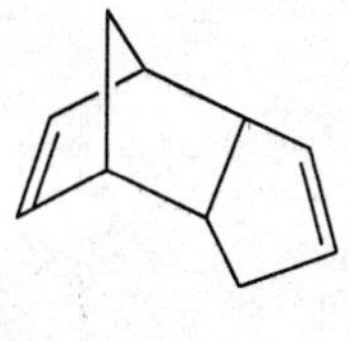

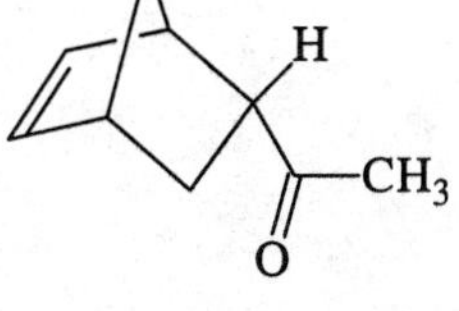

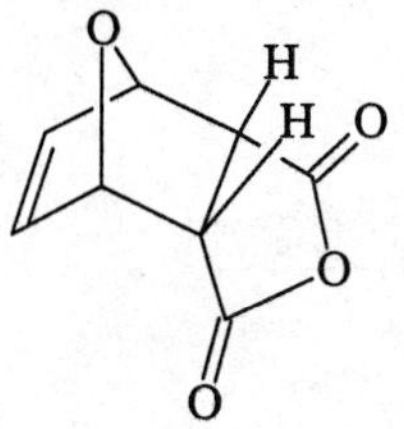

Conjugated Dienes: *Diels-Alder Reactions - Synthesis II*

Using a 4 + 2 cycloaddition reaction, suggest a synthesis for each of the molecules shown below, paying very careful attention to stereochemistry.

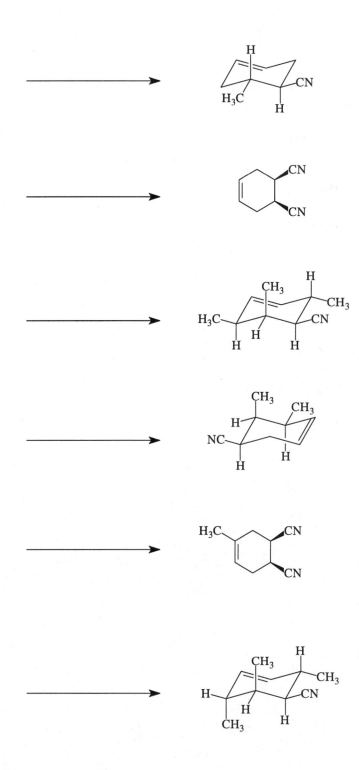

Conjugated Dienes: *Diels-Alder Reactions - Synthesis III*

Beginning with the diene shown on the left, and using at least one 4 + 2 cycloaddition reaction, suggest a synthesis for each of the molecules shown below. Two or more synthetic steps are likely to be required

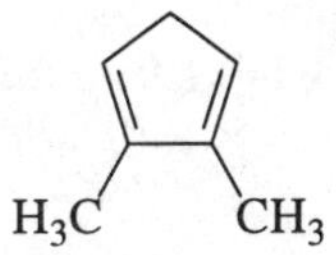

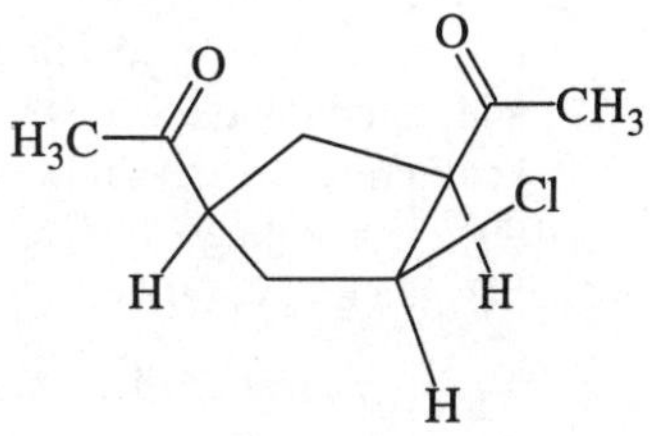

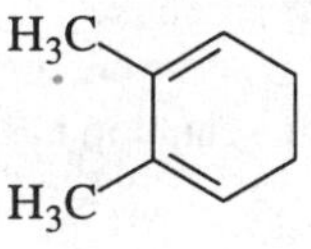

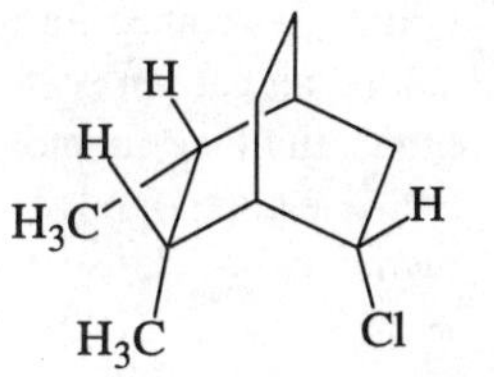

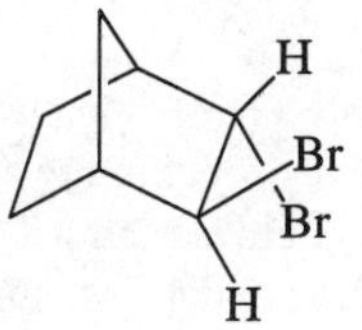

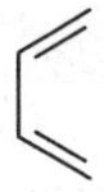

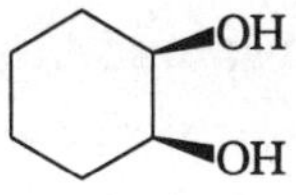

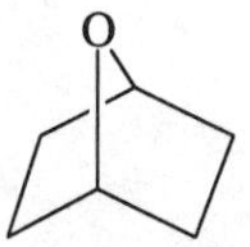

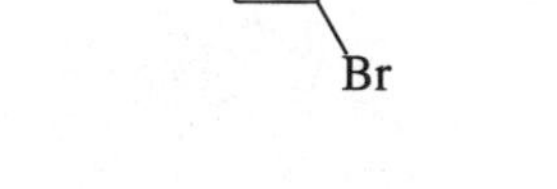

Arenes:

Aromatic Systems and the 4n+2 Rule

Benzene, having the molecular formula C_6H_6, would be consistent with a structure such as cyclohexatriene, having three conjugated double bonds in a six-membered ring. The compound is, however, far more stable than would be predicted for a triene, based on the heat of hydrogenation (the energy evolved when one mole of compound is reacted with H_2 in the presence of Pt or Pd catalyst). Within a sequence of simple cycloalkenes, the inclusion of additional double bonds is generally associated with an increase in the heat of hydrogenation of approximately 25 kcal/mole; benzene, however, has a heat of hydrogenation which is less than that of cyclohexadiene. Further, benzene does not undergo "typical" alkene reactions; it will not react with Br_2 to form a dibromide, nor will it react with halogen acids (i.e., HCl) to give alkyl halides.

The rationalization for the unusual reactivity of benzene which is generally accepted today is that the conjugated π-system forms a continuous molecular orbital above and below the plane of the ring, and that this planar, continuous π-system containing six electrons has unusual stability. The delocalization of the electrons is typically shown by writing resonance forms in which the double bonds in benzene compounds can be shown to be distributed equally among all carbon centers (shown below for dibromobenzene). Remember that resonance forms represent structural limits and that the molecule is "never" one form or the other, but is a hybrid of both. To show this, the bonding in benzene compounds is often written as a circle within the ring, although this type of structural representation has its own drawbacks, as we will see when we consider substitution reactions.

The resonance description of benzene explains the geometry of the molecule but does not explain the unusual stability of benzene and its derivatives. The stability of benzene is suggested to arise from the fact that the conjugated π-system is **planar** and **contains 4n + 2 electrons** (with $n = 1$), and it is suggested that all compounds having planar, conjugated 1 systems containing $4n + 2$ electrons will share this stability. This property, described originally by Hückel, is referred to as **aromaticity**. Aromaticity, and unusual stability, will therefore be associated with molecules having 6 ($n = 1$), 10 ($n = 2$), 14 ($n = 3$), 18 ($n = 4$), etc., electrons. Unshared pairs of electrons on heteroatoms within the ring can also be counted to achieve aromaticity.

Electrophilic Aromatic Substitution

One of the most important reactions of arenes is electrophilic aromatic substitution, in which an electrophile reacts with the ring, forming a new bond to a ring carbon with the loss of one hydrogen. In general, these reactions require a **Lewis acid catalyst**, as shown below for the reaction of bromine with benzene, catalyzed by $FeBr_3$. The role of the $FeBr_3$ is to complex the bromine to form a bromonium cation-like species (which is often simply referred to as Br^+) which is the actual electrophilic agent.

This electrophile first forms a loose complex with the π-cloud, which rearranges to a cationic sigma-complex, in which the electrophile is directly bonded to a ring carbon. Since the ring is a conjugated system, the cationic charge which forms on the adjacent carbon is delocalized over the ring, with partial positive charge developing on the carbons which are *ortho-* and *para-* to the position where the electrophile bonded. Loss of H^+ from the sigma-complex regenerates the aromatic π-system (with its associated stability), and gives bromobenzene and HBr as the final products.

...the driving force for loss of H⁺ is the restoration of the aromatic π system.

Chlorination proceeds by a similar mechanism; for iodination, $I_2/CuCl_2$ is typically utilized to generate the electrophilic I⁺ cation. Aryl fluorides cannot be prepared by this direct method, but can be prepared using thallium trifluoroacetate to form an intermediate aryl-thallium compound, which then reacts with fluoride anion to give the desired product.

Arenes can also be nitrated by a similar mechanism using a mixture of nitric and sulfuric acids to generate the electrophile NO_2^+, which adds to the ring to form a sigma complex, and looses a proton to give the nitro compound. Fuming sulfuric acid (H_2SO_4 saturated with SO_3) to generate SO_3H^+, which is an electrophilic agent, yielding the aryl sulfonic acid as the product.

Perhaps the most notable (and useful) example of electrophilic aromatic substitution is the introduction of alkyl groups using the **Friedel-Crafts** reaction. In this reaction, a Lewis acid complexes with an alkyl halide to give a species with electrophilic character on the carbon of the alkyl halide. This then reacts by the standard mechanism to give an intermediate sigma-complex, and the alkylated benzene as the final product.

There are several limitations of the Friedel-Crafts alkylation reaction; summarizing, only alkyl halides can be utilized (not aryl- or vinyl halides); the ring must be activated, since the electrophile is generally less reactive than those encountered previously; multiple substitutions are possible, and perhaps most important, since the carbon of the alkyl halide has carbocation character, rearrangements often occur. In general this means that an alkyl halide such as 1-bromopropane is not suitable in this reaction, since it would be prone to rearrange to the more stable isopropyl carbocation.

A derivative of the Friedel-Crafts alkylation is the **Friedel-Crafts acylation** reaction in which the arene is converted to an aryl ketone. The electrophile in this reaction is an acylium ion-like species which is formed by reaction of an acid halide, or an acid anhydride, with the Lewis acid. Unlike the alkylation reaction, rearrangements do not occur (the acylium cation is a very stable, resonance-stabilized carbocation), although an activated ring is still required.

Nucleophilic Aromatic Substitution

Arenes having strongly electron-withdrawing groups on the ring, and at least one potential leaving group (typically a halogen) can undergo substitution to strong nucleophiles, as shown below for 1-fluoro-2,4-dinitrobenzene.

This substitution, however, cannot be occurring by a simple S_N1 or S_N2 mechanisms since the pathway for displacement in the S_N2 mechanism is blocked by the ring, and the S_N1 mechanism would involve the generation of an unstable aryl cation on a ring which is already extremely electron deficient. The reaction, instead, involves an **addition-elimination pathway**, in which the nucleophile first adds to the ring at the carbon bearing the leaving group to form an anionic complex. The electrons from the nucleophile are generally delocalized by the other electron-withdrawing groups on the ring, and then are utilized to allow the halogen to leave as an anion, restoring the stable aromatic system.

...addition

...elimination

Nitro-groups are especially good at promoting this type of reaction, since the intermediate anionic charge can be delocalized onto the oxygens of the nitro group as relatively stable oxyanions.

Reactions of Aryl Side-Chains

Arenes having an alkyl side-chain with at least one benzylic hydrogen will undergo oxidation in the presence of neutral MnO_4^- anion to give the corresponding benzoic acid. Note that in the example given on the following page, the same product (benzoic acid) is produced by all three reactions, with the remaining carbons appearing as secondary oxidation products. As with all reactions involving MnO_4^-, the reaction involves radical intermediates and side reactions are common.

[Reaction scheme: toluene (PhCH₃), propylbenzene (PhCH₂CH₂CH₃), and cyclopentylbenzene each react with MnO₄⁻/H₂O, heat to give benzoic acid (PhCOOH).]

Alkyl-substituted arylsulfonic acids undergo a somewhat brutal reaction know as "alkali fusion" in which the sulfonic acid residue is replaced by a hydroxyl group, yielding a substituted phenol. Because of the extreme reaction conditions, the reaction is limited to simple compounds, but is a useful pathway to forming phenols.

Since benzyl radicals are quite stable (being resonance-stabilized by the adjacent ring), free radical bromination occurs quite rapidly on alkyl benzenes having at least one benzylic hydrogen. The reaction conditions employed often utilize NBS (N-bromosuccinimide) in CCl_4 in the presence of a "radical initiator" to generate the bromine radical.

Since arenes are resistant to catalytic reduction, alkene side-chains can be specifically reduced to the alkane without reducing the ring. If you want to reduce the ring, high temperatures and pressure are required when standard catalysts are utilized (Pt or Pd), although Rh will catalyze the reduction under very mild conditions.

Catalytic reduction will also reduce aryl nitro groups to the corresponding amine (or, specific reduction can be accomplished using acidic $SnCl_2$). Likewise, aryl ketones are smoothly reduced catalytically to give the corresponding alkane. This latter reaction is quite often useful since an alkyl chain which would be prone to rearrangement in a Friedel-Crafts alkylation can be introduced using an acylation, and then simple reduced to the alkane.

Benzene Derivatives: *Nomenclature*

Give the proper IUPAC name for each of the compounds shown below.

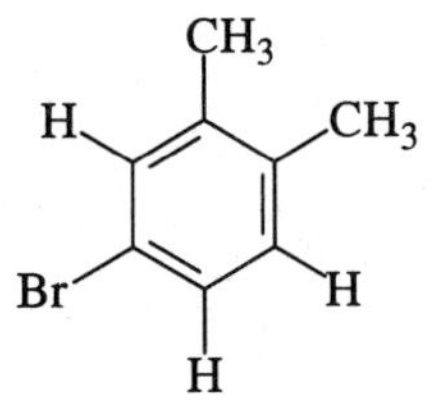

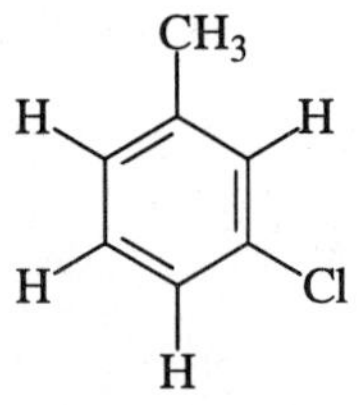

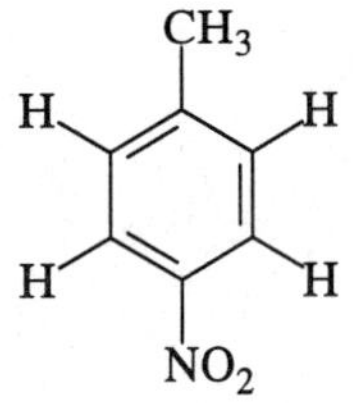

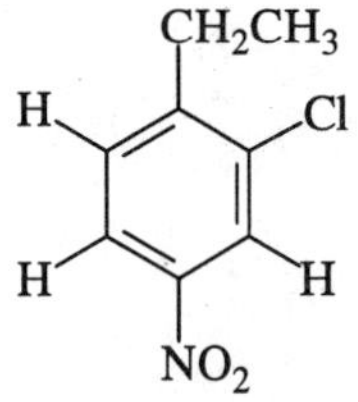

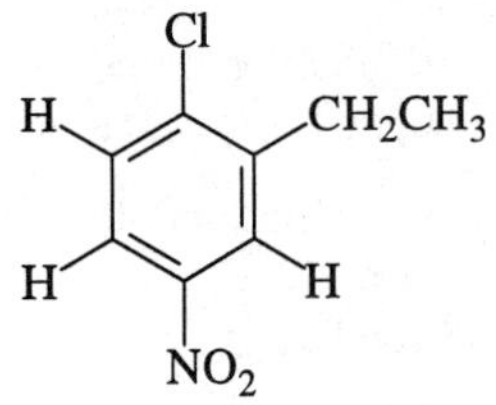

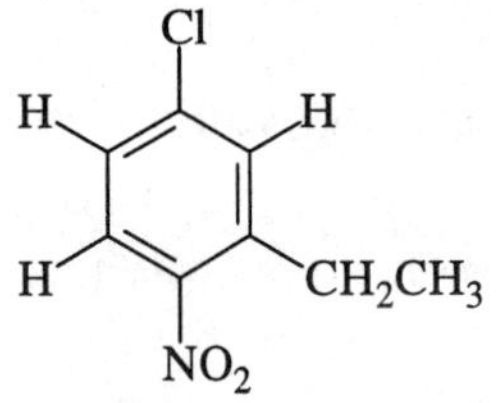

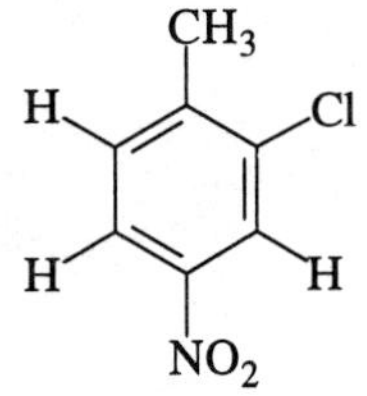

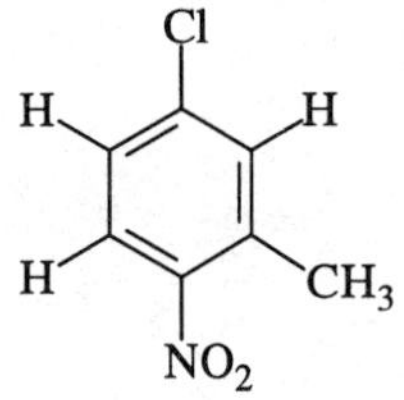

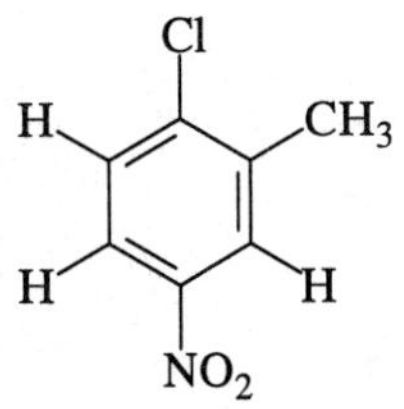

Benzene Derivatives: *Nomenclature*

Draw the structure corresponding to each of the following IUPAC names.

1-bromo-2-chlorobenzene

1-ethyl-2,4-dinitrobenzene

para-chlorotoluene

1,3,5-tribromobenzene

1-bromo-3,5-dimethylbenzene

ortho-bromopropylbenzene

1-fluoro-2,4-dinitrobenzene

2-methyl-6-phenylheptane

2,3,5-trinitrophenol

(1-bromoethyl)benzene

para-(bromomethyl)-toluene

Arenes: *Aromaticity*

Determine if each of the compounds shown below are expected to be aromatic, based on the Hükel criteria.

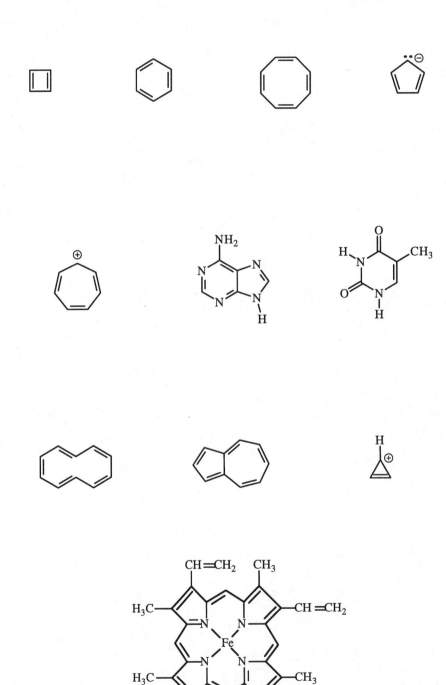

An Overview of Electrophilic Aromatic Substitution

Reagents	Product	Reaction
$Br_2/FeBr_3$	PhBr	Bromination
$Cl_2/FeCl_3$	PhCl	Chlorination
$I_2/CuCl_2$	PhI	Iodination
HNO_3/H_2SO_4	$PhNO_2$	Nitration
SO_3/H_2SO_4	$PhSO_3H$	Sulfonation
$RCl/AlCl_3$	PhR	Friedel-Crafts Alkylation
R−C(=O)−Cl, $AlCl_3$	PhC(=O)R	Friedel-Crafts Acylation
R−C(=O)−O−C(=O)−R, $AlCl_3$	PhC(=O)R	Friedel-Crafts Acylation
H_2/Rh;C	cyclohexane	Catalytic Hydrogenation
Li, NH_3	1,4-cyclohexadiene	Birch Reduction

2,4-dinitrochlorobenzene + Nucleophile: i.e., RSH → 2,4-dinitro-SR-benzene — Nucleophilic Aromatic Substitution

PhBr + NH_2^-/NH_3 (via benzyne intermediate) → $PhNH_2$ — Addition to Benzyne Intermediate

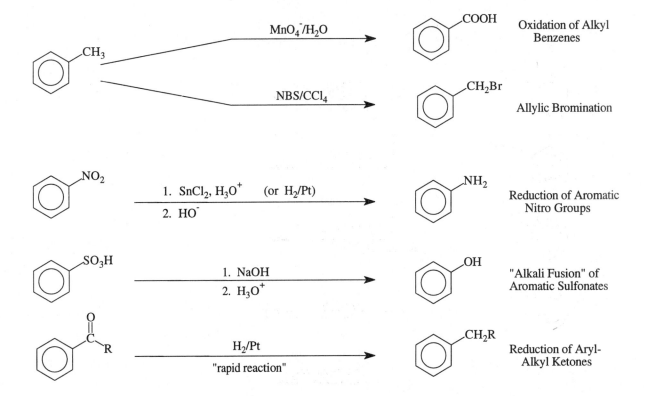

Benzene Derivatives: *Electrophilic Aromatic Substitution*

Give the major product for each of the following electrophilic aromatic substitution reactions.

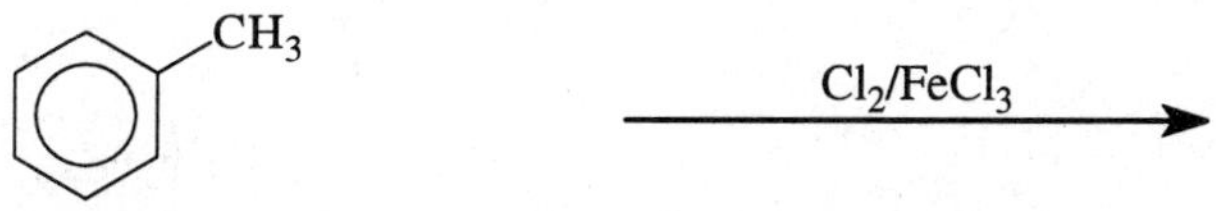

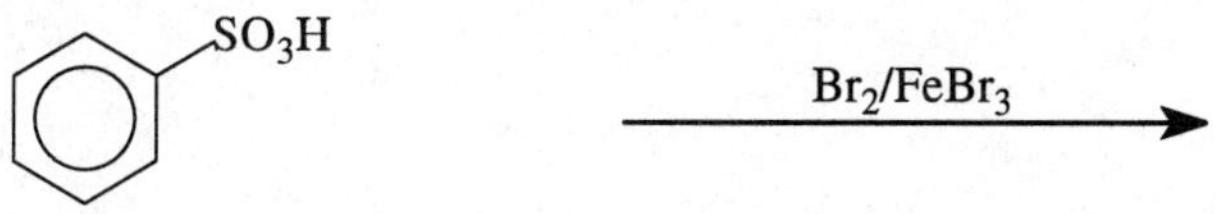

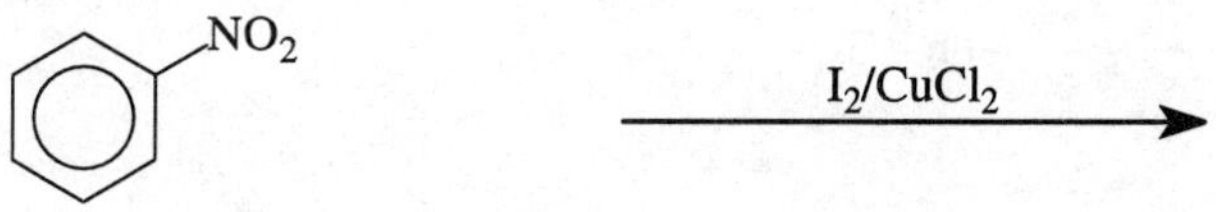

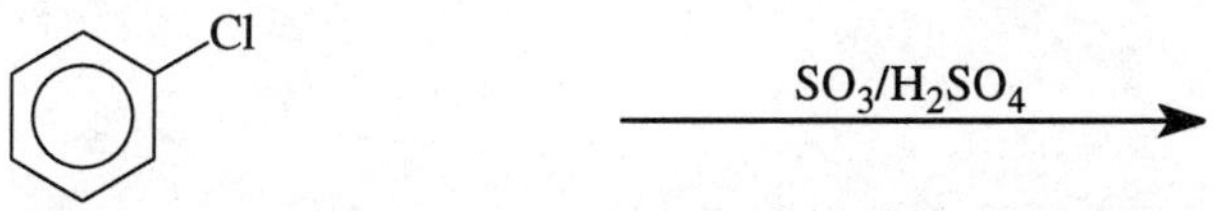

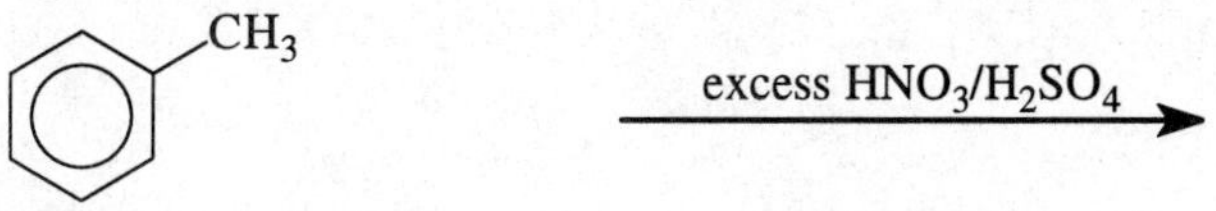

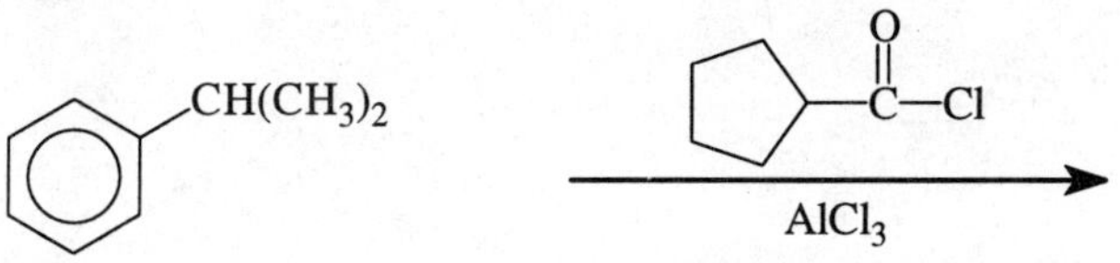

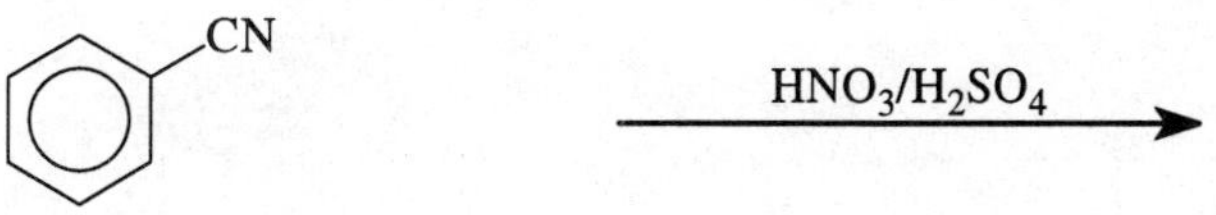

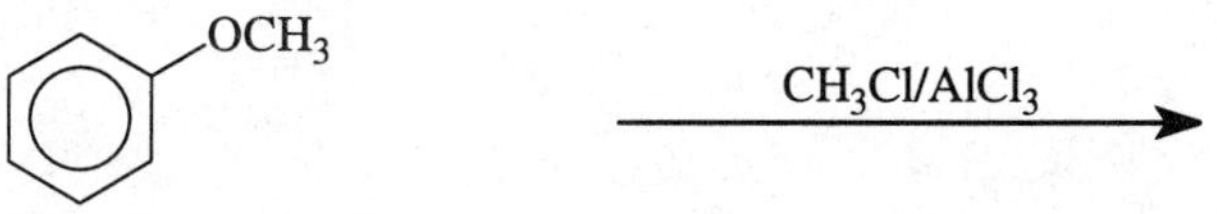

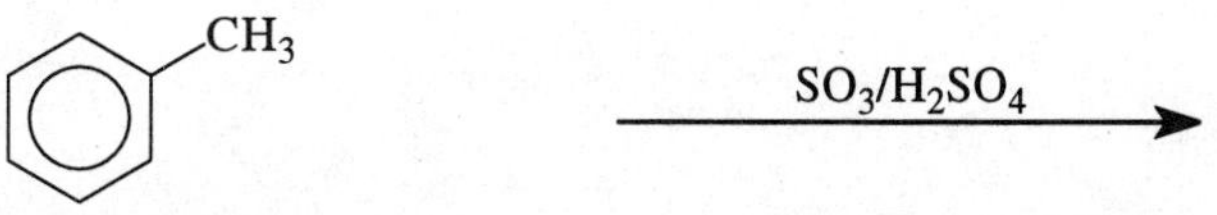

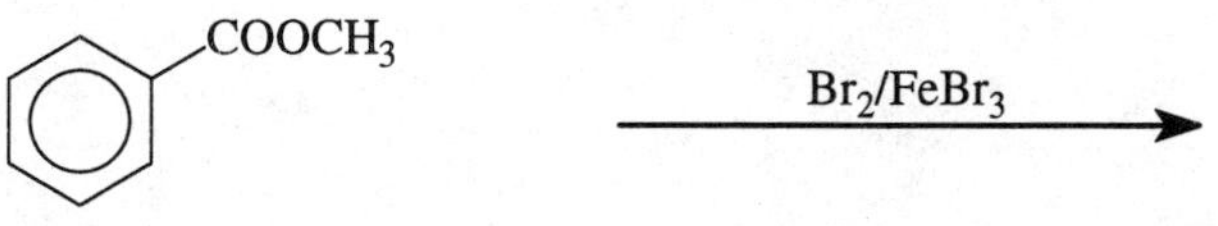

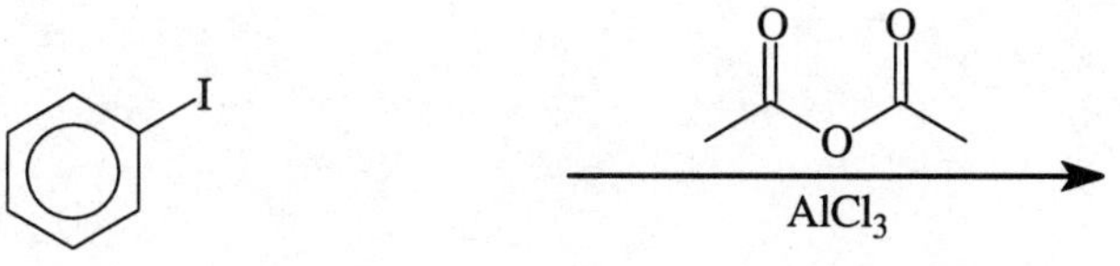

PhCH₃ + Cl₂/FeCl₃ →

PhSO₃H + Br₂/FeBr₃ →

PhNO₂ + I₂/CuCl₂ →

PhCl + SO₃/H₂SO₄ →

PhCH₃ + excess HNO₃/H₂SO₄ →

PhCH(CH₃)₂ + cyclopentyl-C(O)Cl / AlCl₃ →

PhCN + HNO₃/H₂SO₄ →

PhOCH₃ + CH₃Cl/AlCl₃ →

PhCH₃ + SO₃/H₂SO₄ →

PhCOOCH₃ + Br₂/FeBr₃ →

PhI + (CH₃CO)₂O / AlCl₃ →

Benzene Derivatives: *"Other Reactions"*

Give the major products for each of the following reactions.

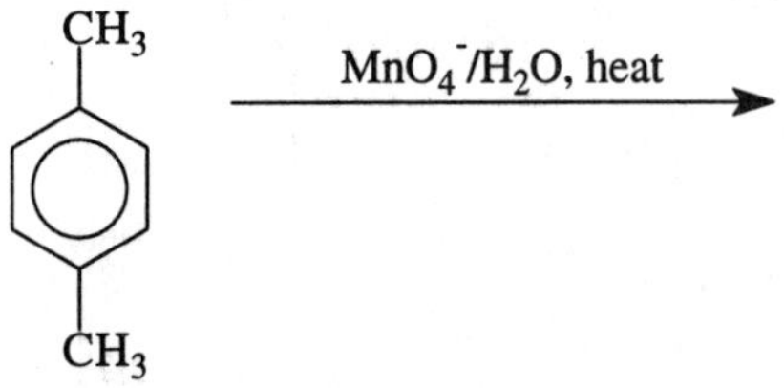

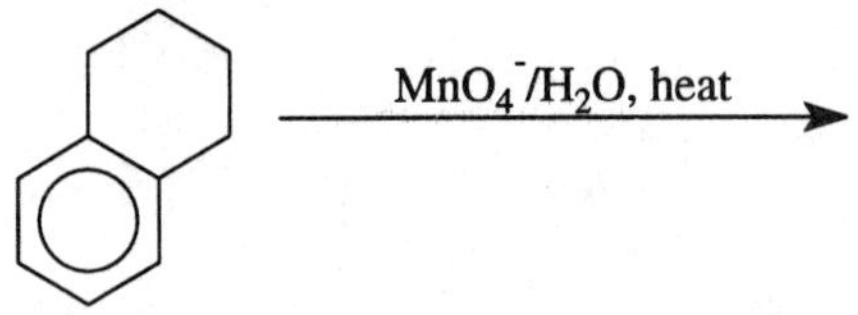

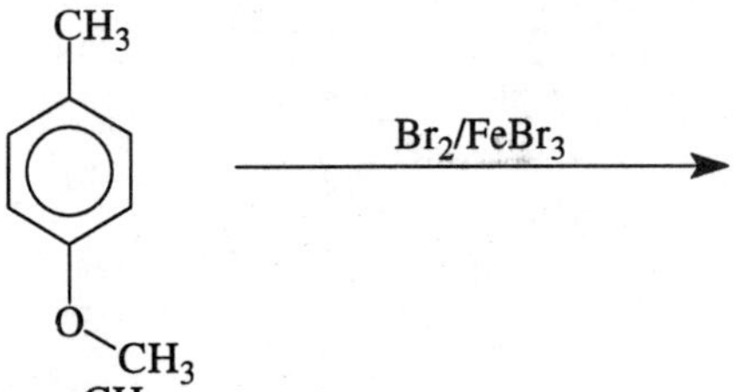

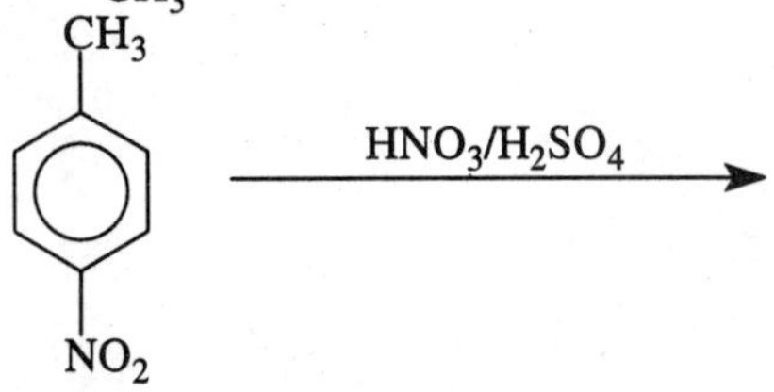

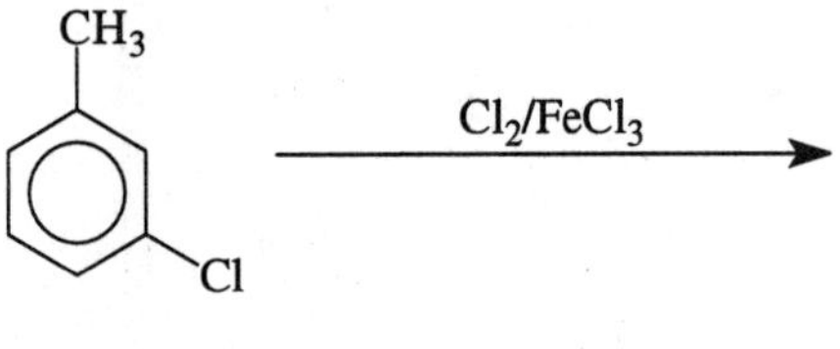

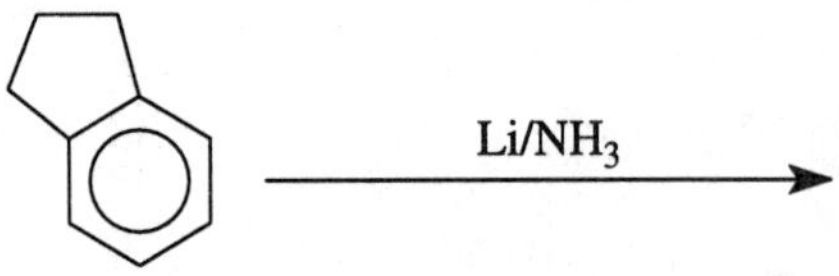

1,4-dimethylbenzene (p-xylene) $\xrightarrow{\text{MnO}_4^-/\text{H}_2\text{O, heat}}$

1,2,3,4-tetrahydronaphthalene $\xrightarrow{\text{MnO}_4^-/\text{H}_2\text{O, heat}}$

4-methylanisole (with CH$_3$ para to OCH$_3$) $\xrightarrow{\text{Br}_2/\text{FeBr}_3}$

4-nitrotoluene $\xrightarrow{\text{HNO}_3/\text{H}_2\text{SO}_4}$

3-chlorotoluene $\xrightarrow{\text{Cl}_2/\text{FeCl}_3}$

indane $\xrightarrow[\text{"radical initiator"}]{\text{NBS/CCl}_4}$

indane $\xrightarrow{\text{Li/NH}_3}$

Benzene Derivatives: *Synthesis*

Beginning with the compound shown on the left, suggest a synthesis for each of the compounds shown below. Clearly show all intermediates and reaction conditions. Several steps are required for each synthetic conversion.

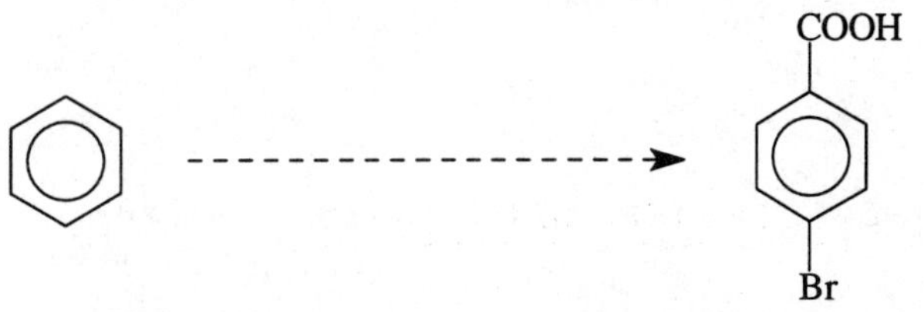

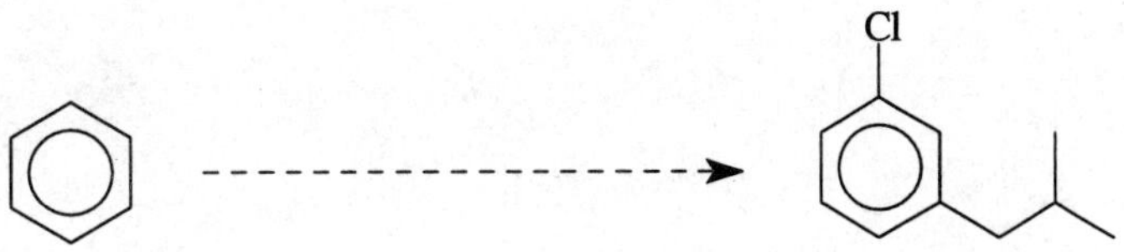

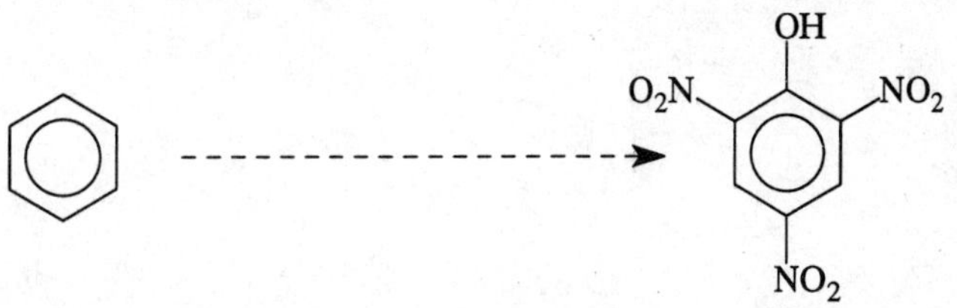

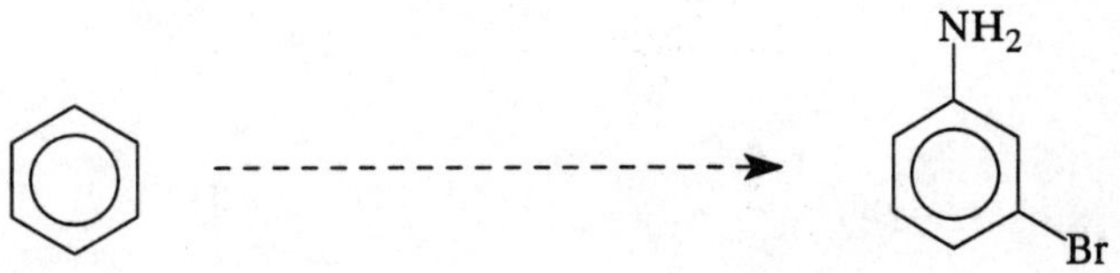

benzene ⇢ 4-bromobenzoic acid (COOH para to Br)

benzene ⇢ 3-chloro-isobutylbenzene (Cl meta to isobutyl group)

benzene ⇢ 2,4,6-trinitrophenol (OH with NO$_2$ at 2, 4, 6 positions)

benzene ⇢ 3-bromoaniline (NH$_2$ meta to Br)

Benzene Derivatives: *Synthesis, continued*

Beginning with the compound shown on the left, suggest a synthesis for each of the compounds shown below. Clearly show all intermediates and reaction conditions. Several steps are required for each synthetic conversion.

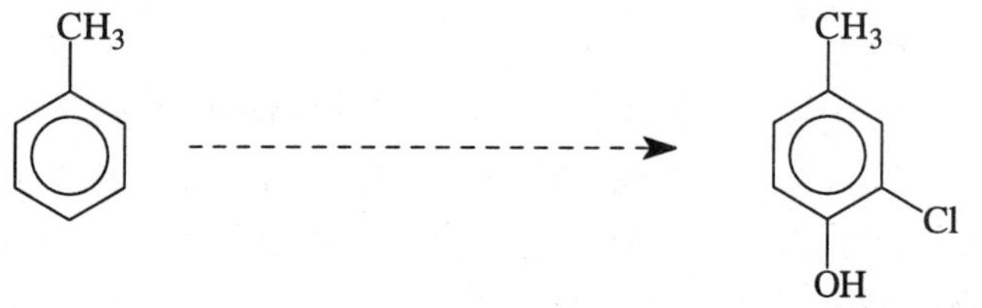

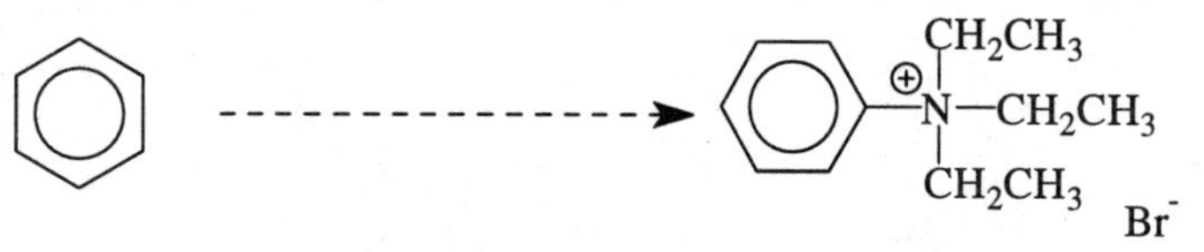

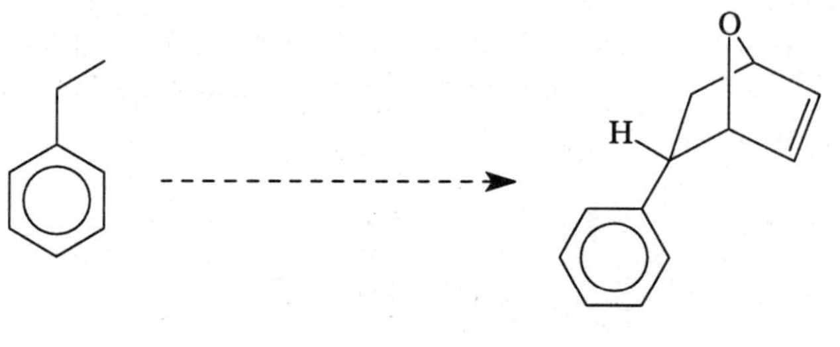

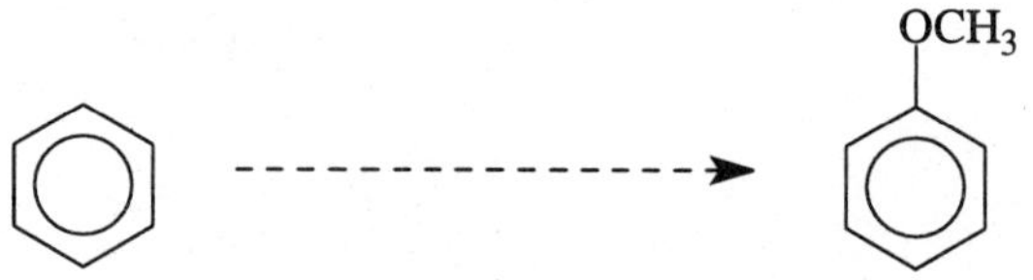

Alcohols, Ethers & Phenols

Nomenclature

Simple alcohols are named as derivatives of the parent alkane, using the suffix **-ol**, using the following simple rules:

1. Select the longest continuous carbon chain, containing the hydroxyl group, and derive the parent name by replacing the **-e** ending with **-ol**.

2. Number the carbon chain, beginning at the end nearest to the hydroxyl group.

3. Number the substituents and write the name, listing substituents alphabetically.

Simple ethers are named either by identifying the two organic residues and adding the word ether, or, if other functionality is present, the ether residue is named as an alkoxy substituent.

Reactions that Yield Alcohols

In previous chapters, we have encountered a number of reactions which yield alcohols; these are, briefly:

Hydration of Alkenes: Simple acid-catalyzed hydration of alkenes is a stepwise reaction involving a carbocation intermediate. Rearrangements will often occur and hydroxide anion will bond to the most stable carbocation center in the molecule (Markovnikov's Rule).

Oxymercuration of Alkenes: Oxymercuration of alkenes is a stepwise reaction involving a bridged mercurinium ion intermediate. In unsymmetrical alkenes, the alkene carbon which would form the most stable carbocation will bear more of the positive charge and will be attacked by hydroxide anion (or water) to give the addition intermediate; rearrangements do not occur, but the orientation follows Markovnikov's Rule. In a second step, BH_4^- is used to remove the mercury and give the final product.

Hydroboration of an Alkene, with Oxidative Work-up: Reaction of an alkene with BH_3 results in the *syn*-addition of the boron and a hydrogen across the double bond. Rearrangements do not occur and the hydrogen will bond to the carbon of the alkene which would form the most stable carbocation center (overall anti-Markovnikov's addition). The driving force for the regiochemistry may actually be more steric than electronic, but viewing the reaction as a concerted, but polar transition state, easily rationalizes the observed product distribution.

Formation of 1,2-Diols from Alkenes: Reaction of an alkene with MnO_4^- or OsO_4 results in the formation of a *cis*-1,2-diol. The reaction involves the formation of an intermediate cyclic ester, which decomposes to give the diol. For OsO_4, reaction with HSO_3^- is necessary to decompose the intermediate ester and form the final product.

Preparation of Alcohols from Aldehydes and Ketones:

Reduction Reactions: Reduction of simple aldehydes and ketones with BH_4^- yields the corresponding alcohol directly. The reaction works well for simple compounds, but reaction of BH_4^- with α,β–unsaturated aldehydes and ketones can result in significant reduction of the double bond.

A much more powerful reductant is $LiAlH_4$, which will reduce aldehydes, ketones, esters, carboxylic acids and nitriles. Some sample reactions are shown below:

[Reaction schemes showing:
1. CH₃O-CO-CH=CH-CH₃ → HO-CH₂-CH=CH-CH₃ (1. LiAlH₄, ether; 2. H₃O⁺)
2. Camphor → borneol (1. LiAlH₄, ether; 2. H₃O⁺)
3. PhCH₂COOH → PhCH₂CH₂OH (1. LiAlH₄, ether; 2. H₃O⁺)]

As seen in the first example, the reduction of carboxylate esters results in the addition of two moles of hydride to the carbonyl carbon, with loss of the alcohol portion of the ester, forming the corresponding primary alcohol.

Addition of Grignard Reagents to Carbonyl Compounds: The reaction of an alkyl, aryl or vinyl halide with magnesium metal in ether solvent, produces an organometallic complex of uncertain structure, but which behaves as if it has the structure R-Mg-X and is commonly referred to as a **Grignard Reagent**.

$$R\text{—}X + Mg \xrightarrow{\text{ether}} \overset{\delta-\ \ \delta+}{R\text{—}MgX}$$

R = 1°, 2°, or 3° alkyl, aryl or vinyl

X = Cl, Br or I

[Mechanism diagram showing Grignard addition to ketone forming alkoxide with MgX⁺ counterion]

The "R" group in this complex (alkyl, aryl or vinyl), acts as if it was a stabilized carbanion and Grignard reagents react with water and other compounds containing acidic hydrogens to give hydrocarbons (just as would be expected for a well-behaved, highly basic carbanion). In the absence of acidic hydrogens, the Grignard reagent can function as a powerful nucleophile, and is most often used in addition reactions involving carbonyl compounds, as shown above. The product of these addition reactions is typically a secondary or tertiary alcohol (primary alcohols can be formed by reaction with formaldehyde).

Carboxylate esters also react with Grignard reagents, undergoing the addition of **two moles** of reagent to give the final product. The reason for this is clearly seen in the example below, where the product of the first mole of addition is the simple carbonyl compound, which rapidly adds a second mole of reagent.

[Mechanism diagram showing ester + R-MgX → tetrahedral intermediate → ketone + R-MgX → tertiary alkoxide]

Strained ethers also react with Grignard reagents to give alcohols, as shown below for reaction with epoxides; the regiochemistry of the reaction with epoxides is generally dictated by steric factors.

Reactions of Alcohols

Conversion to Alkyl Chlorides by Reaction with HCl: Tertiary alcohols, or alcohols which can lose the hydroxyl group to form a stable carbocation, can undergo an S_N1 substitution reaction with HCl gas dissolved in ether to give the corresponding alkyl chloride. Again, the reaction is limited to alcohols that can from stable carbocations.

Conversion to Alkyl Bromides by Reaction with PBr₃: Primary and secondary alcohols react with PBr_3 to form an intermediate phosphite ester which undergoes S_N2 attack by bromide anion to yield the alkyl bromide with inversion of configuration (the stereochemical inversion is simply a result of the S_N2 displacement).

Conversion to Alkyl Chlorides by Reaction with SOCl₂: Primary and secondary alcohols react with $SOCl_2$ in polar solvents (i.e., pyridine) to form an intermediate sulfite ester which undergoes S_N2 attack by chloride anion to yield the alkyl chloride with inversion of configuration (the stereochemical inversion is simply a result of the S_N2 displacement). If the reaction is performed in a non-polar solvent such as benzene, an unusual S_Ni mechanism occurs involving frontside attack, and yielding retention of stereochemistry. This reaction is unusual, but is often useful if you desire to control the stereochemical course of a synthesis.

Dehydration of Tertiary Alcohols: Tertiary alcohols, or alcohols which can lose the hydroxyl group to form a stable carbocation, can undergo an acid-catalyzed E1 elimination reaction to form the corresponding alkene. Again, the reaction is limited to alcohols that can from stable carbocations.

Dehydration of Secondary and Tertiary Alcohols with POCl₃: Secondary and tertiary alcohols react with $POCl_3$ to form a dichlorophosphate ester, which undergoes an E2 elimination reaction to form the corresponding alkene. Since an E2 elimination is occurring, the hydrogen abstracted must be anti- and coplanar with the oxygen on the leaving group (antarafacial).

Oxidation of Alcohols with Pyridinium Chlorochromate: Primary and secondary alcohols are smoothly oxidized by pyridinium chlorochromate (PCC) in CH_2Cl_2 to form aldehydes and ketones, respectively. The PCC oxidation of primary alcohols to give aldehydes is a very useful reaction, since aldehydes are difficult to prepare and are easily over-oxidized to the carboxylic acid. Primary and secondary alcohols are oxidized by CrO_3/H_2SO_4 (**Jones Reagent**) to form carboxylic acids and ketones, respectively; sodium dichromate in acetic acid ($Na_2Cr_2O_7$) can also be used.

Conversion to Silyl Ethers: Alcohols react with chlorotrimethylsilane to form trimethylsilyl ethers which are stable to many reactions which occur in aprotic medium, but can be readily cleaved by reaction with aqueous acid, regenerating the alcohol. This reaction is often utilized to "protect" an alcohol during a synthesis, such as that shown below (in the synthesis shown, the Grignard reagent would react with the acidic proton on the alcohol, destroying the reagent).

Ethers, Synthesis & Reactions

Synthesis of Ethers:

S_N2 Displacement Reactions: Unhindered primary and secondary alkyl halides react with simple (unhindered) alkoxides by an S_N2 mechanism yielding ethers (the Williamson Ether Synthesis).

Oxymercuration of Alkenes in the Presence of Alcohols: Oxymercuration of alkenes is a stepwise reaction involving a bridged mercurinium ion intermediate. In unsymmetrical alkenes, the alkene carbon which would form the most stable carbocation will bear more of the positive charge and, in alcohols, will be attacked by alkoxide anion (or the alcohol) to give the addition intermediate; rearrangements do not occur, but the orientation follows Markovnikov's Rule. In a second step, BH_4^- is used to remove the mercury and give the final product.

Formation of Epoxides by Oxidation of Alkenes: Alkenes undergo partial oxidation with peracids to form epoxides. A stable and useful reagent for this reaction is the magnesium salt of monoperoxyphthalate (MMPP).

Formation of Epoxides by an Internal S_N2 Reaction in Halohydrins: Halohydrins (prepared by the addition of HOX to an alkene) undergo an internal S_N2 reaction in the presence of strong base (NaOH) to give epoxides.

Reactions of Ethers

Cleavage of Ethers with HI: Ethers undergo cleavage in the presence of aqueous HI to give the corresponding alkyl iodide. Attack will be at the least hindered carbon and E1 reactions, with carbocation intermediates, are common with ethers with groups which can form stable carbocations.

Ring-Opening Reactions of Epoxides: The three-membered ring of epoxides is highly strained and undergoes ring-opening reactions with a variety of nucleophiles, as shown on the following pages; for reactions involving acid catalysis, the first step involves protonation of the ether oxygen to make it a better leaving group, followed by nucleophilic attack.

Phenols

Reactions of Phenols

Phenols, like simple alcohols, will form an anion which will undergo an S_N2 reaction with alkyl halides (or alkyl groups with "good leaving groups") to give ethers. They will also react with activated carbonyl compounds to undergo acyl transfer reactions; thus aryl esters are readily formed by the reaction of phenols with acid halides or acid anhydrides. As with aryl amines, the ring of phenols is electron-rich and will rapidly react with Br_2 to give a tri-substituted product.

The last reaction shown above is a specific and unusual reaction of phenols, that is the direct reaction with CO_2 to give carboxylation at the *ortho*-postion. This reaction is called the Kolbe-Schmitt carboxylation, and is important since the product is salicylic acid, which is widely used in pharmaceuticals.

Another reaction which is highly specific for phenols is oxidation with Fremy's salt: potassium nitrosodisulfonate, to give the *para*-quinone as product.

An Overview of Alcohol Reactions

RCH_2-OH
- $\xrightarrow{PCC,\ CH_2Cl_2}$ $R-\overset{\overset{O}{\|}}{C}-H$
- $\xrightarrow{CrO_3,\ H_2SO_4}$ $R-\overset{\overset{O}{\|}}{C}-OH$

$R\underset{\underset{R'}{|}}{CH}-OH$ $\xrightarrow{PCC/CH_2Cl_2 \text{ or } CrO_3,\ H_2SO_4}$ $R-\overset{\overset{O}{\|}}{C}-R'$

$R\underset{\underset{R'}{|}}{CH}-OH$ R' = alkyl, aryl or H

- $\xrightarrow{PBr_3}$ $R\underset{\underset{R'}{|}}{CH}-Br$ S_N2 mechanism, inversion of configuration
- $\xrightarrow{SOCl_2/pyridine}$ $R\underset{\underset{R'}{|}}{CH}-Cl$ S_N2 mechanism, inversion of configuration
- $\xrightarrow{SOCl_2/benzene}$ $R\underset{\underset{R'}{|}}{CH}-Cl$ S_Ni mechanism, retention of configuration
- $\xrightarrow{CH_3-C_6H_4-SO_2Cl}$ $R\underset{\underset{R'}{|}}{CH}-OTos$
- $\xrightarrow{(CH_3)_3SiCl}$ $R\underset{\underset{R'}{|}}{CH}-OSi(CH_3)_3$

$R\underset{\underset{R'}{|}}{\overset{\overset{R''}{|}}{C}}-OH$ R' = alkyl, aryl or H

- $\xrightarrow{HX}$ $R\underset{\underset{R'}{|}}{CH}-X$ 3° alcohols only, carbocation mechanism
- $\xrightarrow{H_3O^+}$ alkene   dehydration, tertiary alcohols only
- $\xrightarrow{POCl_3,\ pyridine}$ alkene   dehydration, 2° and 3° alcohols only

$R\underset{\underset{R'}{|}}{CH}-Br$ R' = alkyl, aryl or H

- $\xrightarrow{1.\ (H_2N)_2C=S;\ 2.\ H_2O,\ NaOH}$ $R\underset{\underset{R'}{|}}{CH}-SH$ S_N2 mechanism, inversion of configuration
- $\xrightarrow{HS^-}$ $R\underset{\underset{R'}{|}}{CH}-SH$ S_N2 mechanism, inversion of configuration

An Overview of Reactions that Yield Alcohols

Reactions that Yield Alcohols

Substrate	Reagents	Product	Notes
alkene	H^+/H_2O	alcohol (H, OH added)	carbocation rearrangements
alkene	1. BH_3, THF 2. H_2O_2, HO^-	alcohol (syn)	*anti*-Markovnikov regiochemistry Syn addition
alkene	1. OsO_4 2. $NaHSO_3$, H_2O or $KMnO_4$, HO^-/H_2O	diol (syn)	Syn addition
alkene	1. $Hg(OAc)_2$, H_2O 2. $NaBH_4$	alcohol	Markovnikov regiochemistry *trans* addition
$R\text{-}\underset{\underset{O}{\|\|}}{C}\text{-}H$	1. $NaBH_4$ or $LiAlH_4$ 2. H_3O^+	$RCH_2\text{-}OH$	
$R\text{-}\underset{\underset{O}{\|\|}}{C}\text{-}R$	1. $NaBH_4$ or $LiAlH_4$ 2. H_3O^+	$RCH(R')\text{-}OH$	
$R\text{-}\underset{\underset{O}{\|\|}}{C}\text{-}OR'$	1. $LiAlH_4$ 2. H_3O^+	$RCH_2\text{-}OH$	
$R\text{-}\underset{\underset{O}{\|\|}}{C}\text{-}OH$	1. $LiAlH_4$ 2. H_3O^+	$RCH_2\text{-}OH$	
$H\text{-}\underset{\underset{O}{\|\|}}{C}\text{-}H$	1. **R**MgX, ether 2. H_3O^+	$RCH_2\text{-}OH$	
$R\text{-}\underset{\underset{O}{\|\|}}{C}\text{-}H$	1. **R**MgX, ether 2. H_3O^+	$RCH(R)\text{-}OH$	
$R\text{-}\underset{\underset{O}{\|\|}}{C}\text{-}R'$	1. **R**MgX, ether 2. H_3O^+	$RC(R)(R')\text{-}OH$	
$R\text{-}\underset{\underset{O}{\|\|}}{C}\text{-}OR'$	1. **R**MgX, ether 2. H_3O^+	$RC(R)(R)\text{-}OH$	

Alcohols & Thiols: *Nomenclature*

Provide proper IUPAC names for each of the molecules shown below.

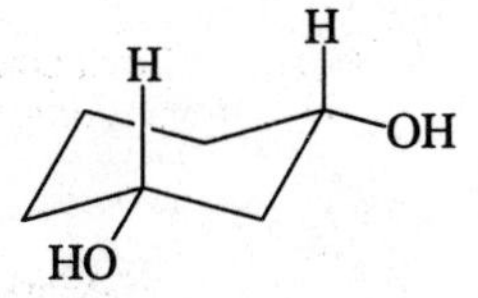

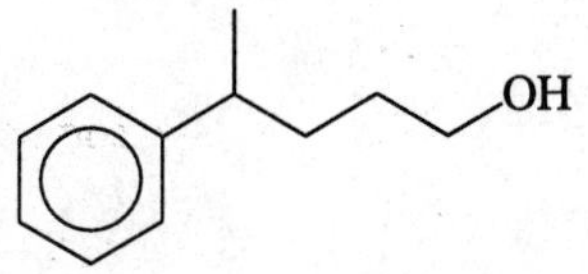

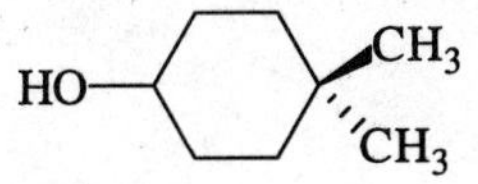

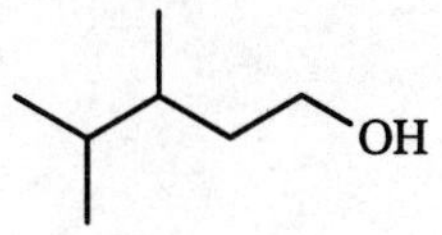

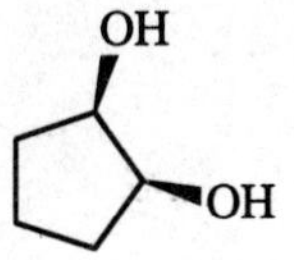

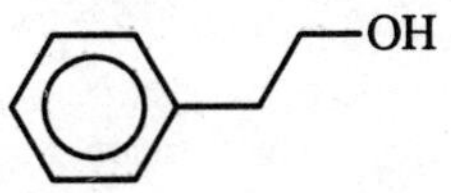

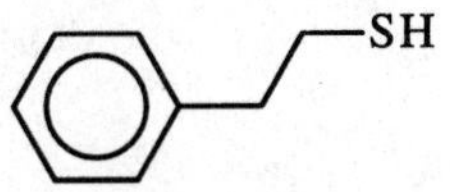

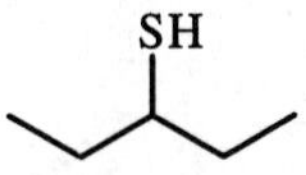

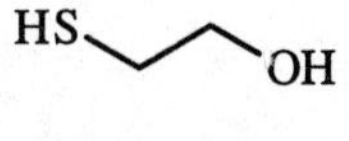

Alcohols: *Preparation of Alcohols by Reduction of Carbonyl Groups*

For each of the reactions shown below, draw the structure of the major organic product. For the synthesis problems, show the structure of the carbonyl compound necessary to product the desired product. Clearly show stereochemistry when appropriate.

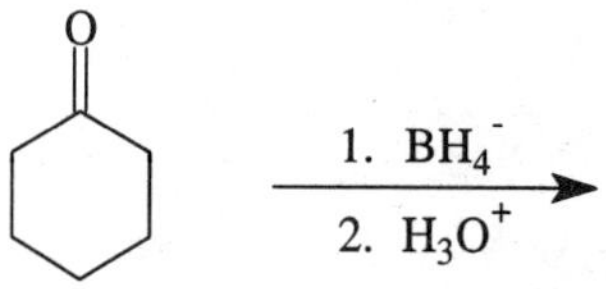

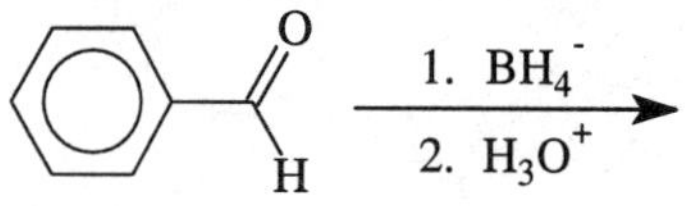

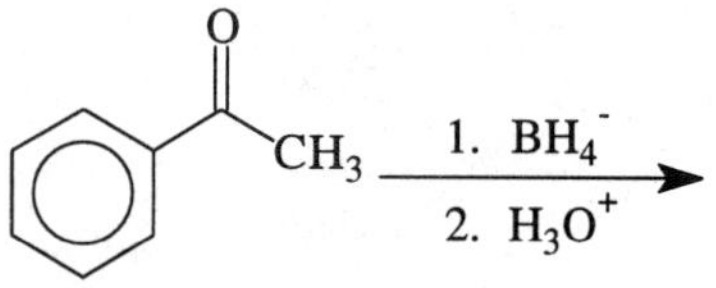

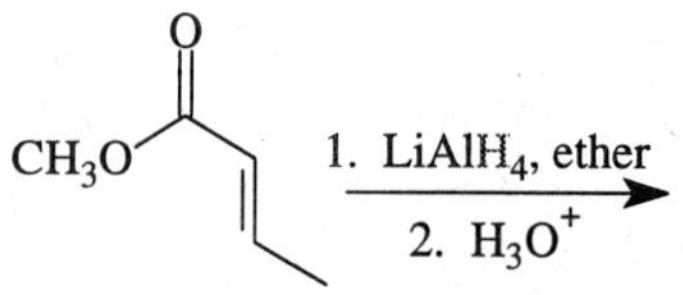

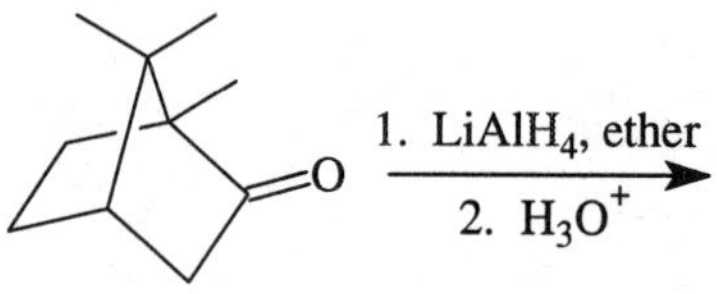

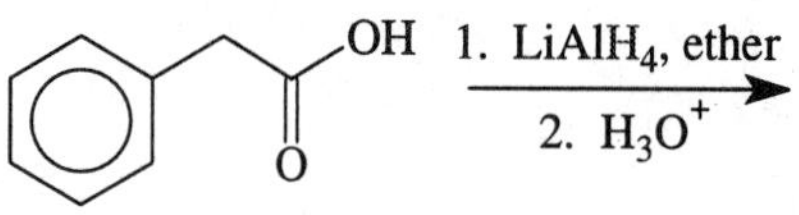

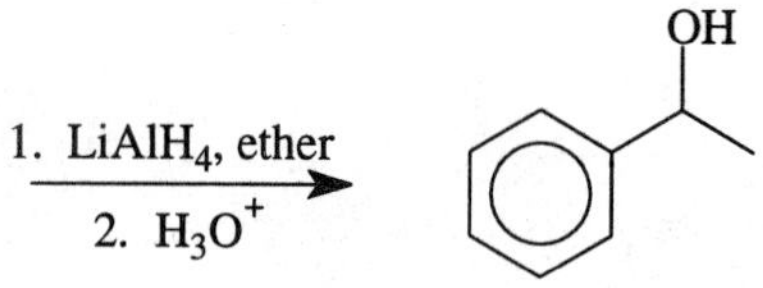

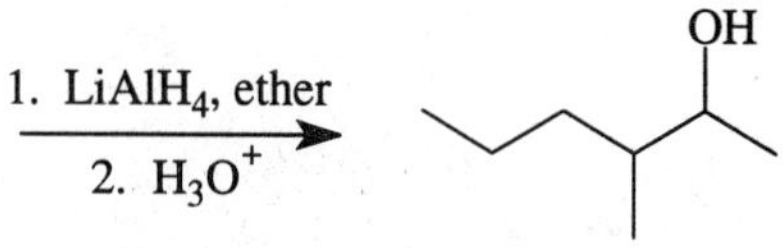

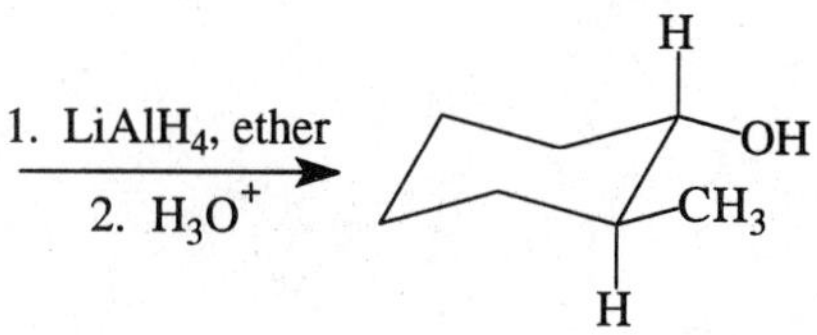

cyclohexanone $\xrightarrow{\text{1. BH}_4^-\;\; 2.\; H_3O^+}$

benzaldehyde $\xrightarrow{\text{1. BH}_4^-\;\; 2.\; H_3O^+}$

acetophenone $\xrightarrow{\text{1. BH}_4^-\;\; 2.\; H_3O^+}$

methyl crotonate (CH$_3$O-CO-CH=CH-CH$_3$) $\xrightarrow{\text{1. LiAlH}_4,\text{ ether}\;\; 2.\; H_3O^+}$

camphor $\xrightarrow{\text{1. LiAlH}_4,\text{ ether}\;\; 2.\; H_3O^+}$

phenylacetic acid $\xrightarrow{\text{1. LiAlH}_4,\text{ ether}\;\; 2.\; H_3O^+}$

$\xrightarrow{\text{1. LiAlH}_4,\text{ ether}\;\; 2.\; H_3O^+}$ 1-phenylethanol (PhCH(OH)CH$_3$)

$\xrightarrow{\text{1. LiAlH}_4,\text{ ether}\;\; 2.\; H_3O^+}$ 3-methyl-2-hexanol

$\xrightarrow{\text{1. LiAlH}_4,\text{ ether}\;\; 2.\; H_3O^+}$ trans-2-methylcyclohexanol (H, OH, CH$_3$, H shown on cyclohexane)

Alcohols: *Preparation by the Grignard Reaction*

For each of the reactions shown below, draw the structure of the major organic product. Clearly show stereochemistry when appropriate.

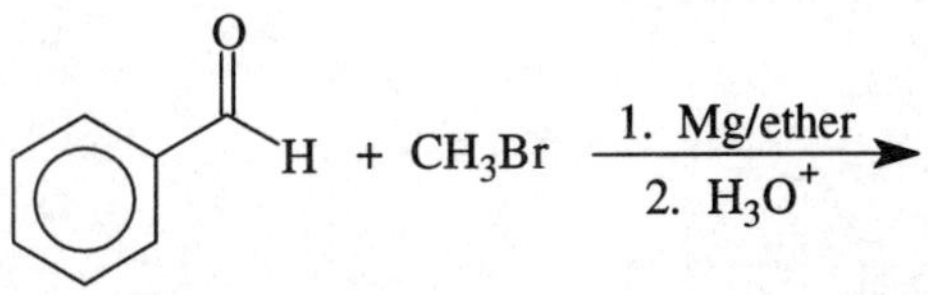

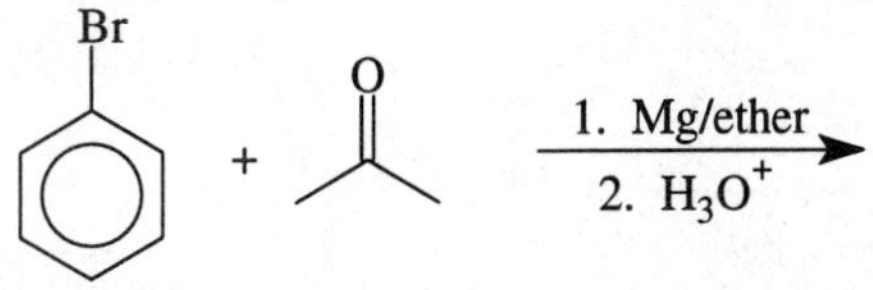

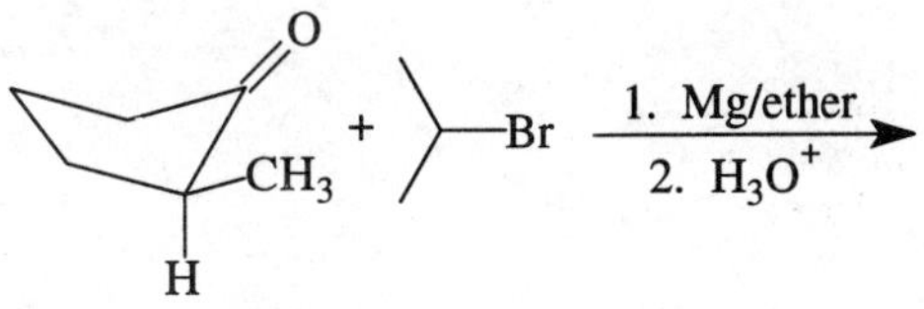

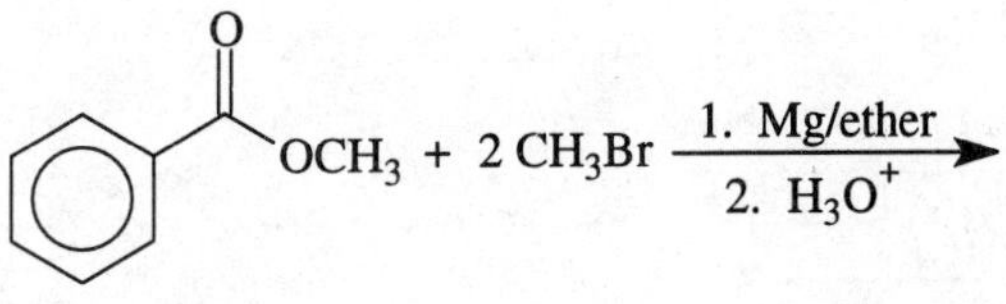

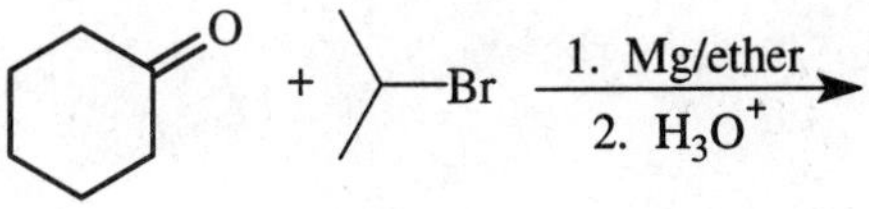

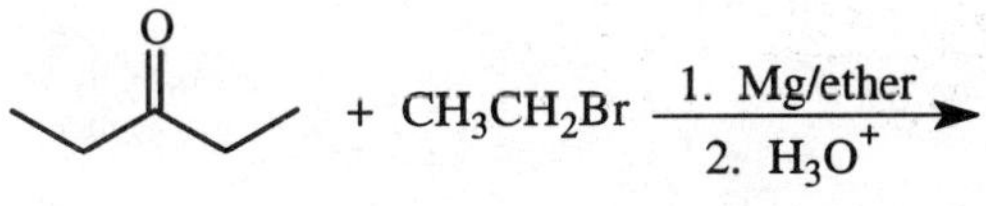

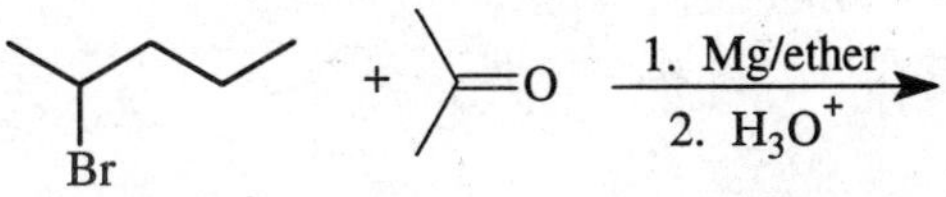

PhCHO + CH₃Br $\xrightarrow{\text{1. Mg/ether} \quad \text{2. H}_3\text{O}^+}$

PhBr + (CH₃)₂C=O $\xrightarrow{\text{1. Mg/ether} \quad \text{2. H}_3\text{O}^+}$

(2-methylcyclohexanone, with CH₃ and H shown) + (CH₃)₂CHBr $\xrightarrow{\text{1. Mg/ether} \quad \text{2. H}_3\text{O}^+}$

PhC(O)OCH₃ + 2 CH₃Br $\xrightarrow{\text{1. Mg/ether} \quad \text{2. H}_3\text{O}^+}$

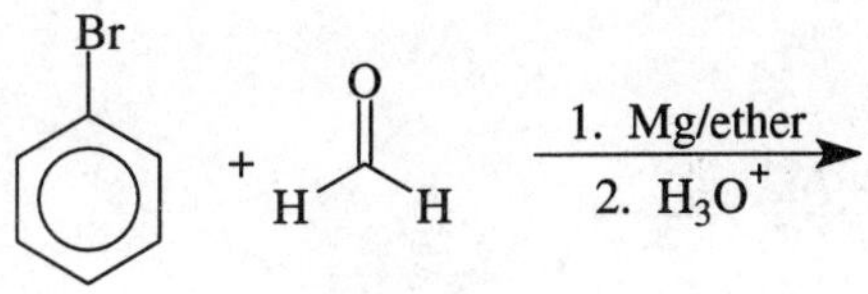

PhBr + HCHO $\xrightarrow{\text{1. Mg/ether} \quad \text{2. H}_3\text{O}^+}$

cyclohexanone + (CH₃)₂CHBr $\xrightarrow{\text{1. Mg/ether} \quad \text{2. H}_3\text{O}^+}$

2-butanone + CH₃CH₂Br $\xrightarrow{\text{1. Mg/ether} \quad \text{2. H}_3\text{O}^+}$

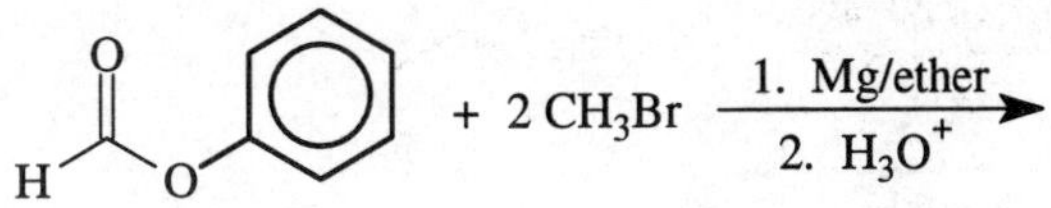

HC(O)OPh + 2 CH₃Br $\xrightarrow{\text{1. Mg/ether} \quad \text{2. H}_3\text{O}^+}$

sec-butyl bromide + (CH₃)₂C=O $\xrightarrow{\text{1. Mg/ether} \quad \text{2. H}_3\text{O}^+}$

Alcohols: *Reactions*

For each of the reactions shown below, draw the structure of the major organic product. Clearly show all stereochemistry, if appropriate.

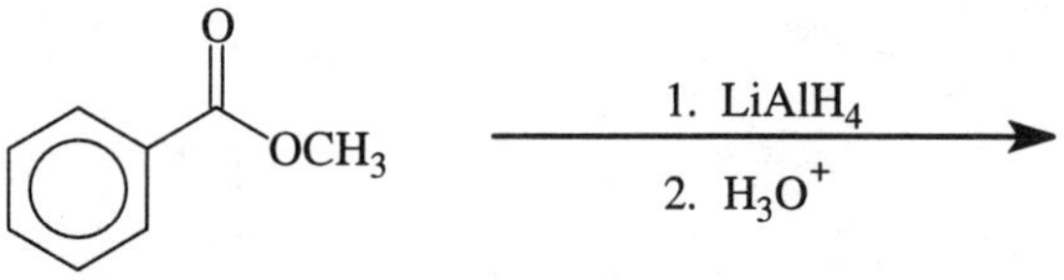

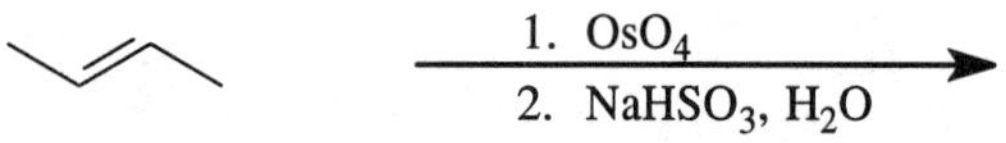

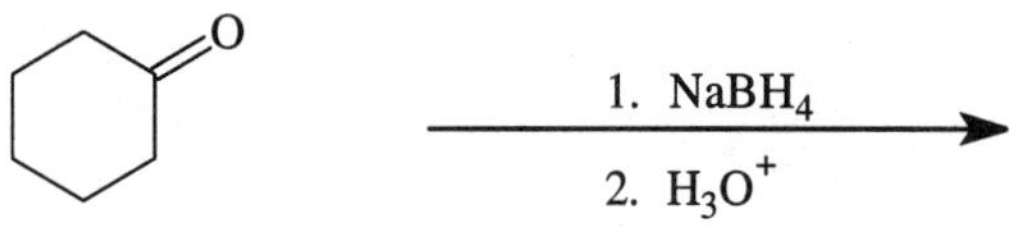

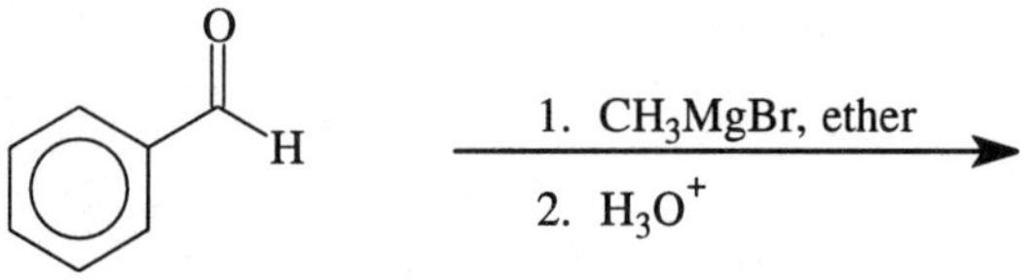

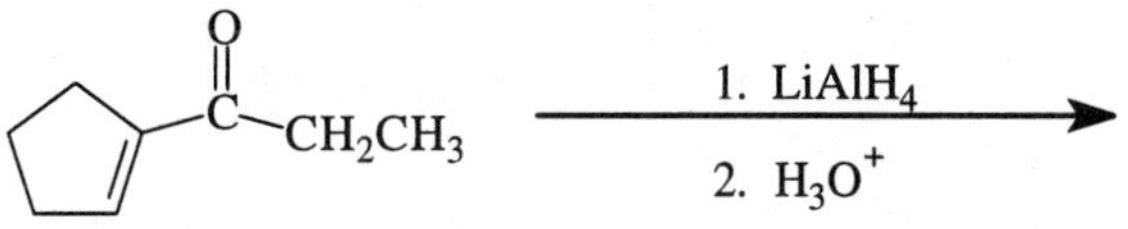

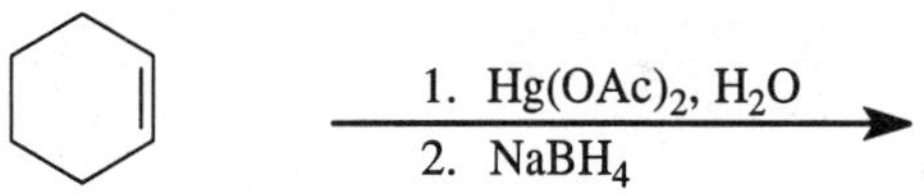

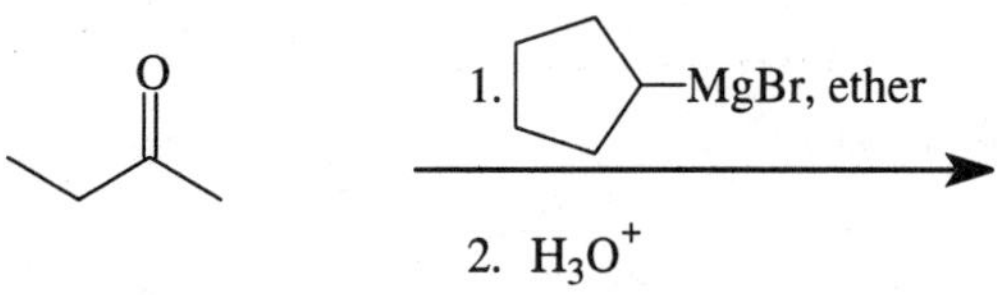

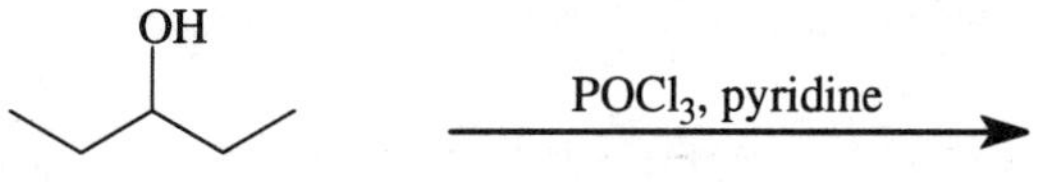

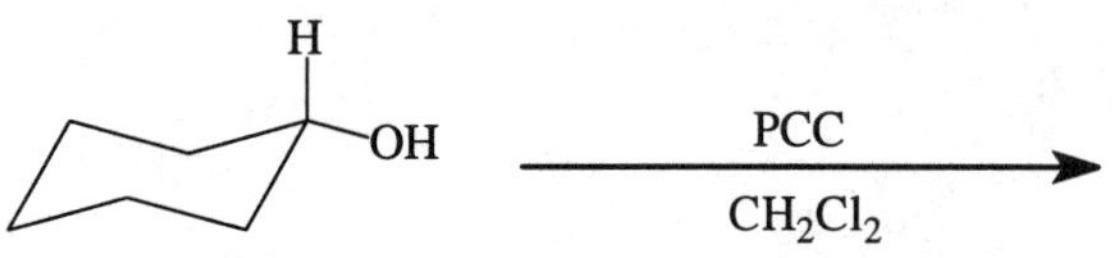

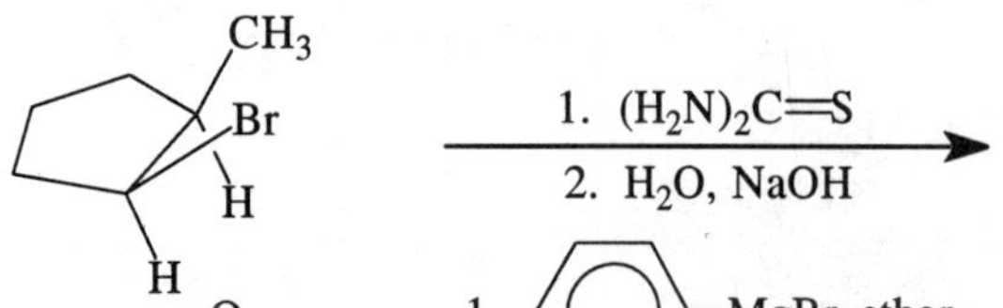

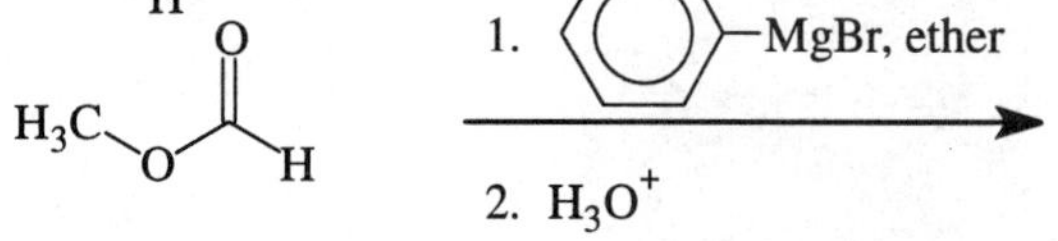

1-methylcyclopentene → 1. BH₃, THF; 2. H₂O₂, HO⁻

methyl benzoate → 1. LiAlH₄; 2. H₃O⁺

propene (CH₃CH=CHCH₃ trans-2-butene) → 1. OsO₄; 2. NaHSO₃, H₂O

cyclohexanone → 1. NaBH₄; 2. H₃O⁺

benzaldehyde → 1. CH₃MgBr, ether; 2. H₃O⁺

1-(1-cyclopentenyl)propan-1-one → 1. LiAlH₄; 2. H₃O⁺

cyclohexene → 1. Hg(OAc)₂, H₂O; 2. NaBH₄

2-butanone → 1. cyclopentyl-MgBr, ether; 2. H₃O⁺

2-pentanol → POCl₃, pyridine

trans-4-substituted cyclohexanol (H, OH) → PCC, CH₂Cl₂

1-bromo-2-methylcyclopentane (with stereochem) → 1. (H₂N)₂C=S; 2. H₂O, NaOH

methyl formate (H₃C–O–CHO) → 1. phenyl-MgBr, ether; 2. H₃O⁺

Alcohols: *Synthesis-I*

Suggest a simple synthesis for each of the compounds shown below, beginning with the starting material shown on the left. Clearly show all reagents and reaction conditions.

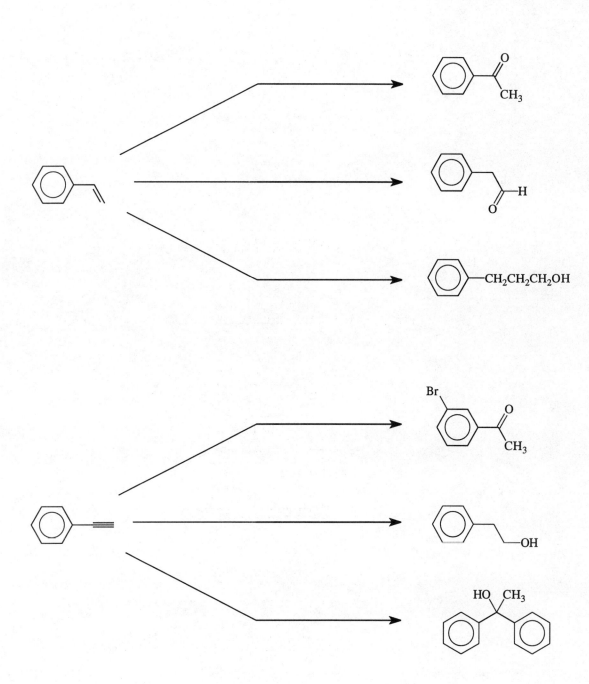

Alcohols: *Synthesis-II*

Suggest a simple synthesis for each of the molecules shown below, beginning with any one of the compounds shown below. Clearly show all required reagents and reaction conditions.

Starting Materials:

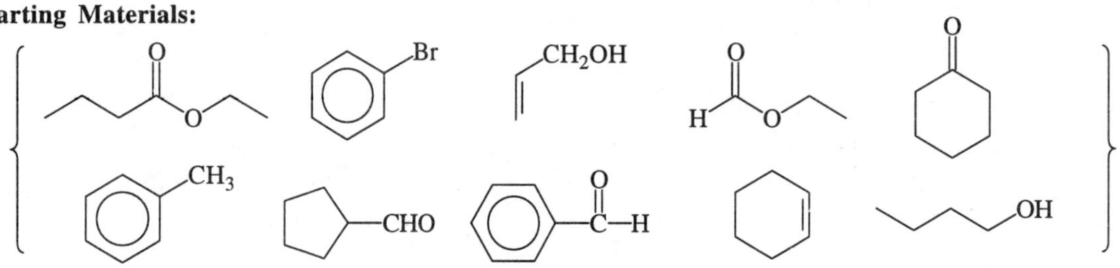

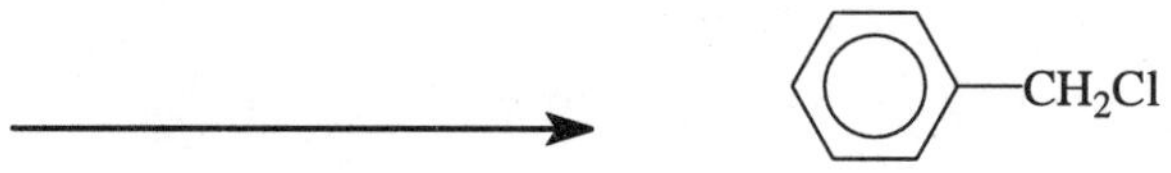

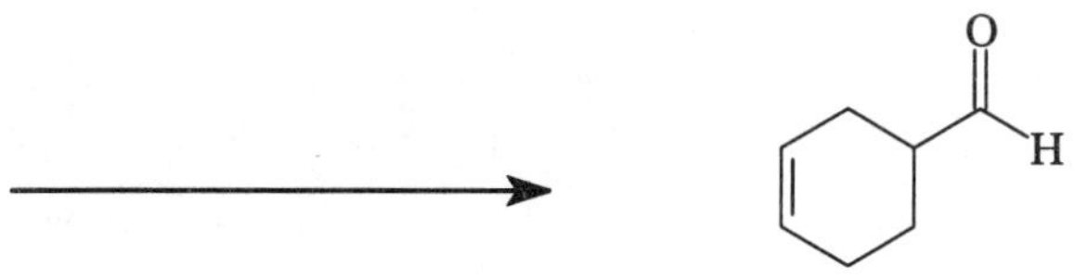

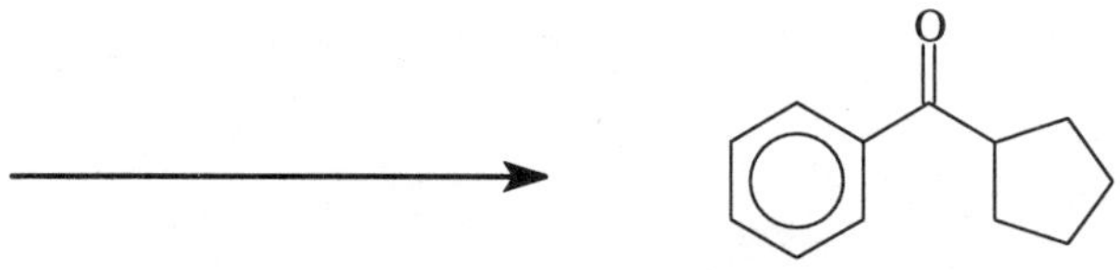

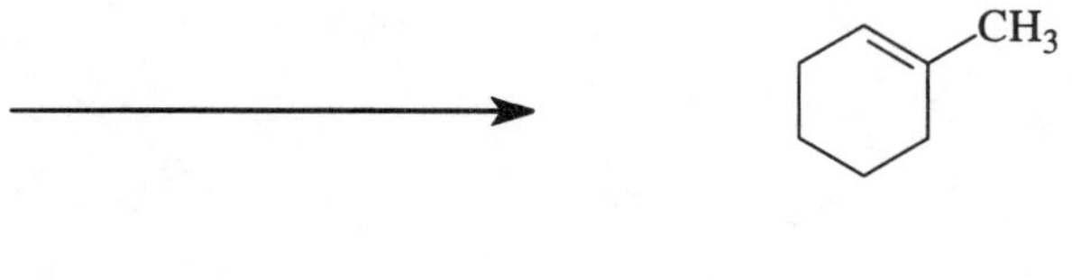

{ ethyl butyrate, bromobenzene, allyl alcohol (CH₂=CH-CH₂OH), ethyl formate, cyclohexanone, toluene, cyclopentanecarbaldehyde, benzaldehyde, cyclohexene, 1-butanol }

⟶ 3-bromobenzyl alcohol (Br-C₆H₄-CH₂OH, meta)

⟶ benzyl chloride (C₆H₅-CH₂Cl)

⟶ cyclohex-3-ene-1-carbaldehyde

⟶ phenyl cyclopentyl ketone

⟶ 4-heptanol (di-n-propyl carbinol)

⟶ 1-methylcyclohexene

Ethers: *Preparation and Reactions*

For each of the reactions shown below, draw the structure of the major organic product. Clearly show stereochemistry when appropriate.

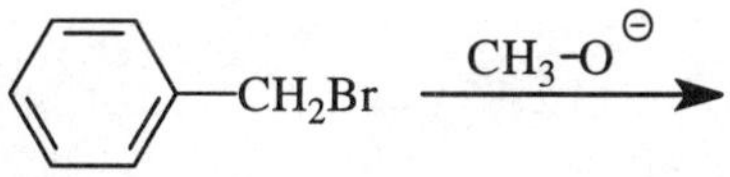

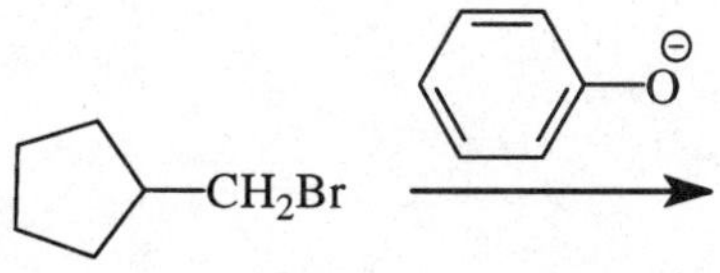

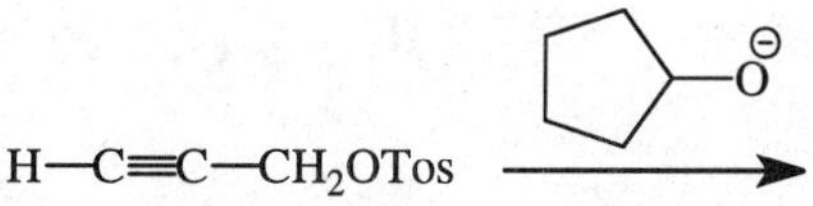

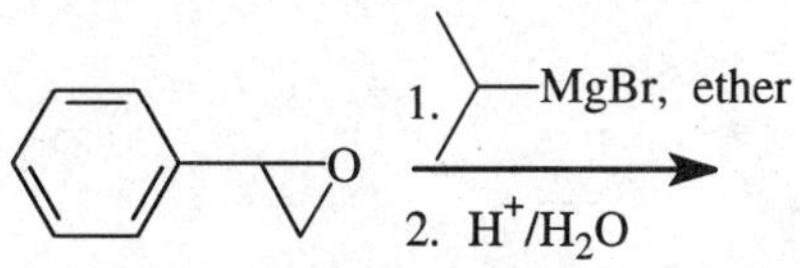

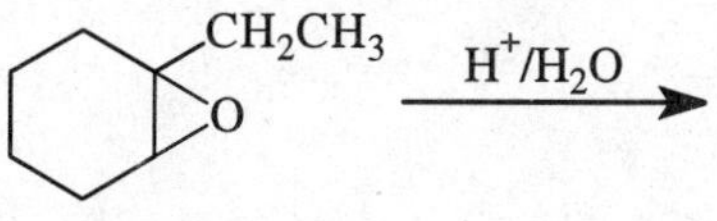

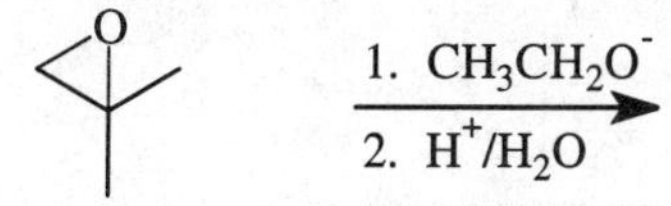

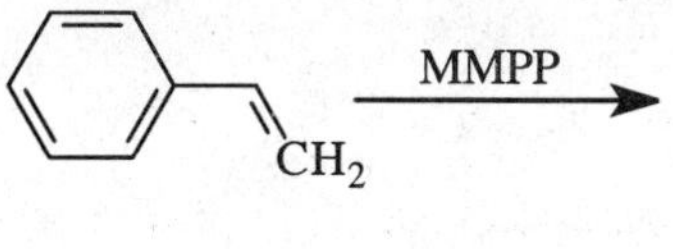

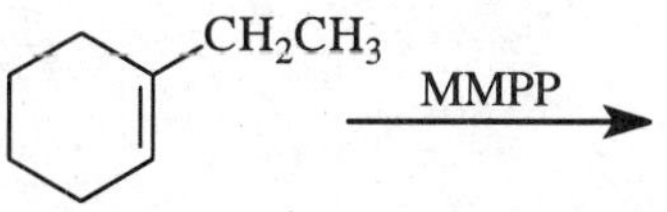

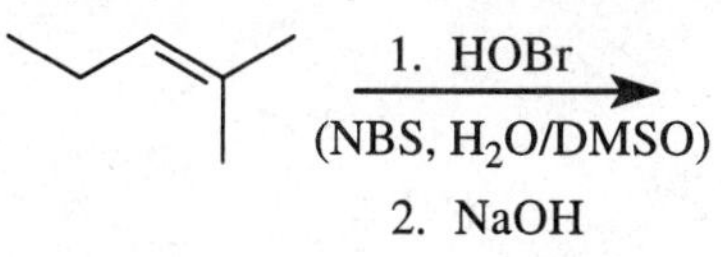

Ph–CH$_2$Br $\xrightarrow{\text{CH}_3\text{–O}^{\ominus}}$

(cyclopentyl)–CH$_2$Br $\xrightarrow{\text{PhO}^{\ominus}}$

H–C≡C–CH$_2$OTos $\xrightarrow{\text{cyclopentyl-O}^{\ominus}}$

Ph–(epoxide) $\xrightarrow[\text{2. H}^+/\text{H}_2\text{O}]{\text{1. (CH}_3)_2\text{CH–MgBr, ether}}$

1-ethyl-1,2-epoxycyclohexane $\xrightarrow{\text{H}^+/\text{H}_2\text{O}}$

2-methyl-1,2-epoxypropane $\xrightarrow[\text{2. H}^+/\text{H}_2\text{O}]{\text{1. CH}_3\text{CH}_2\text{O}^-}$

Ph–CH=CH$_2$ $\xrightarrow{\text{MMPP}}$

1-ethylcyclohexene $\xrightarrow{\text{MMPP}}$

2-methyl-2-pentene $\xrightarrow[\text{2. NaOH}]{\text{1. HOBr (NBS, H}_2\text{O/DMSO)}}$

Ethers: *Synthesis Problems*

Beginning with the compound shown on the left, suggest a synthesis for each of the following compounds. Multiple steps are required. Clearly show all intermediates and reaction conditions.

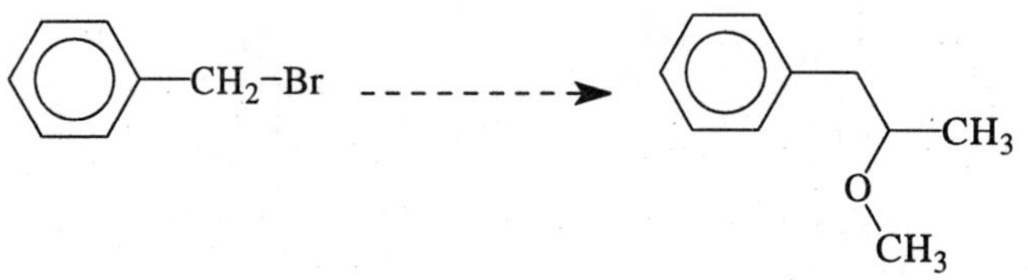

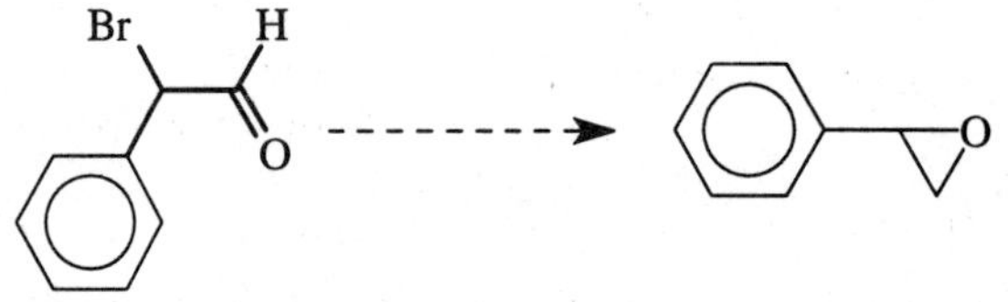

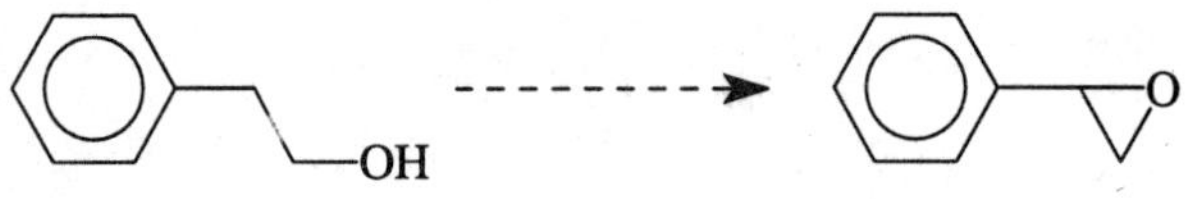

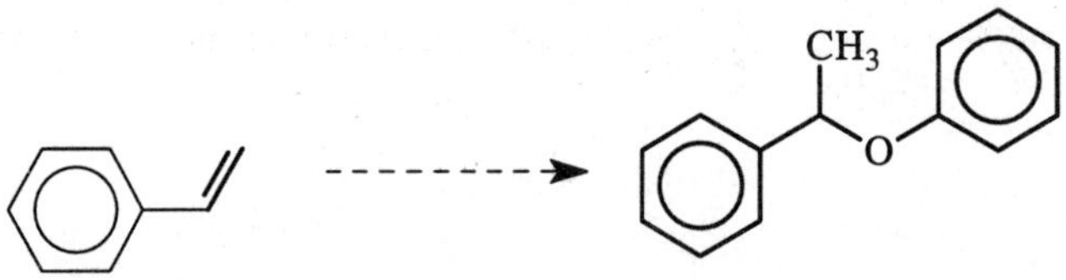

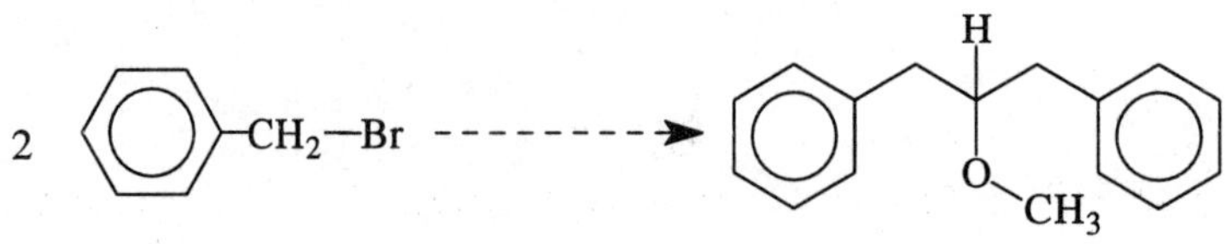

Ph–CH₂–Br ------→ Ph–CH₂–CH(CH₃)–O–CH₃

Ph–CHBr–CHO ------→ Ph–(epoxide)

Ph–CH₂–CH₂–OH ------→ Ph–(epoxide)

Ph–CH=CH₂ ------→ Ph–CH(CH₃)–O–Ph

2 Ph–CH₂–Br ------→ Ph–CH₂–CH(OCH₃)–CH₂–Ph

Aldehydes & Ketones

Nomenclature

Simple aldehydes and ketones are named using the standard rules of nomenclature which we have used in the past with the following specific changes:

1. Aldehydes are named by replacing the terminal -e of the parent alkane with the suffix **-al**; the suffix for ketones is **-one**.

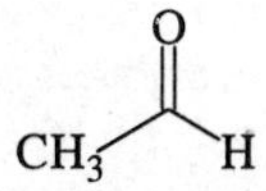

Ethanal (Acetaldehyde) 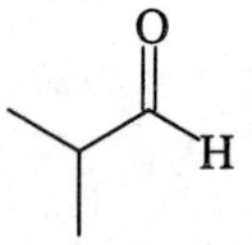   2-methylpropanal

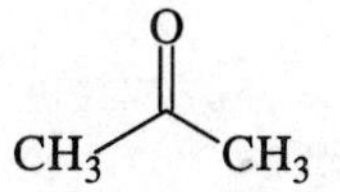

2-propanone (Acetone) 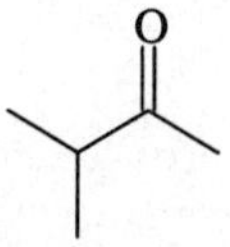   2-methyl-2-butanone

2. The parent chain selected must contain the carbonyl group.

3. Number the carbon chain, beginning at the end nearest to the carbonyl group.

4. Number the substituents and write the name, listing substituents alphabetically.

5. When an aldehyde is a substituent on a ring, it is referred to as a **carbaldehyde** group.

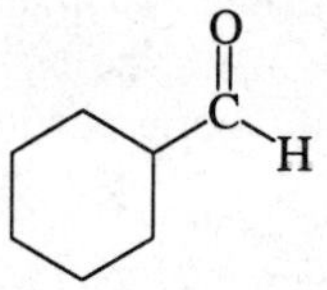

Cyclohexanecarbaldehyde 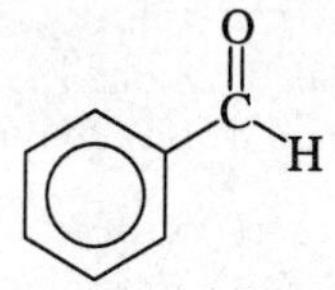   Benzenecarbaldehyde
(Benzaldehyde)

6. When the -COR group becomes a substituent on another chain, it is referred to as an **acyl** group and the name is formed using the suffix **-yl**.

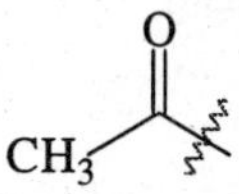

Acetyl 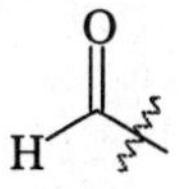   Formyl

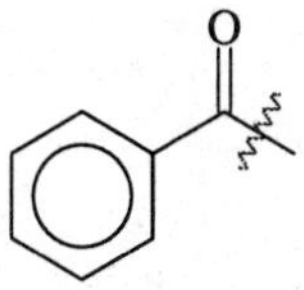

Benzoyl

7. When the carbonyl group becomes a substituent on another chain, it is referred to as an oxo group.

5-oxohexanal

Reactions that Yield Aldehydes & Ketones

Oxidation of Alcohols with Pyridinium Chlorochromate: Primary and secondary alcohols are smoothly oxidized by pyridinium chlorochromate (PCC) in CH_2Cl_2 to form aldehydes and ketones, respectively. The PCC oxidation of primary alcohols to give aldehydes is a very useful reaction, since aldehydes are difficult to prepare and are easily over-oxidized to the carboxylic acid.

Oxidation of Alcohols with "Jones Reagent": Primary and secondary alcohols are oxidized by CrO_3/H_2SO_4 (Jones Reagent) to form carboxylic acids and ketones, respectively; sodium dichromate in acetic acid ($Na_2Cr_2O_7$) can also be used.

Ozonolysis of Alkenes: Simple alkenes are oxidized by O_3 to form an intermediate ozonide, which undergoes dissolving metal reduction with Zn/H_3O^+ to produce aldehydes and ketones.

Oxidation of Alkenes: Simple alkenes are oxidized by MnO_4^- to produce aldehydes and ketones. Terminal alkenes yield CO_2, while alkene carbons bearing one hydrogen form the corresponding carboxylic acid.

Hydration of Alkynes: Acid-catalyzed hydration of alkynes in the presence of Hg^{+2} yields an intermediate enol, which rapidly equilibrates with the corresponding carbonyl compound. The regiochemistry of the reaction is "Markovnikov"; that is, hydroxide anion will bond to the most stable potential carbocation center of the alkyne.

Hydroboration of an Alkyne, with Oxidative Work-up: Reaction of an alkyne with BH_3 results in the *syn*-addition of the boron and a hydrogen across the triple bond. Rearrangements do not occur and the hydrogen will bond to the carbon of the alkyne which would form the most stable carbocation center (overall anti-Markovnikov's addition). The driving force for the regiochemistry may actually be more steric than electronic, but viewing the reaction as a concerted, but polar transition state, easily rationalizes the observed product distribution. On oxidative work-up, the borane is converted to the enol, which rapidly equilibrates with the corresponding carbonyl compound.

Friedel-Crafts Acylation: Arenes react with acid halides and acid anhydrides in the presence of $AlCl_3$ to form aryl ketones. This is an example of electrophilic aromatic substitution, and the reaction does not proceed well on rings which are strongly deactivated.

Reactions of Aldehydes & Ketones

The Grignard Reaction: The reaction of an alkyl, aryl or vinyl halide with magnesium metal in ether solvent, produces an organometallic complex of uncertain structure, but which behaves as if it has the structure R-Mg-X and is commonly referred to as a **Grignard Reagent**. The "R" group in this complex (alkyl, aryl or vinyl), acts as if it was a stabilized carbanion and reacts with aldehydes and ketones to yield secondary or tertiary alcohols (primary alcohols can be formed by reaction with formaldehyde).

Hydration of Aldehydes & Ketones: The hydration of carbonyl compounds is an equilibrium process and the extent of that equilibrium generally parallels the reactivity of the parent aldehyde or ketone towards nucleophilic substitution; aldehydes are more reactive than ketones and are more highly hydrated at equilibrium.

Formation of Cyanohydrins: The reaction of carbonyl compounds with HCN is an equilibrium process and, again, the extent of that equilibrium generally parallels the reactivity of the parent aldehyde or ketone towards nucleophilic substitution.

Reaction with Amines: The reaction of carbonyl compounds with amines involves the formation of an intermediate carbinolamine which undergoes dehydration to form an immonium cation which can loose a proton to form the neutral imine.

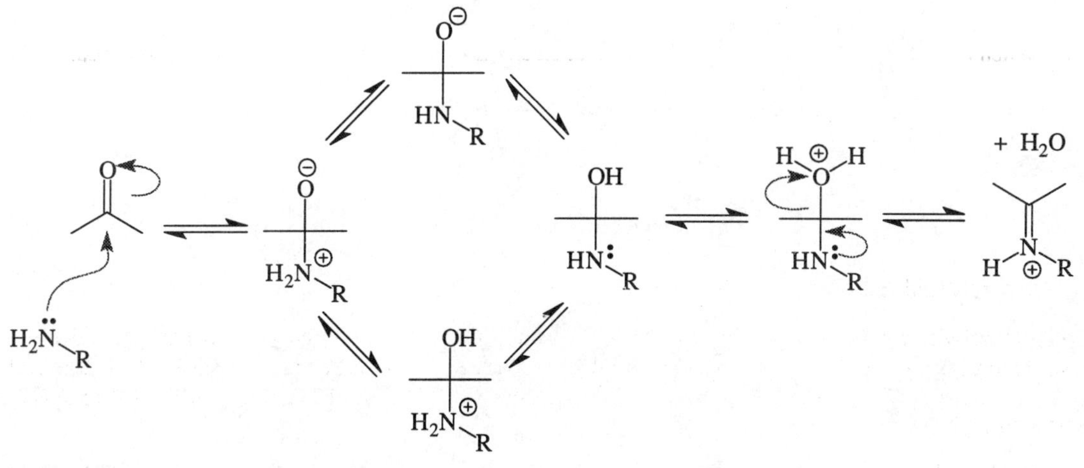

Some examples of common imine-forming reactions are given below:

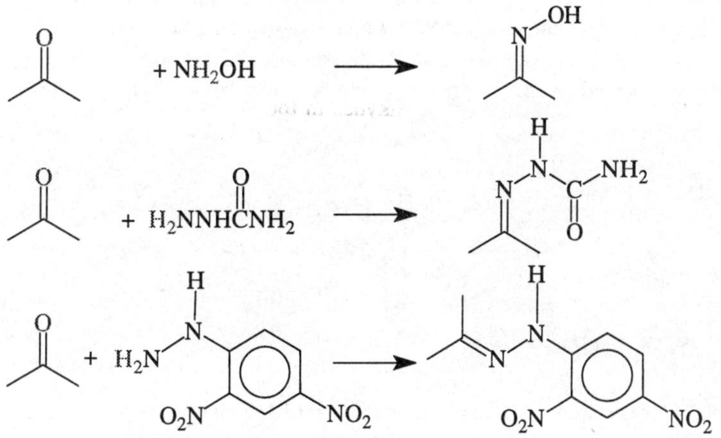

Imines formed from secondary amines can loose a proton from the α-carbon to form an enamine. Because of resonance, enamines maintain a partial carbanion character on the α-carbon and can be utilized as nucleophiles, as will be discussed in the section on "alpha alkylations".

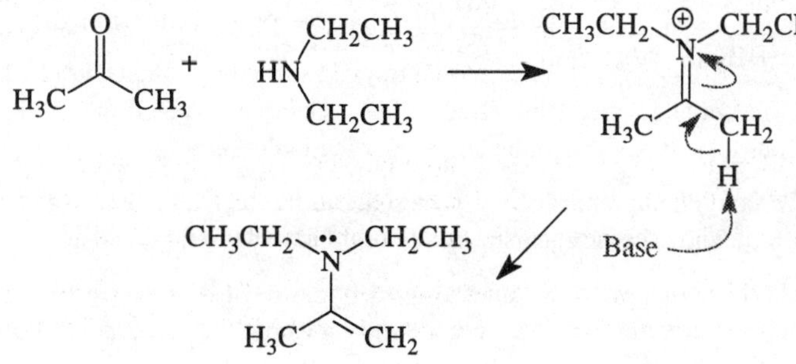

Ketal and Acetal Formation: Ketones and aldehydes react with excess alcohol in the presence of acid to give ketals and acetals, respectively. The mechanism of acetal formation involves equilibrium protonation, attack by alcohol, and then loss of a proton to give the neutral hemiacetal (or hemiketal). The hemiacetal undergoes protonation and loss of water to give an oxocarbonium ion, which undergoes attack by another mole of alcohol and loss of a proton to give the final product; note that acetal (or ketal) formation is an equilibrium process.

The Wittig Reaction: Ketones and aldehydes react with phosphorus ylides to form alkenes. Phosphorus ylides are prepared by an S_N2 reaction between an alkyl halide and triphenylphosphine, followed by deprotonation by a strong base such as *n*-butyllithium. The mechanism of the Wittig reaction involves nucleophilic addition to give an intermediate **betaine**, which decomposes to give the alkene and triphenylphosphine oxide. The Wittig reaction works well to prepare mono- di- and tri-substituted alkenes; tetra-substituted alkenes cannot be prepared by this method.

Oxidation & Reduction of Aldehydes and Ketones

Preparation of Alcohols by Reduction of Aldehydes and Ketones: Reduction of simple aldehydes and ketones with BH_4^- yields the corresponding alcohol directly. The reaction works well for simple compounds, but reaction of BH_4^- with α,β–unsaturated aldehydes and ketones can result in significant reduction of the double bond. A much more powerful reductant is $LiAlH_4$, which will reduce aldehydes, ketones, esters, carboxylic acids and nitriles. Although the reduction of esters with $LiAlH_4$ proceeds to produce the alcohol, reduction of carboxylate esters by diisobutylaluminum hydride (DIBAH) stops at the aldehyde.

Wolff-Kishner Reduction: The imine formed from an aldehyde or ketone on reaction with hydrazine (NH_2NH_2) is unstable in base, and undergoes loss of N_2 to give the corresponding hydrocarbon.

Clemmensen Reduction: Carbonyl compounds can also be reduced by the Clemmensen reduction using zinc-mercury amalgam in the presence of acid; the mechanism most likely involves free radicals.

The Formation of Thioketal and Thioacetals: Ketones and aldehydes react with excess thiol in the presence of acid to give thioketals and thioacetals, respectively. These compounds are smoothly reduced by Raney-Nickel to give the corresponding hydrocarbons.

Oxidation of Aldehydes by Silver Oxide: Reaction of simple aldehydes with aqueous Ag_2O in the presence of NH_3 yields the corresponding carboxylic acid and metallic silver. The silver is generally deposited in a thin metallic layer which forms a reflective "mirror" on the inside surface of the reaction vessel. The formation of this mirror forms the basis of a qualitative test for aldehydes, called the **Tollens Test**.

Oxidation of Aldehydes to form Carboxylic Acids: Reaction of simple aldehydes with acidic MnO_4^-, or CrO_3/H_2SO_4 yields the corresponding carboxylic acid. Aldehydes oxidize very easily and it is often difficult to prevent oxidation, even by atmospheric oxygen.

Oxidation of Ketones: Ketones are more resistant to oxidation, but can be cleaved with acidic MnO_4^- to yield carboxylic acids.

Reactions of Aldehydes & Ketones; Conjugate Additions

Conjugated ketones and aldehydes can undergo a reaction which is analogous to the 1,4-addition reaction of dienes, in which a nucleophile adds to the terminal carbon of the double bond. As in dienes, resonance throughout the conjugated system places cationic character on both carbons #1 and 3.

Addition of Amines: Amines add to conjugated ketones and aldehydes to give the conjugate addition product, almost exclusively.

Addition of HCN: HCN, or preferably $(CH_3CH_2)_2AlCN$, will also add to conjugated ketones and aldehydes to exclusively give the conjugate addition product.

Addition of Alkyl and Aryl Groups: Alkyl, aryl and vinyl groups can be added to conjugated ketones using the appropriate dialkylcopperlithium reagent.

Reactions of Aldehydes & Ketones

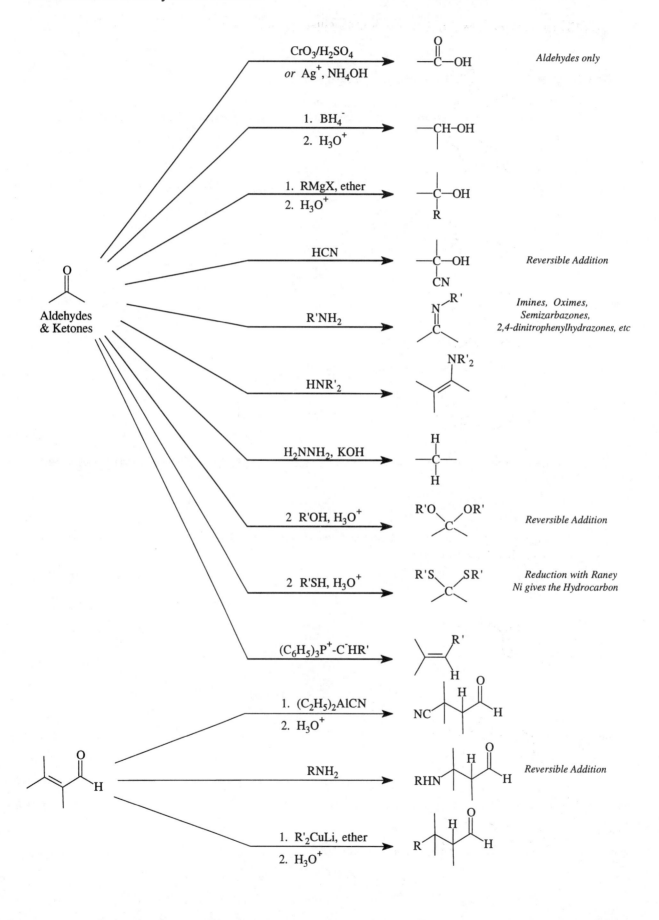

Aldehydes & Ketones: *Nomenclature*

Give the proper IUPAC name for each of the following compounds:

Aldehydes & Ketones: *Selected Reactions-I*

For each of the reactions shown below, draw the structure of the major organic product. Clearly show stereochemistry, if appropriate.

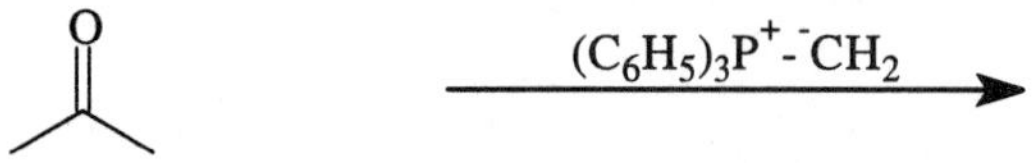

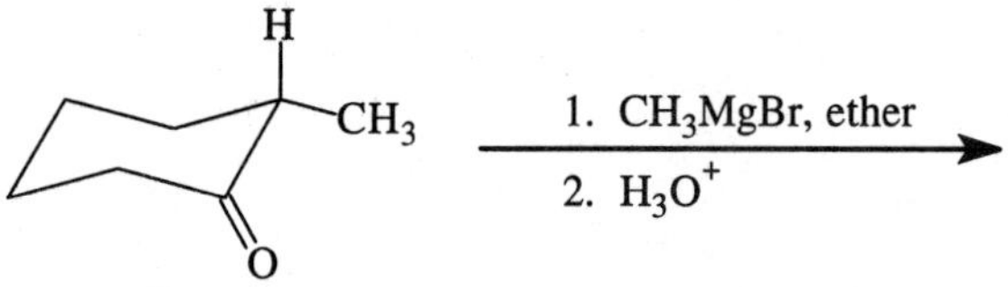

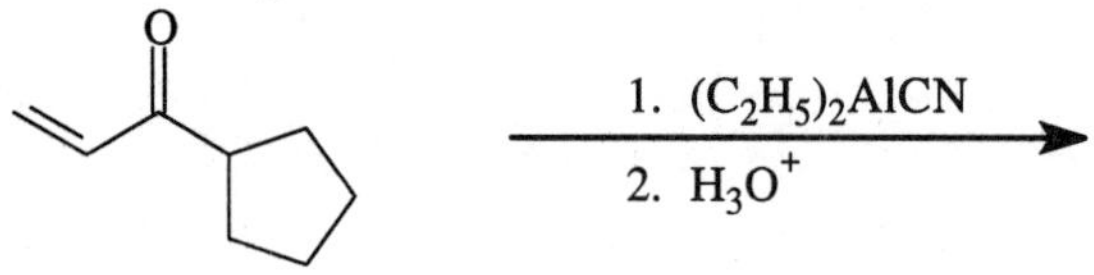

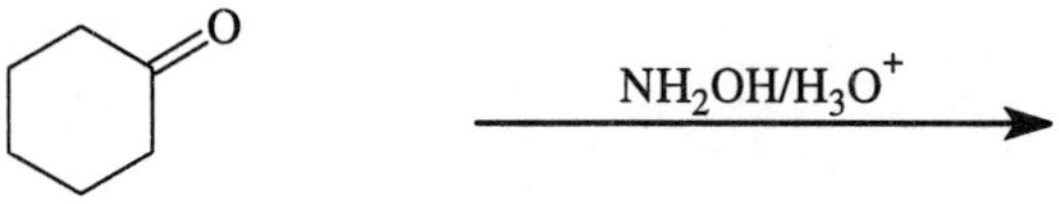

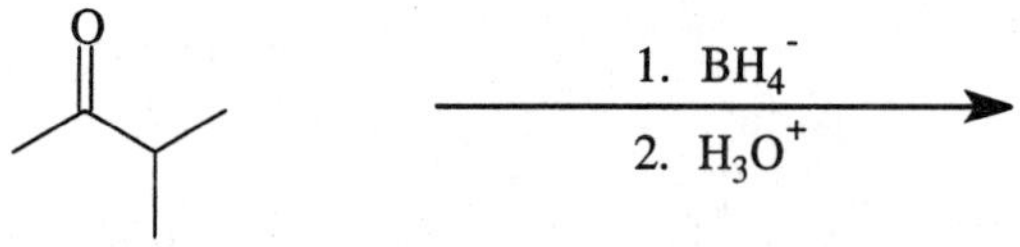

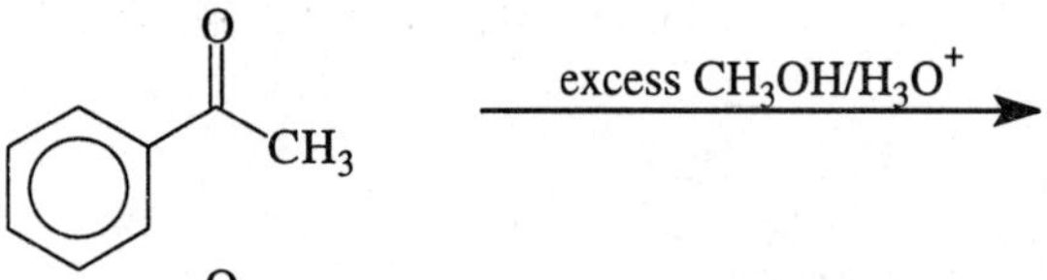

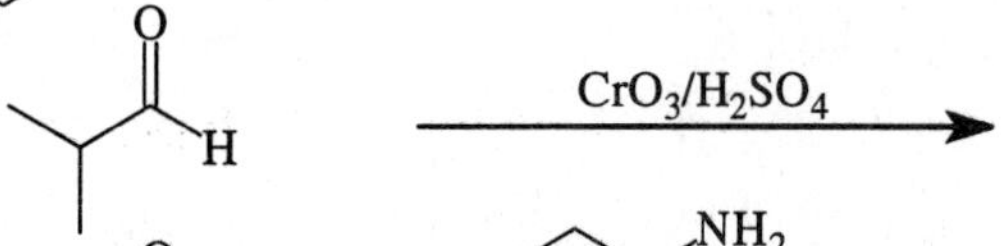

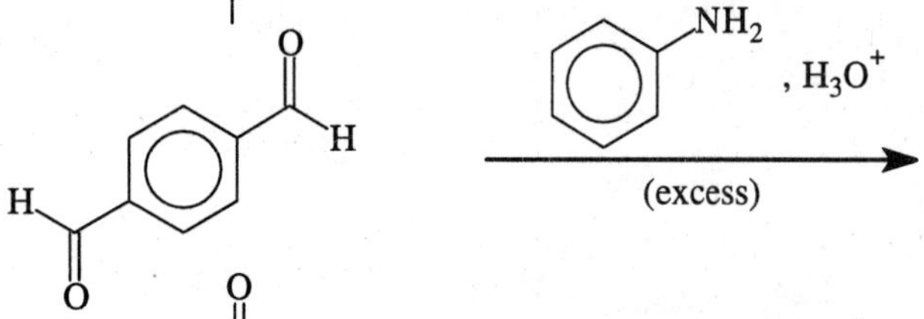

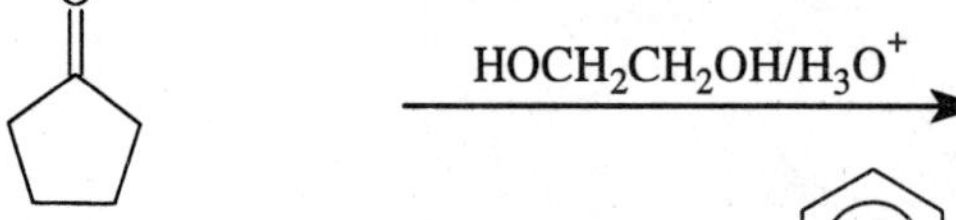

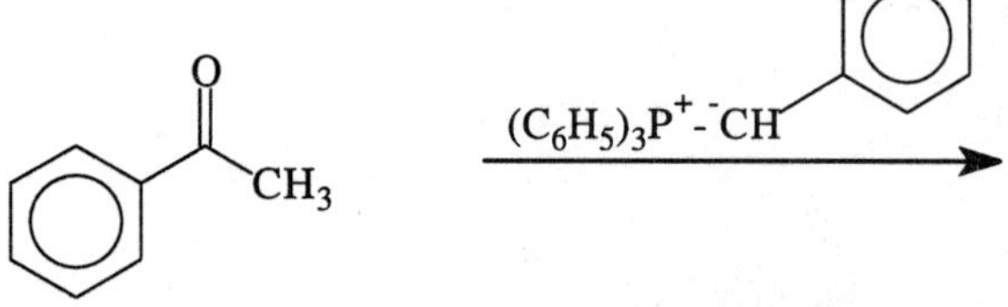

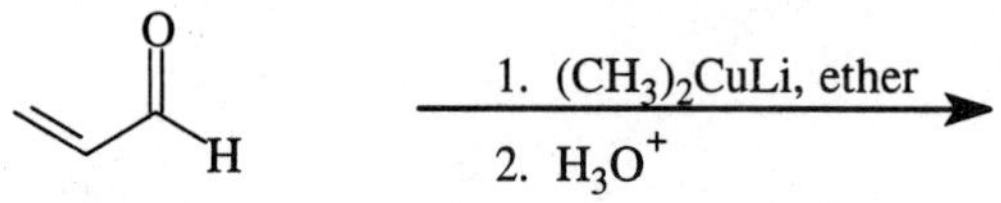

acetone + $(C_6H_5)_3P^+{}^-CH_2$ →

trans-2-methylcyclohexanone (H up, CH₃ down)
1. CH_3MgBr, ether
2. H_3O^+ →

1-cyclopentyl-2-propen-1-one
1. $(C_2H_5)_2AlCN$
2. H_3O^+ →

cyclohexanone + NH_2OH/H_3O^+ →

3-methyl-2-butanone
1. BH_4^-
2. H_3O^+ →

acetophenone + excess CH_3OH/H_3O^+ →

isobutyraldehyde + CrO_3/H_2SO_4 →

terephthalaldehyde + aniline ($PhNH_2$), H_3O^+ (excess) →

cyclopentanone + $HOCH_2CH_2OH/H_3O^+$ →

acetophenone + $(C_6H_5)_3P^+{}^-CH(C_6H_5)$ →

acrolein (CH₂=CH-CHO)
1. $(CH_3)_2CuLi$, ether
2. H_3O^+ →

Aldehydes & Ketones: *Selected Reactions-II*

For each of the reactions shown below, draw the structure of the major organic product. Clearly show stereochemistry, if appropriate.

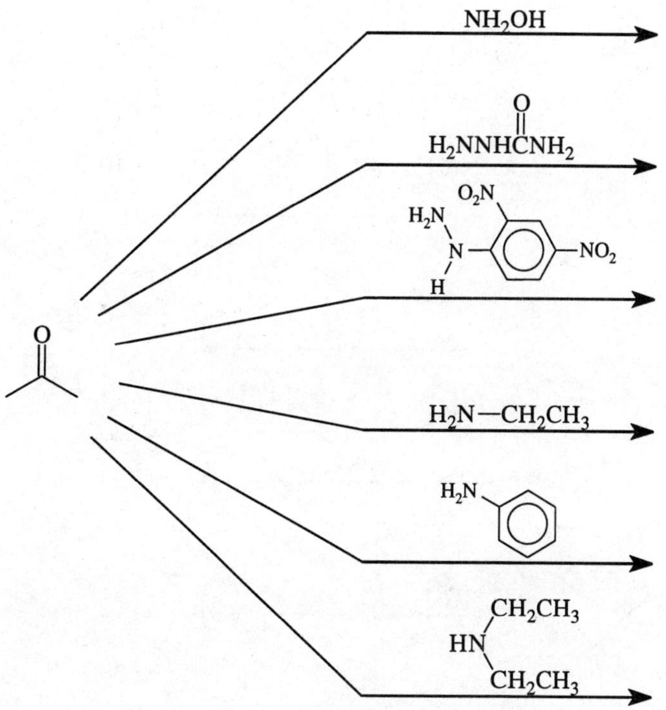

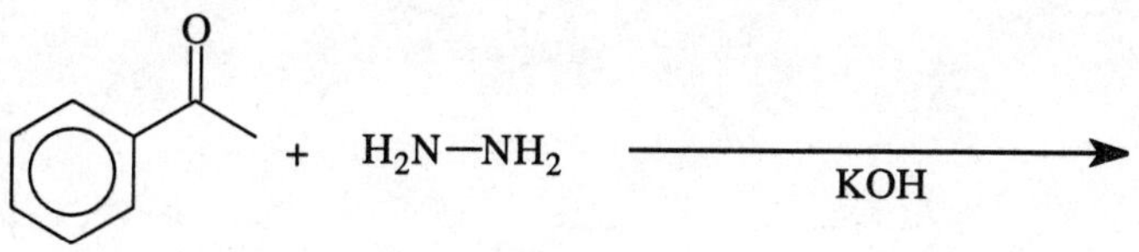

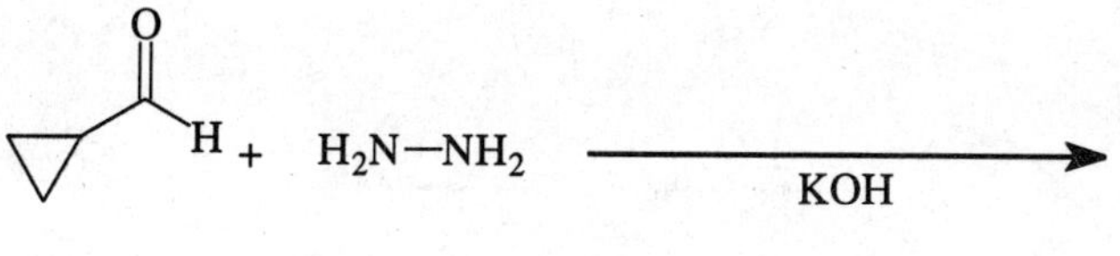

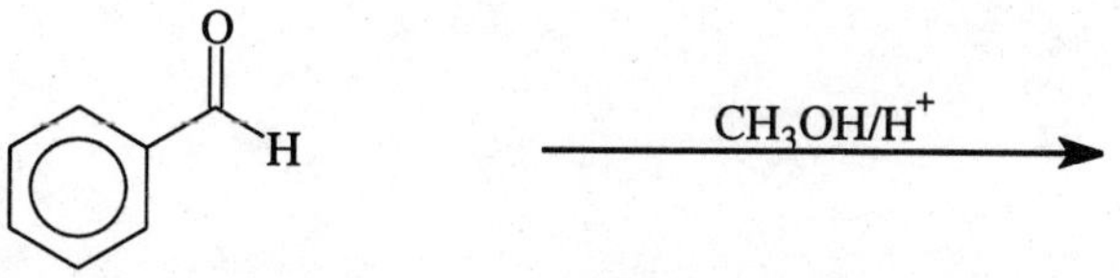

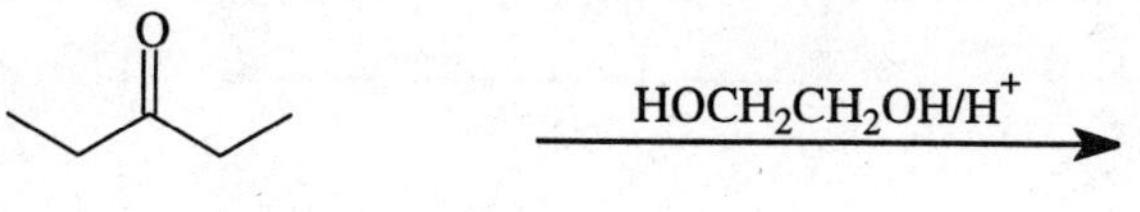

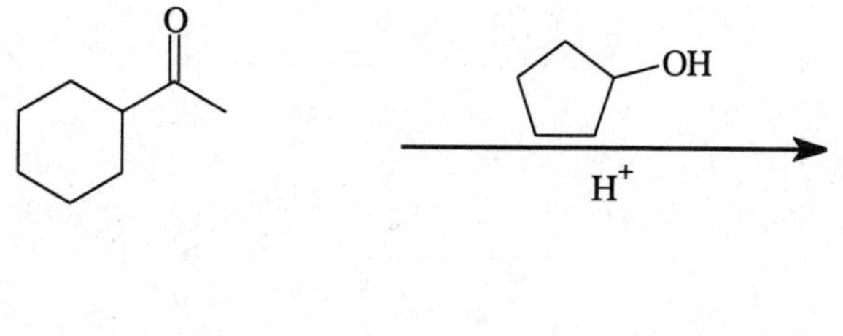

(acetone) + NH$_2$OH →

(acetone) + H$_2$NNHCNH$_2$ (with C=O) →

(acetone) + 2,4-dinitrophenylhydrazine →

(acetone) + H$_2$N–CH$_2$CH$_3$ →

(acetone) + H$_2$N–C$_6$H$_5$ (aniline) →

(acetone) + HN(CH$_2$CH$_3$)$_2$ →

PhC(O)CH$_3$ + H$_2$N–NH$_2$ $\xrightarrow{\text{KOH}}$

cyclopropyl-CHO + H$_2$N–NH$_2$ $\xrightarrow{\text{KOH}}$

PhCHO $\xrightarrow{\text{CH}_3\text{OH/H}^+}$

3-pentanone $\xrightarrow{\text{HOCH}_2\text{CH}_2\text{OH/H}^+}$

cyclohexyl methyl ketone + cyclopentanol $\xrightarrow{\text{H}^+}$

Aldehydes & Ketones: *Synthesis-I*

Suggest a synthesis for each of the compounds shown below, beginning with the compound shown on the left. Clearly show all required reagents and intermediates.

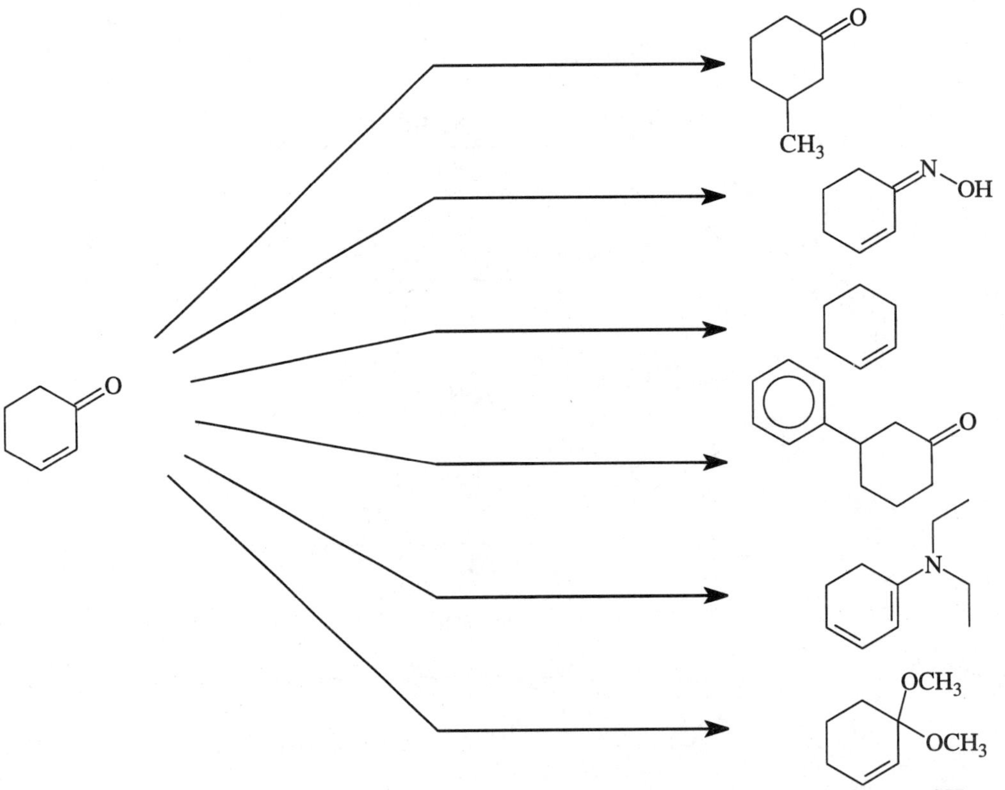

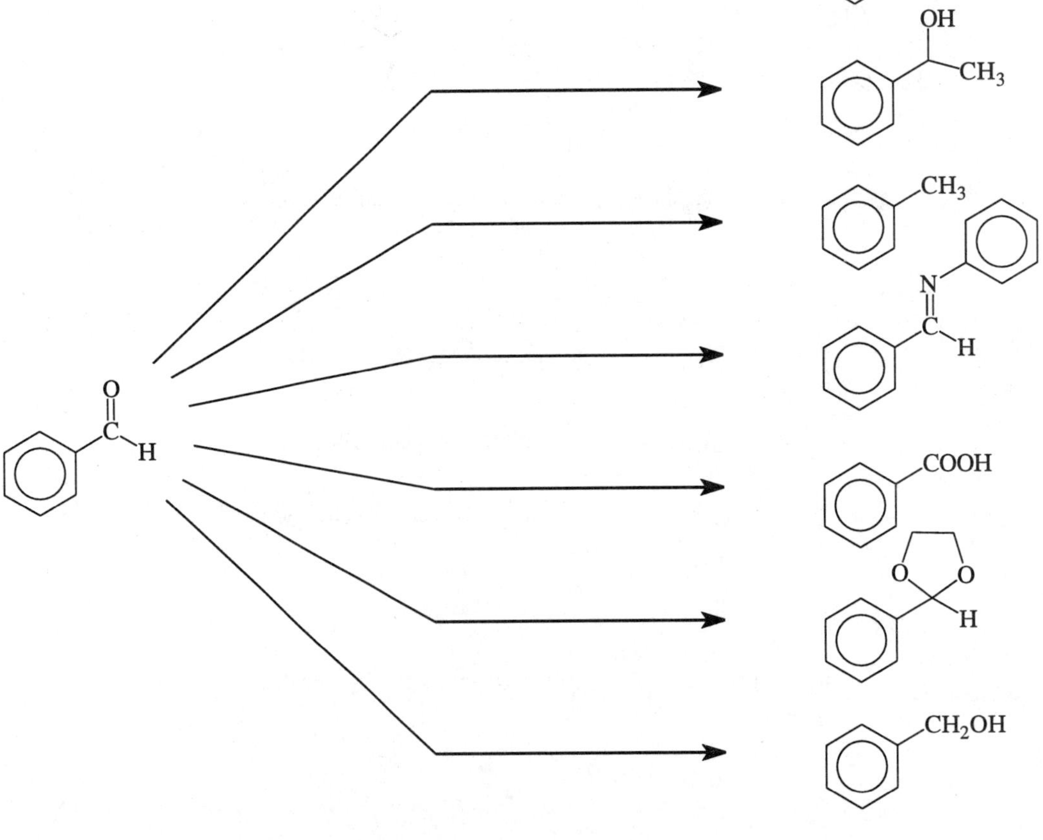

Aldehydes & Ketones: *Synthesis-II*

Suggest a synthesis for each of the compounds shown below, beginning with the starting material shown on the left. Clearly show all intermediates and reaction conditions. Multiple steps are likely to be required.

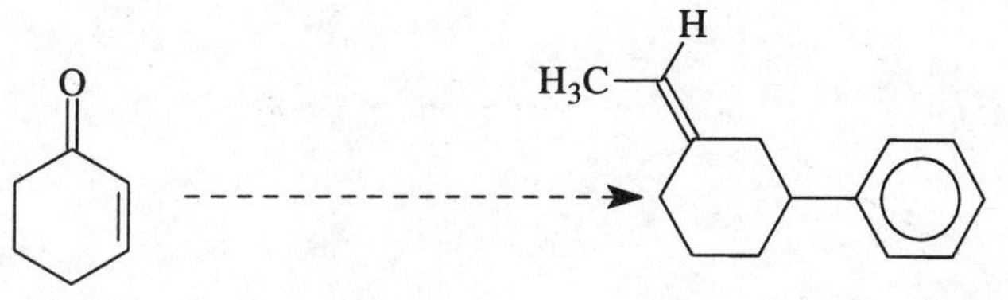

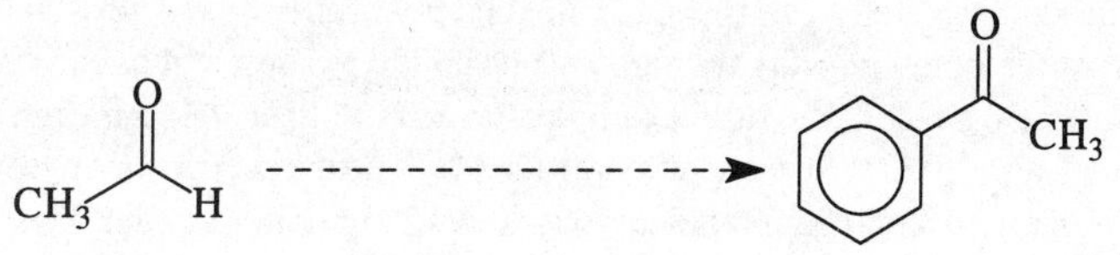

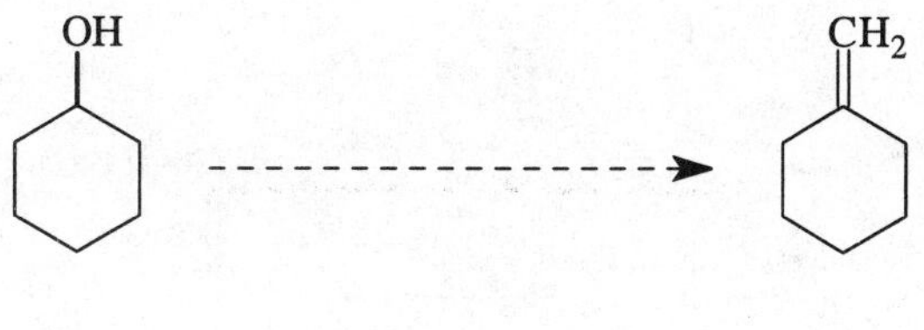

Carboxylic Acids

Nomenclature:

Simple carboxylic acids are named as derivatives of the parent alkane, using the suffix **-oic** acid

1. Select the longest continuous carbon chain, containing the carboxylic acid group, and derive the parent name by replacing the -e ending with -oic acid.
2. Number the carbon chain, beginning at the end nearest to the carboxylic acid group.
3. Number the substituents and write the name, listing substituents alphabetically.
4. Carboxylic acid substituents attached to rings are named using the suffix **-carboxylic acid**.

Several simple examples are shown below:

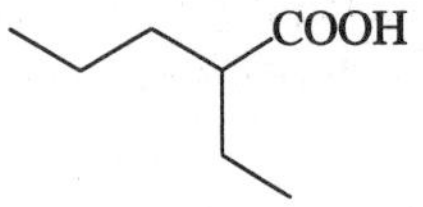

2-ethylpentanoic acid

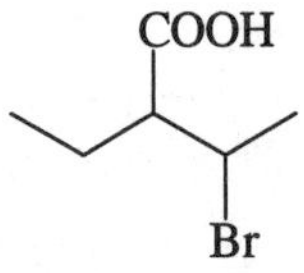

3-bromo-2-ethylbutanoic acid

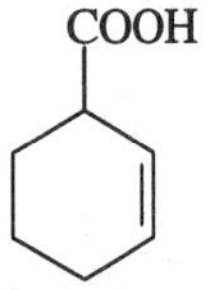

2-cyclohexenecarboxylic acid

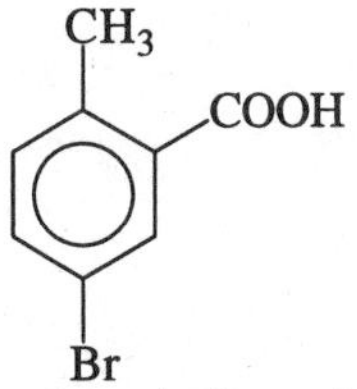

5-bromo-2-methylbenzoic aicd

In the first example, the parent chain is a pentane and the carboxylic acid group is assigned as carbon #1. On the pentane parent, there is an ethyl group in position #2; hence the name, **2-ethylpentanoic acid**. In the second example, there are two potential four-carbon chains; in this case, the chain with the most substituents is selected as parent, (a butanoic acid). Attached to the butanoic acid at carbon #2 is an ethyl group and at carbon #3, a bromine; hence the name **3-bromo-2-ethylbutanoic acid**. In the third example, the carboxylic acid is attached to a cycloalkene ring and will therefore be named as a "carboxylic acid" substituent (rule #4). The parent ring is a cyclohexene; letting the carboxylic acid be carbon #1, the name is **2-cyclohexenecarboxylic acid**. In the last example, the name is based on benzoic acid as the parent. In this case, we simply number the substituents to give the lowest number sequence at the first point of difference and arrange alphabetically; **5-bromo-2-methylbenzoic acid**.

Reactions which Yield Carboxylic Acids

Oxidation of Aromatic Side-Chains with Neutral Permanganate: Warm, neutral permanganate anion will oxidize aromatic side-chains which contain at least one benzylic hydrogen to the corresponding carboxylic acid.

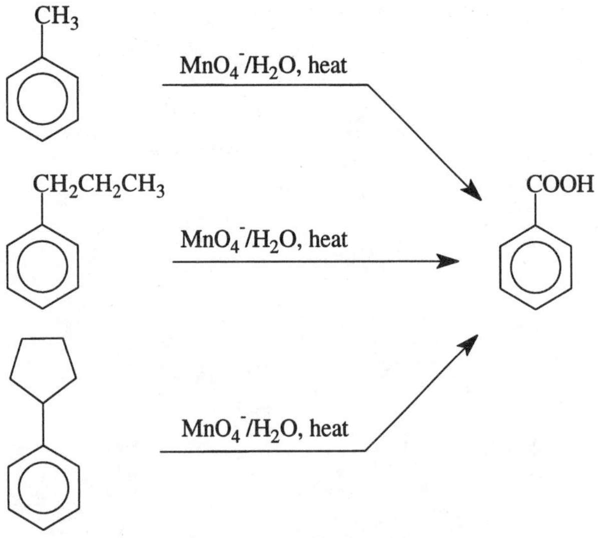

Oxidation of Primary Alcohols: Primary alcohols can be oxidized smoothly to the corresponding carboxylic acid with either CrO_3/H_2SO_4 or sodium dichromate.

$$\text{cyclohexyl-CH}_2\text{OH} \xrightarrow[\substack{\text{or } Na_2Cr_2O_7 \\ CH_3COOH/H_2O}]{\substack{CrO_3, H_2SO_4 \\ \text{(Jones' Reagent)}}} \text{cyclohexyl-COOH}$$

Oxidation of Alkenes: Acidic MnO_4^- will oxidize an alkene bearing at least one alkyl or aryl substituent to the corresponding carboxylic acid. Terminal alkenes are converted to CO_2 under these conditions.

$$\underset{H}{\overset{R}{>}}C=C\underset{H}{\overset{H}{<}} \xrightarrow{KMnO_4\ H_3O^+} \underset{HO}{\overset{R}{>}}C=O \quad O=C=O$$

Oxidation of Aldehydes: Aldehydes are smoothly oxidized to the corresponding carboxylic acid with either CrO_3/H_2SO_4 or sodium dichromate.

$$\text{cyclohexyl-CHO} \xrightarrow[\substack{\text{or } Na_2Cr_2O_7 \\ CH_3COOH/H_2O}]{\substack{CrO_3, H_2SO_4 \\ \text{(Jones' Reagent)}}} \text{cyclohexyl-COOH}$$

Oxidation of Aldehydes with Silver Oxide: Aldehydes (but not ketones) are oxidized by Ag_2O in aqueous ammonia to give the carboxylic acid and metallic silver. This is used as a qualitative test for aldehydes since the silver metal is deposited in a thin film, forming a "silver mirror" (the **Tollens test**).

$$\text{PhCHO} \xrightarrow[NH_3/H_2O]{Ag_2O} \text{PhCOOH} + Ag$$

Oxidation of Ketones: Acidic MnO_4^- will oxidize a ketone to the corresponding carboxylic acid, in this case, splitting the ring. The reaction is slower than the oxidation of alkenes, allowing disubstituted alkene carbons to be oxidized to the ketone, without significant over-oxidation.

$$\text{cyclohexanone} \xrightarrow{MnO_4^-/H^+} \text{HOOC-(CH}_2)_4\text{-COOH}$$

hexanedioic acid

Hydrolysis of Nitriles: Nitriles can be hydrolyzed to the corresponding carboxylic acids. Typically, vigorous conditions are required (heat and concentrated acid).

$$\text{cyclohexyl-CN} \xrightarrow{H_3O^+} \text{cyclohexyl-COOH}$$

Carboxylation of Grignard Reagents: Grignard reagents react with CO_2 to yield carboxylic acids. This is an important method for the preparation of carboxylic acids and yields are generally good.

Reactions of Carboxylic Acids

Carboxylic acids can be reduced to **primary alcohols** with $LiAlH_4$ or with BH_3 followed by work-up with aqueous acid. In the reduction with $LiAlH_4$, an intermediate aldehyde is formed, which is rapidly reduced to give the primary alcohol.

Heavy metal salts of carboxylic acids undergo decarboxylation on heating in organic solvents. In the presence of Br_2, the radical intermediate is trapped as an **alkyl bromide** (the **Hünsdiecker** reaction).

Carboxylic acids can be converted into **acid halides** by reaction with $SOCl_2$, phosgene, or PBr_3.

Carboxylic acids can be converted into **esters** by reaction with the corresponding alcohol in the presence of an acid catalyst (**Fischer esterification**), by alkylation of the carboxylate anion with an alkyl halide in an S_N2 reaction, or by reaction with diazomethane (methyl esters only).

Carboxylic Acids: *Nomenclature*
For each of the compounds shown below, provide an acceptable IUPAC name.

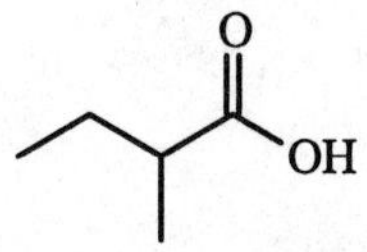

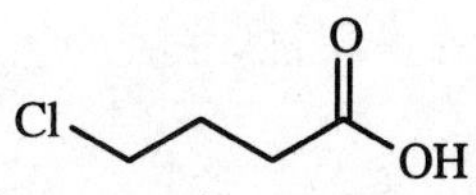

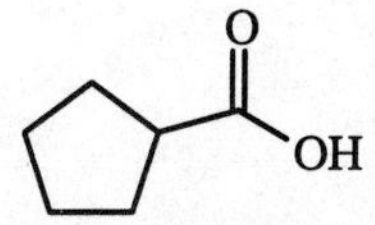

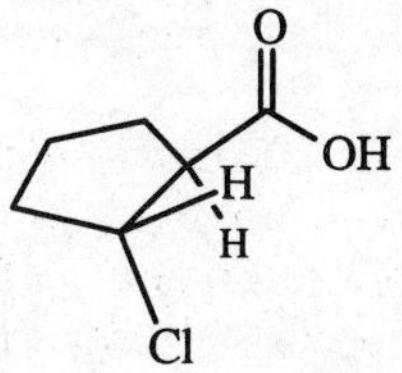

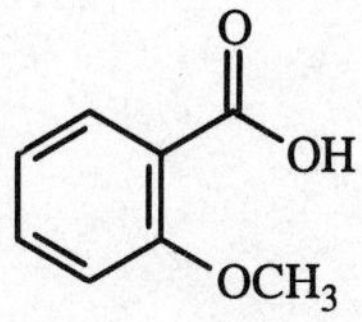

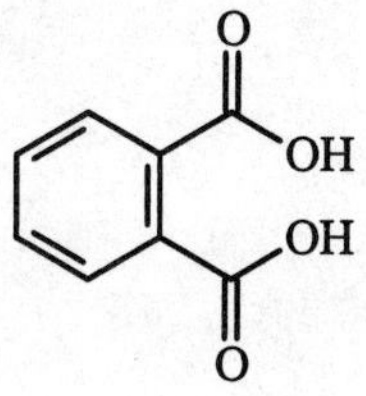

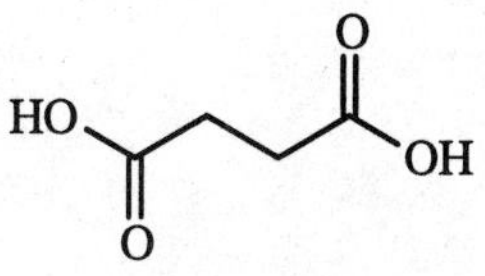

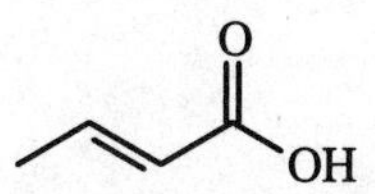

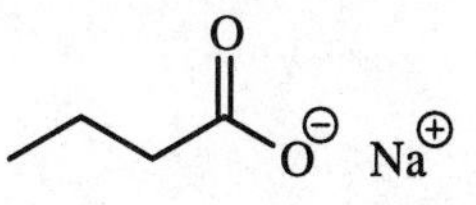

Carboxylic Acids: *Reactions that Yield Carboxylic Acids*

For each of the reactions shown below, draw the structure of the major organic product.

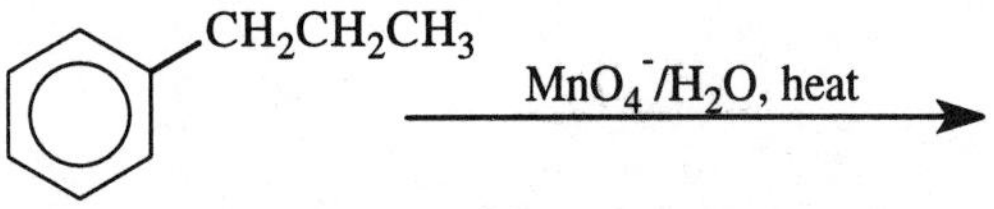

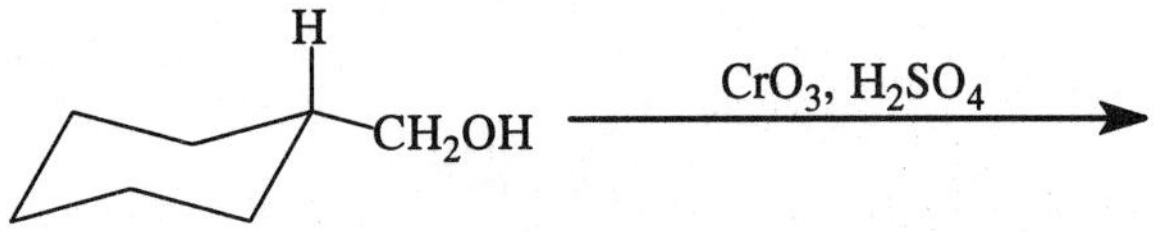

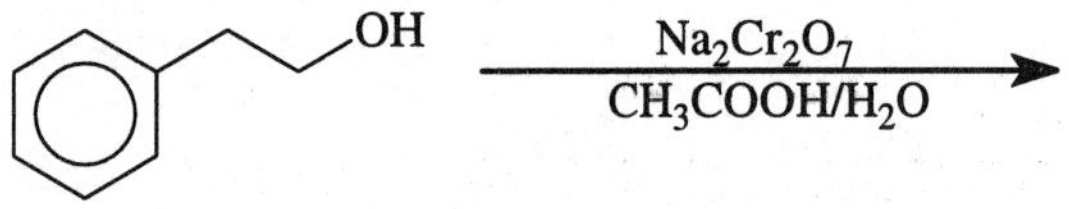

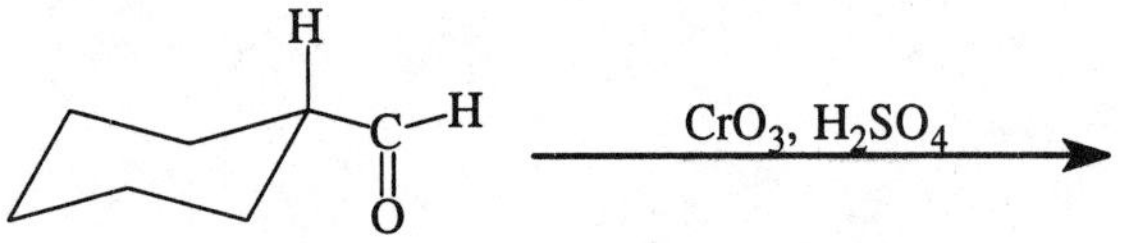

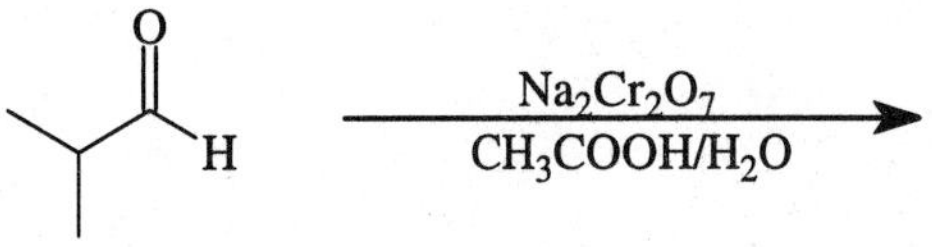

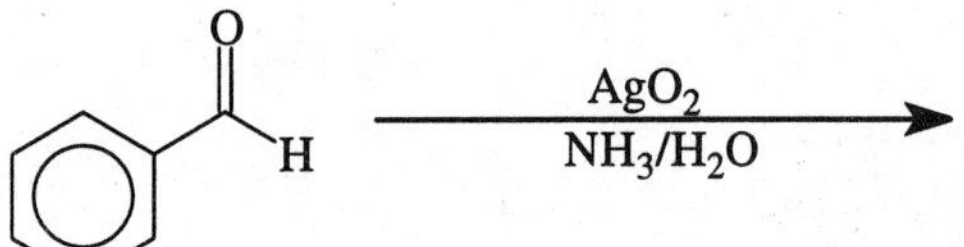

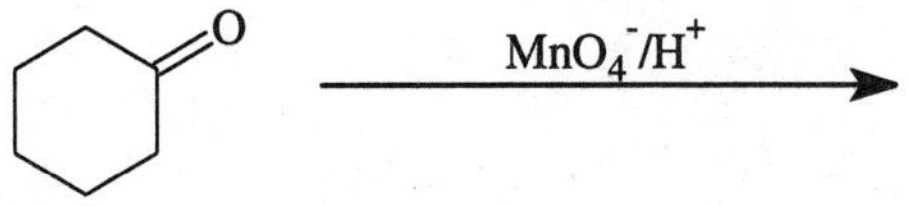

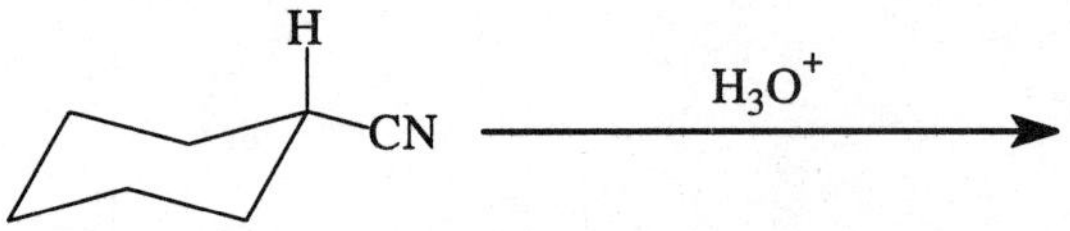

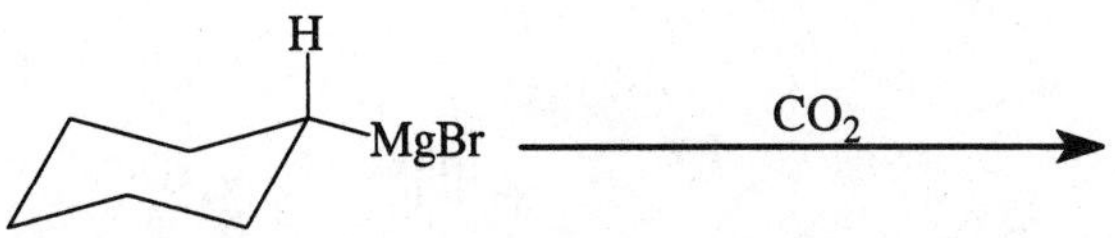

PhCH$_3$ $\xrightarrow{\text{MnO}_4^-/\text{H}_2\text{O, heat}}$

PhCH$_2$CH$_2$CH$_3$ $\xrightarrow{\text{MnO}_4^-/\text{H}_2\text{O, heat}}$

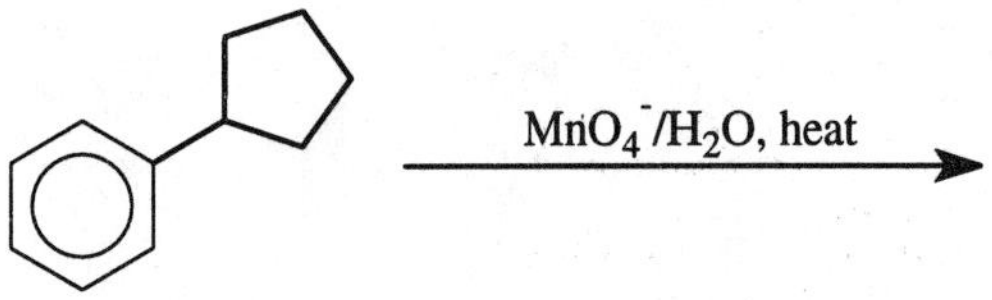

Ph–cyclopentyl $\xrightarrow{\text{MnO}_4^-/\text{H}_2\text{O, heat}}$

cyclohexyl-CH$_2$OH $\xrightarrow{\text{CrO}_3, \text{H}_2\text{SO}_4}$

PhCH$_2$CH$_2$OH $\xrightarrow[\text{CH}_3\text{COOH/H}_2\text{O}]{\text{Na}_2\text{Cr}_2\text{O}_7}$

(CH$_3$)(H)C=C(H)(H) $\xrightarrow{\text{KMnO}_4 \ \text{H}_3\text{O}^+}$

cyclohexyl-CHO $\xrightarrow{\text{CrO}_3, \text{H}_2\text{SO}_4}$

(CH$_3$)$_2$CHCHO $\xrightarrow[\text{CH}_3\text{COOH/H}_2\text{O}]{\text{Na}_2\text{Cr}_2\text{O}_7}$

PhCHO $\xrightarrow[\text{NH}_3/\text{H}_2\text{O}]{\text{AgO}_2}$

cyclohexanone $\xrightarrow{\text{MnO}_4^-/\text{H}^+}$

cyclohexyl-CN $\xrightarrow{\text{H}_3\text{O}^+}$

cyclohexyl-MgBr $\xrightarrow{\text{CO}_2}$

Carboxylic Acids: *Selected Reactions of Carboxylic Acids*

For each of the reactions shown below, draw the structure of the major organic product. Clearly show stereochemistry, if appropriate.

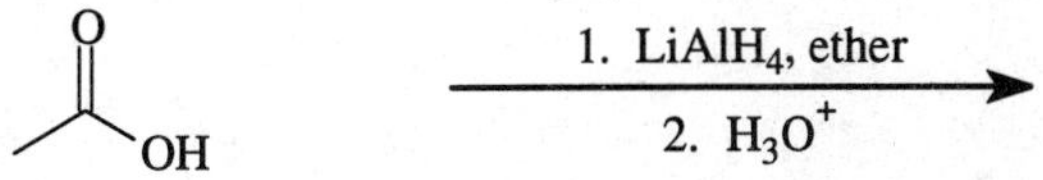

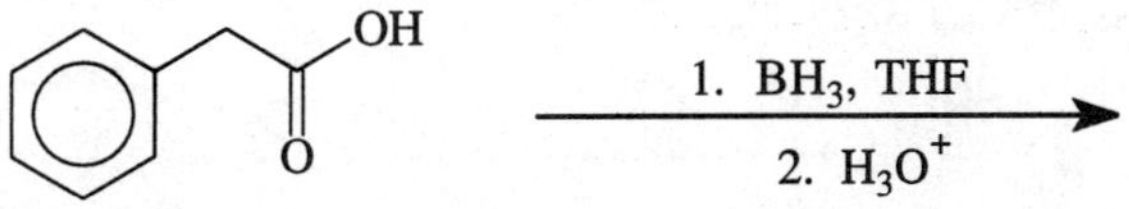

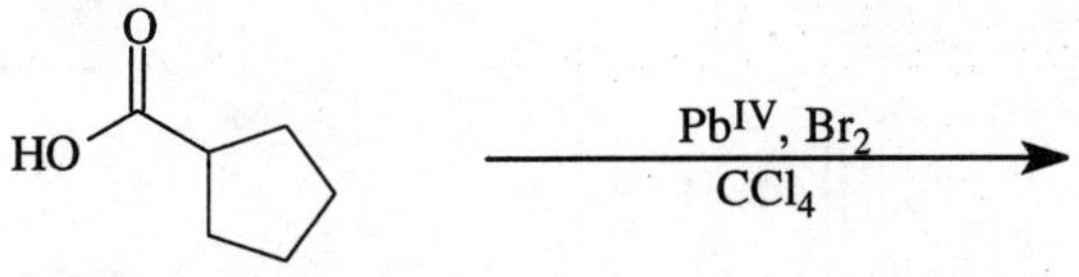

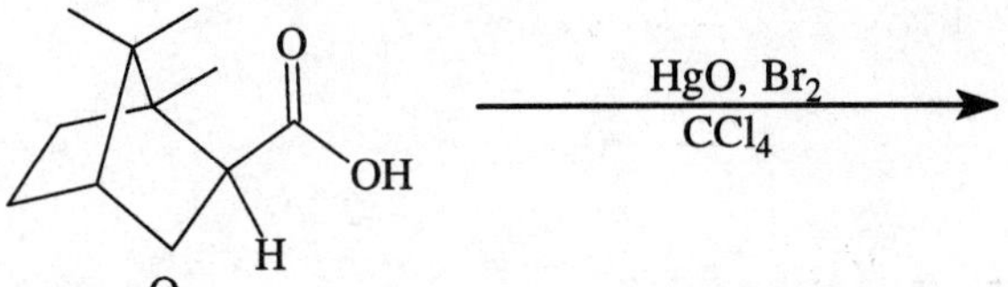

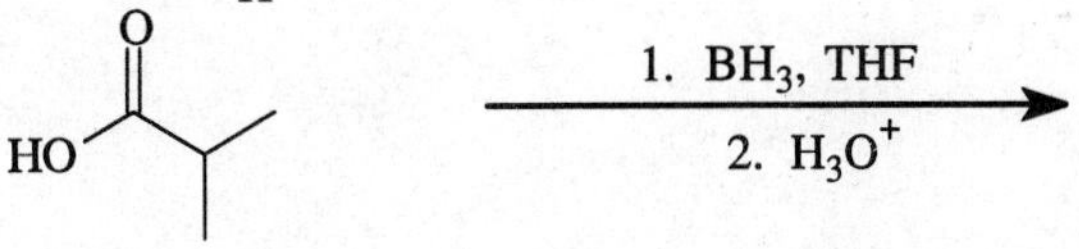

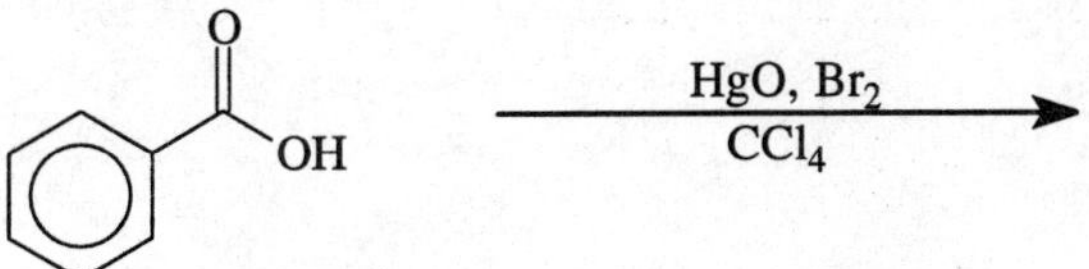

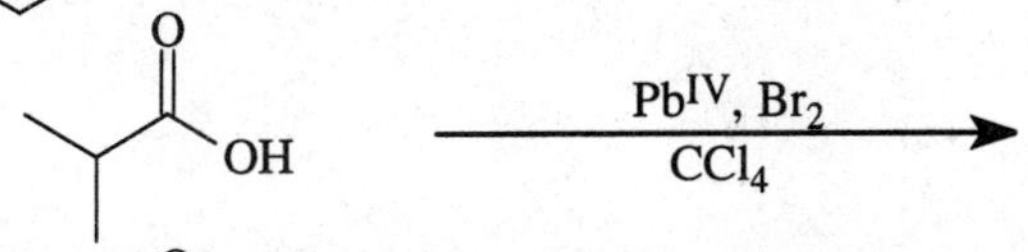

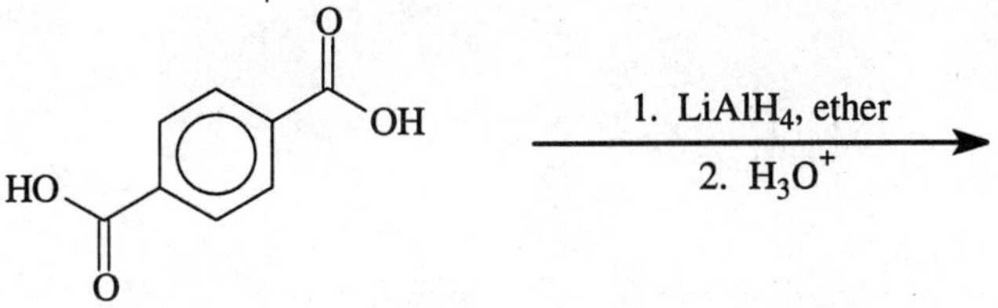

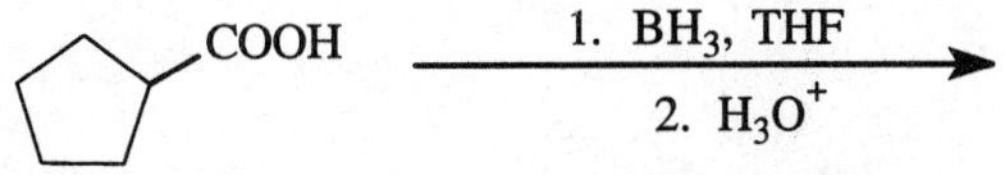

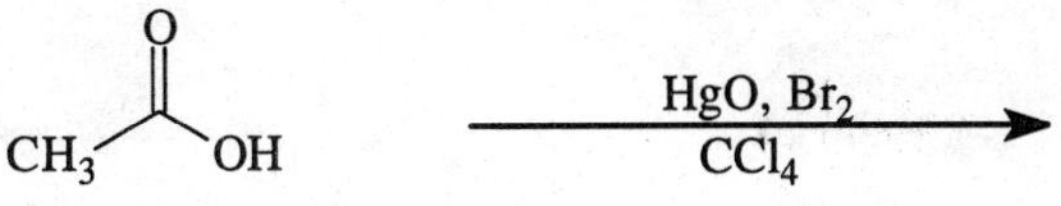

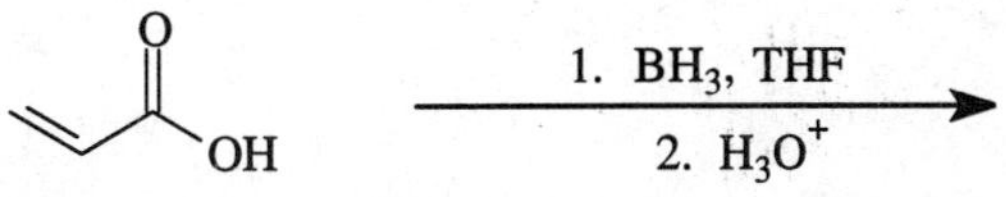

CH₃COOH $\xrightarrow{\text{1. LiAlH}_4,\ \text{ether}}_{\text{2. H}_3\text{O}^+}$

PhCH₂COOH $\xrightarrow{\text{1. BH}_3,\ \text{THF}}_{\text{2. H}_3\text{O}^+}$

cyclopentane-COOH $\xrightarrow{\text{Pb}^{\text{IV}},\ \text{Br}_2}_{\text{CCl}_4}$

camphor-like carboxylic acid $\xrightarrow{\text{HgO, Br}_2}_{\text{CCl}_4}$

(CH₃)₂CHCOOH $\xrightarrow{\text{1. BH}_3,\ \text{THF}}_{\text{2. H}_3\text{O}^+}$

PhCOOH $\xrightarrow{\text{HgO, Br}_2}_{\text{CCl}_4}$

(CH₃)₂CHCOOH $\xrightarrow{\text{Pb}^{\text{IV}},\ \text{Br}_2}_{\text{CCl}_4}$

terephthalic acid (HOOC-C₆H₄-COOH) $\xrightarrow{\text{1. LiAlH}_4,\ \text{ether}}_{\text{2. H}_3\text{O}^+}$

cyclopentyl-COOH $\xrightarrow{\text{1. BH}_3,\ \text{THF}}_{\text{2. H}_3\text{O}^+}$

CH₃COOH $\xrightarrow{\text{HgO, Br}_2}_{\text{CCl}_4}$

CH₂=CHCOOH $\xrightarrow{\text{1. BH}_3,\ \text{THF}}_{\text{2. H}_3\text{O}^+}$

Carboxylic Acids: *Synthesis*

Suggest a synthesis for each of the compounds shown below, beginning with acetaldehyde and/or butanoic acid as starting materials. Clearly show all intermediates and all reaction conditions.

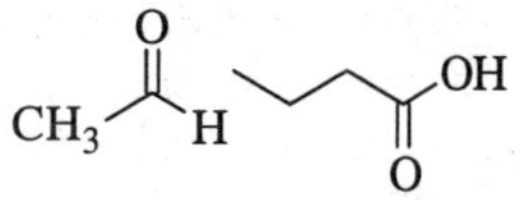

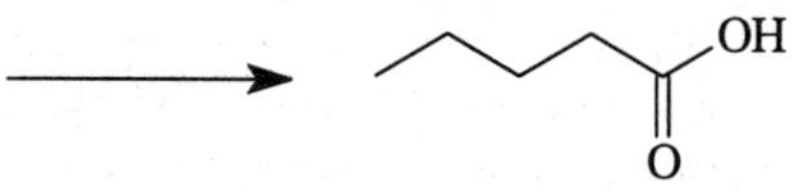

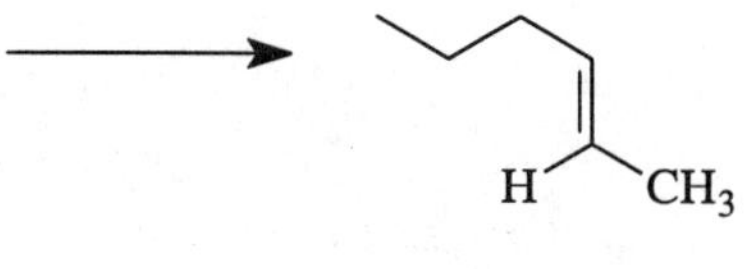

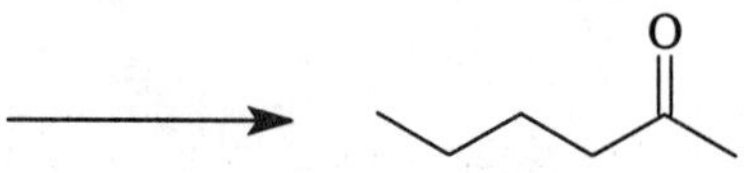

$$CH_3-CHO \qquad \qquad HOOC-CH_2CH_2-COOH$$

⟶ CH₃CH₂CH₂CH₂OH

⟶ CH₃CH₂CH₂CH₂Br

⟶ CH₃CH₂CH₂CH₂COOH

⟶ CH₃CH₂CH₂CH(OH)CH₃

⟶ CH₃CH₂CH₂CH=CHCH₃ (cis)

⟶ CH₃CH₂CH₂CH₂C(O)CH₃

Acyl Compounds (Carboxylic Acid Derivatives)

The group of compounds referred to as **acyl derivatives** include **acid halides, acid anhydrides, esters, carboxylic acids** and **amides; nitriles** can be converted into carboxylic acids by hydrolysis and are often included in this group even though they don't have carbonyl centers and don't undergo acyl transfer reactions. Acyl derivatives can undergo attack by a nucleophile at the carbonyl carbon to give an addition intermediate, which can breakdown to give a different acyl derivative. The order given above generally reflects the inherent reactivity of acyl compounds and acyl transfer reactions can be utilized to convert a given compound into any derivative which is *less reactive* than itself (below it on the reactivity list). Nomenclature of acyl derivatives is similar to that for carboxylic acids, and suffixes are shown below.

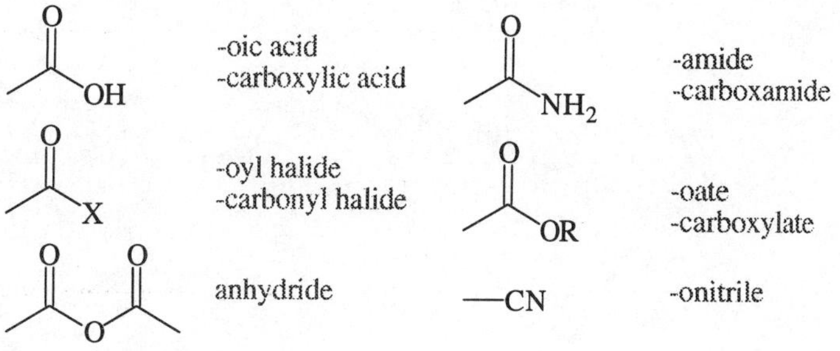

-oic acid
-carboxylic acid

-amide
-carboxamide

-oyl halide
-carbonyl halide

-oate
-carboxylate

anhydride

—CN   -onitrile

The Partitioning of Tetrahedral Intermediates

As described above, the attack of a nucleophile on an acyl derivative results in the formation of a transient tetrahedral intermediate. This intermediate **partitions**, that is, breaks down, by a series of parallel pathways to generate a product which is governed by the exact nature of the intermediate and the individual microscopic rate constants for each potential pathway. For tetrahedral intermediates formed from simply acyl derivatives, it is possible to apply a general set of rules to determine which group around the tetrahedral center is the "best leaving group", and hence, what will be the predominate product of a given reaction. In general, **for tetrahedral intermediates involving anionic leaving groups, you can assume that the best leaving group will be that group which has the strongest conjugate acid.**

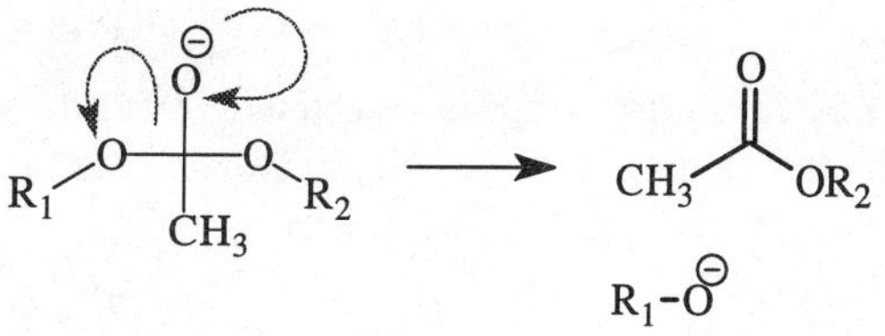

...for anionic leaving groups, in general, the group which has the

strongest conjugate acid (forms the most stable anion) will be the **best leaving group**, that is,

if R_1-OH is a **stronger acid** than R_2-OH, R_1-O⁻ will be **the preferred leaving group**...

For example, consider the intermediate shown below. The potential leaving groups are methoxide and phenoxide. The conjugate acids of methoxide and phenolate anions are methanol, $pK_a = 16$, and phenol, $pK_a = 9$ (approximate pK_a values). The strongest acid is phenol, and the major product formed will be ethyl acetate.

...phenol is the strongest acid, the most stable anion, and is the best leaving group

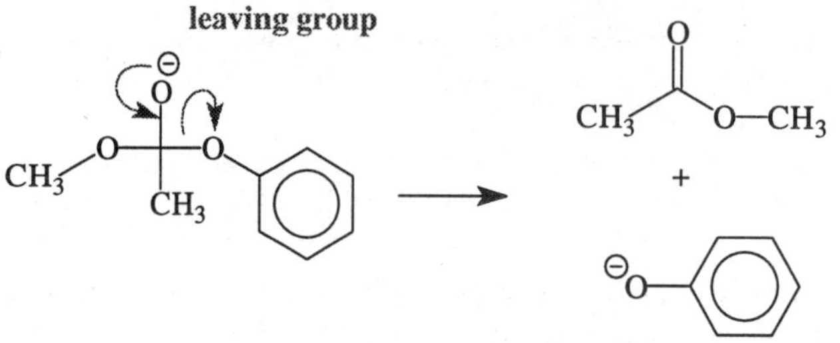

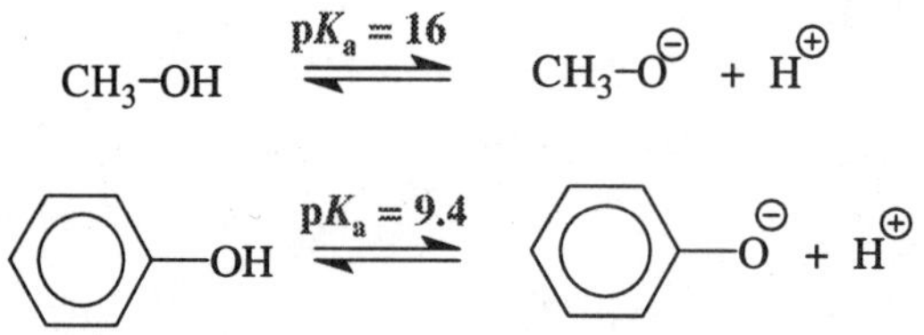

$pK_a = 16$
$CH_3-OH \rightleftharpoons CH_3-O^\ominus + H^\oplus$

$pK_a = 9.4$

Reactions of Carboxylic Acid Derivatives

Acid Halides: Acid halides are the most reactive acyl derivatives, and can be readily converted into carboxylic acids, esters and amides, by simple reaction with the appropriate nucleophile. The reaction involves an addition-elimination mechanism, as shown below:

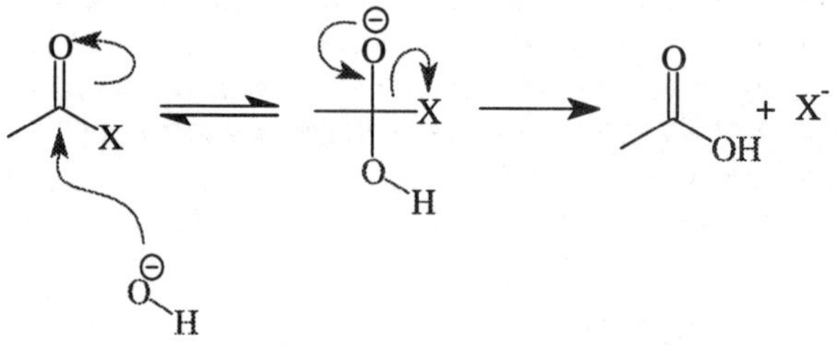

Reaction of acid halides with Grignard reagents, or reduction with $LiAlH_4$, leads to the incorporation of two moles of Grignard (or hydride), in a mechanism which involves an intermediate aldehyde or ketone, as shown below for hydride reduction.

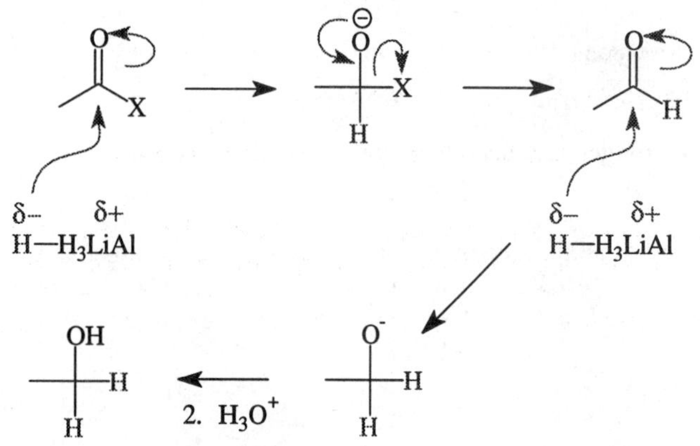

When a bulky reducing agent, such as lithium tri-*tert*-butoxylaluminum hydride is utilized, the reduction or acyl halides can be stopped at the intermediate aldehyde.

[Reaction scheme showing acyl halide reduction with H—LiAl(OC(CH₃)₃)₃ to aldehyde]

Partial alkylation can be achieved with acyl halides by reaction with a dialkylcopperlithium reagent, at low temperatures. This reaction is also useful for α,β-unsaturated aldehydes and ketones since conjugate addition of the alkyl group does not occur at the low temperatures utilized.

[Reaction scheme showing acyl halide + R₂CuLi giving ketone + X⁻]

Acid Anhydrides: Acid halides are the next most reactive acyl derivatives, and can be readily converted into carboxylic acids, esters and amides, by simple reaction with the appropriate nucleophile. As with acid halides, reduction of anhydrides with LiAlH$_4$ results in the addition of two moles of hydride, forming the primary alcohol.

Carboxylic Esters: Carboxylate esters can be readily converted into carboxylic acids and amides, by simple reaction with the appropriate nucleophile. Reaction of esters with Grignard reagents, or reduction with LiAlH$_4$, leads to the incorporation of two moles of Grignard (or hydride), in a mechanism which involves an intermediate aldehyde or ketone, similar to that shown above for acid halides.

Amides: Amides are relatively unreactive in acyl transfer reactions, largely because the electrons from the adjacent nitrogen participate in resonance delocalization with the adjacent carbonyl, making the carbonyl carbon significantly less electropositive.

[Resonance structures of an amide showing the neutral form and the charge-separated form with C=N⁺ and O⁻]

...more single C-O character,
less positive charge on the carbonyl carbon

Amides undergo acid-catalyzed hydrolysis to give carboxylic acids by the standard addition-elimination mechanism, as shown previously. Unsubstituted amides also undergo dehydration in the presence of SOCl$_2$ to give the corresponding nitrile.

$$\underset{NH_2}{\overset{O}{\|}}{\text{C}}\text{—} \xrightarrow{SOCl_2} \text{—C≡N}$$

Nitriles: Nitriles can be hydrolyzed carboxylic acids, converted into ketones by reaction with Grignard reagents, reduced to primary amines with LiAlH$_4$ and partially reduced to aldehydes using DIBAH (diisobutylaluminum hydride), as shown below.

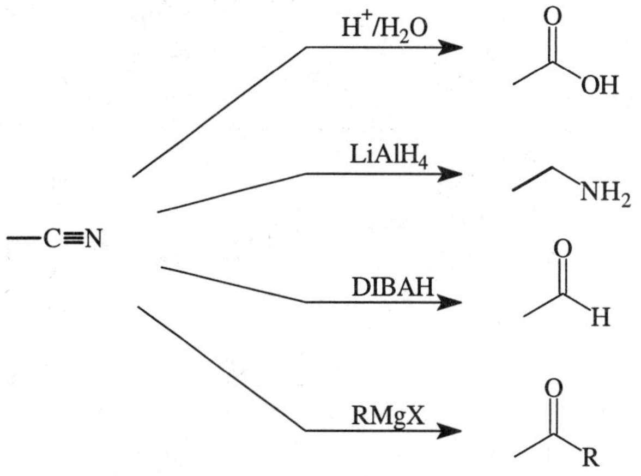

—C≡N

- H$^+$/H$_2$O → carboxylic acid (—COOH)
- LiAlH$_4$ → primary amine (—CH$_2$NH$_2$)
- DIBAH → aldehyde (—CHO)
- RMgX → ketone (—COR)

The reaction with Grignard reagent is typical of these latter reactions and involves nucleophilic attack on the nitrile carbon to give an anionic intermediate which is resistant to further attack, and undergoes hydrolysis to give a ketone, as shown below.

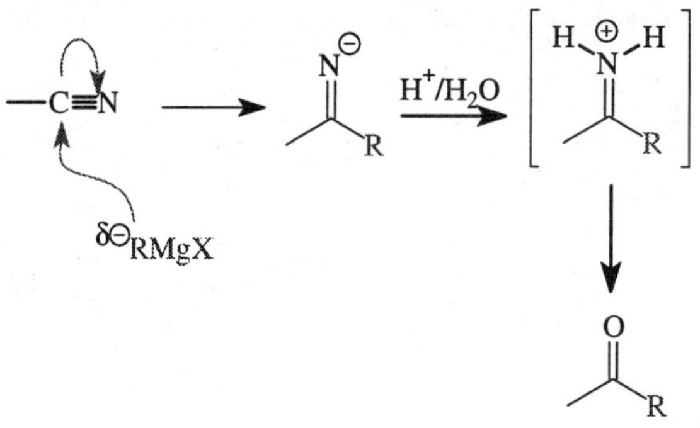

—C≡N + $\delta^{\ominus}$RMgX → imine anion (N$^{\ominus}$=C(R)—) $\xrightarrow{H^+/H_2O}$ [protonated iminium] → ketone (—COR)

Carboxylic Acid Derivatives: *Reactions of Acid Halides, Anhydrides & Esters*

Carboxylic Acid Derivatives: *Reactions of Nitriles.*

—C≡N

- H^+/H_2O → CH₃COOH
- $LiAlH_4$ → CH₃CH₂NH₂
- DIBAH → CH₃CHO
- RMgX → CH₃C(O)R

Carboxylic Acid Derivatives: *Reactions that Yield Acid Halides & Esters*

- PhCOOH —$SOCl_2$→ PhCOCl
- PhCH₂COOH —(COCl)₂→ PhCH₂COCl
- cyclopentyl-COOH —PBr_3→ cyclopentyl-COBr
- PhCOOH —ROH, H^+→ PhCOOR
- PhCH₂COOH —RX→ PhCH₂C(O)—OR
- cyclopentyl-COOH —CH_2N_2→ cyclopentyl-C(O)OCH₃

Carboxylic Acid Derivatives: *Nomenclature*
For each of the compounds shown below, provide an acceptable IUPAC name.

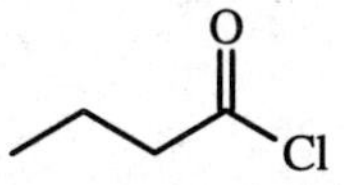

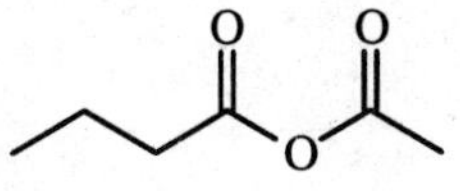

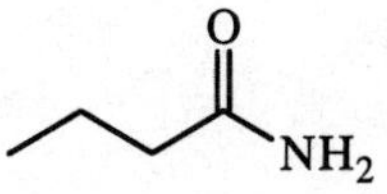

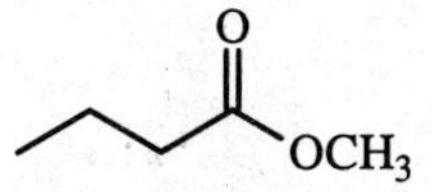

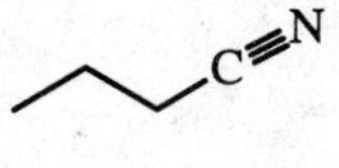

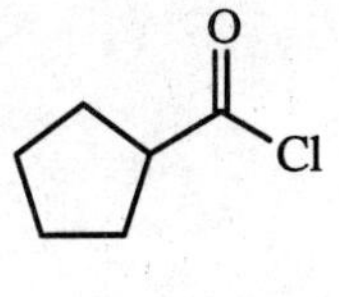

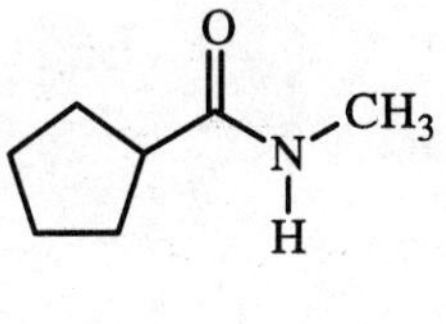

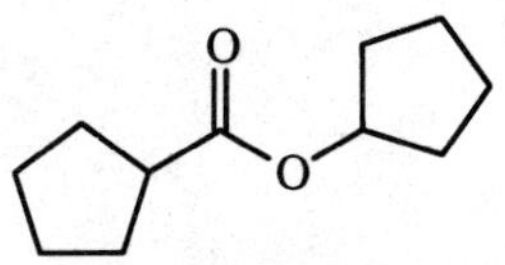

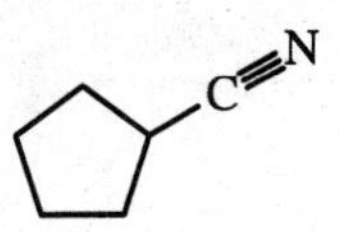

Carboxylic Acid Derivatives: *Tetrahedral Intermediates*

For each of the tetrahedral intermediates shown below, predict the leaving group and draw the expected product.

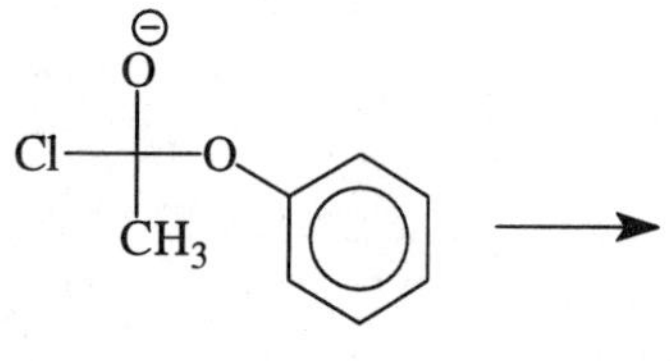

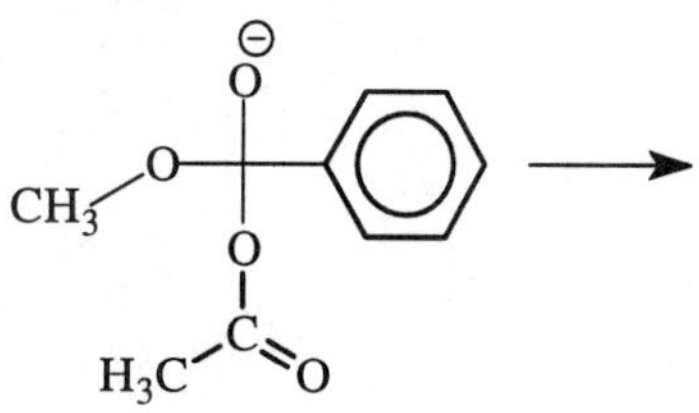

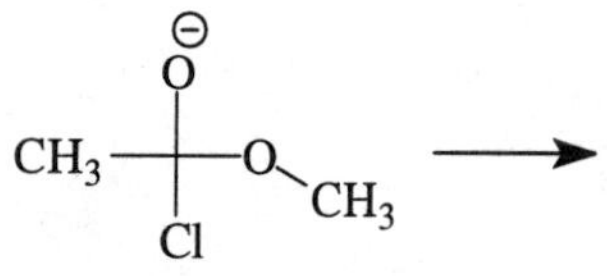

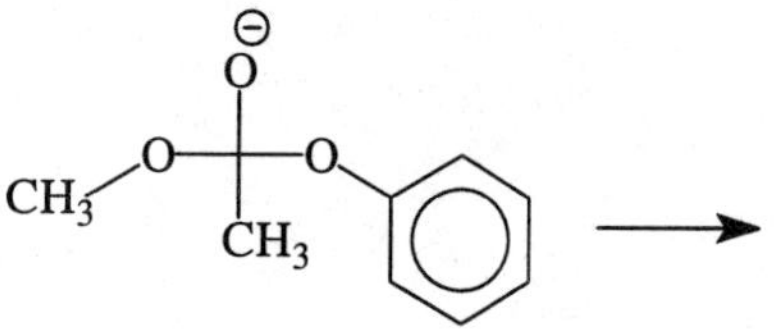

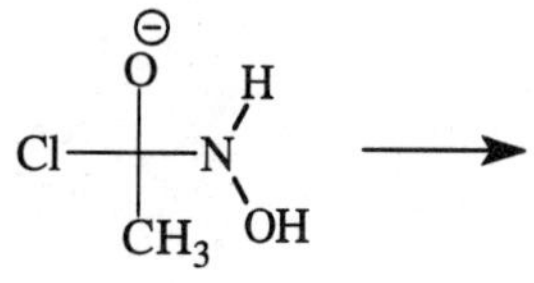

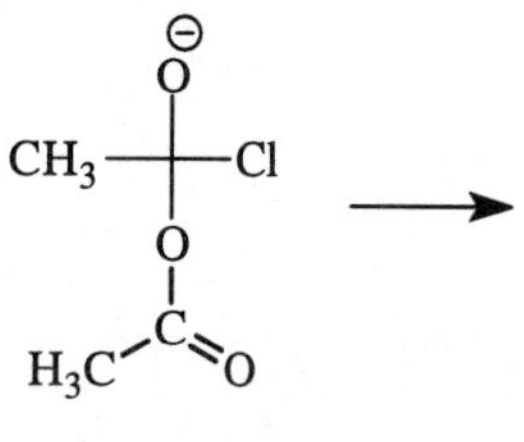

Carboxylic Acid Derivatives: *Reactions that Yield Acyl Derivatives*

For each of the reactions shown below, predict the major organic product.

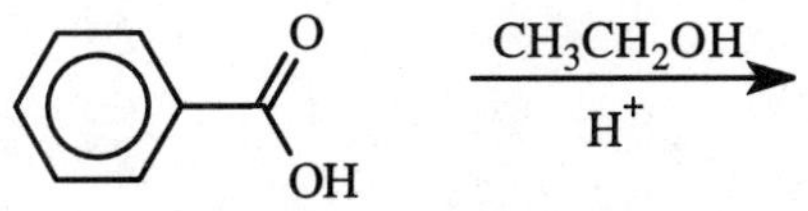

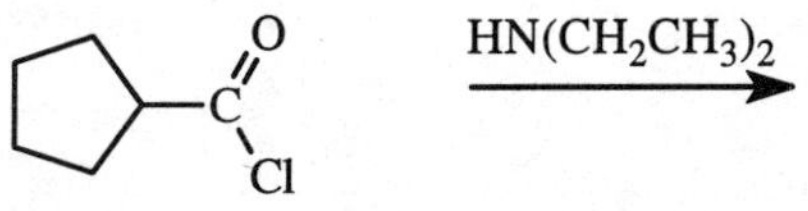

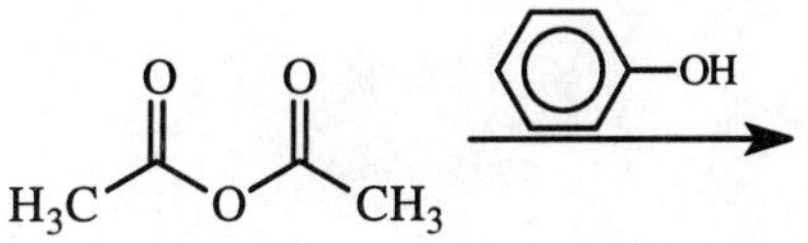

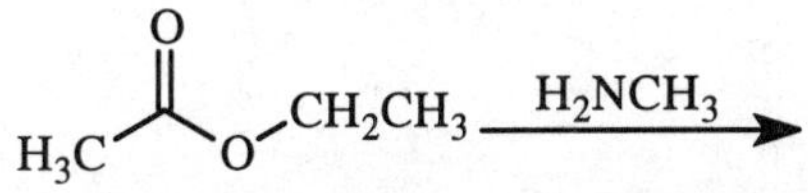

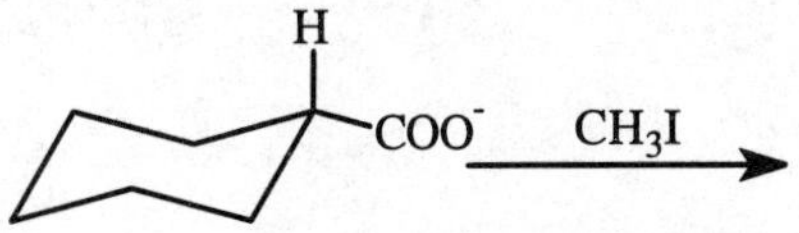

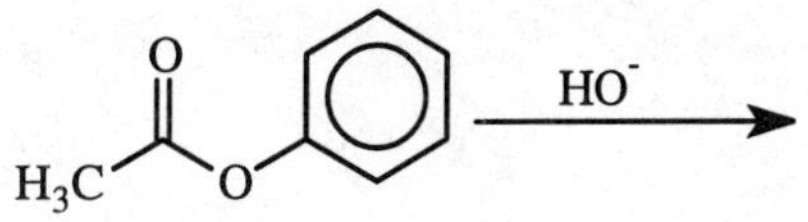

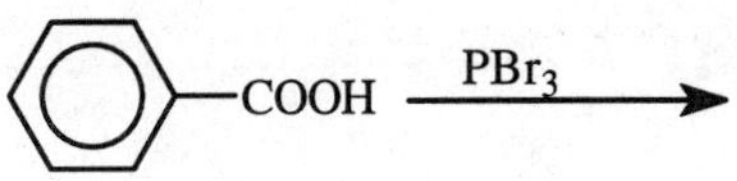

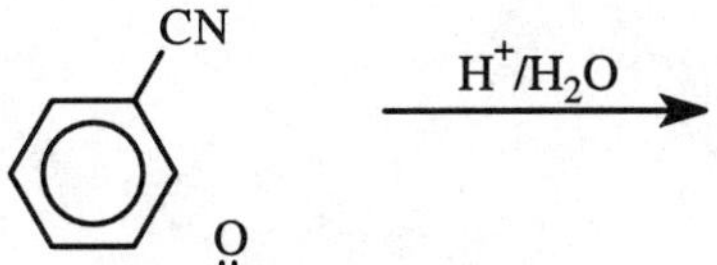

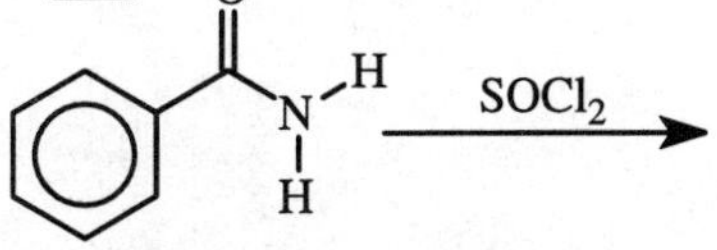

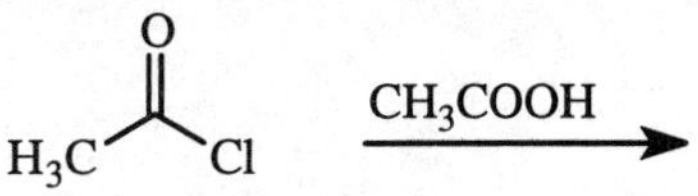

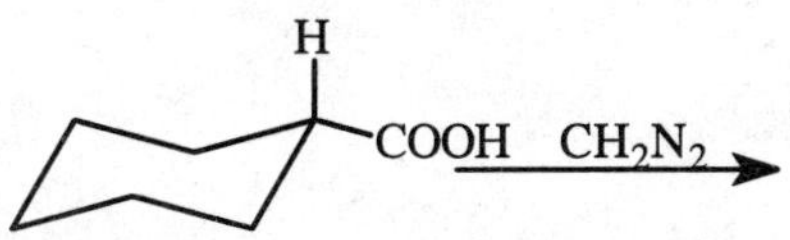

PhCOOH + CH$_3$CH$_2$OH / H$^+$ →

cyclopentyl-COCl + HN(CH$_2$CH$_3$)$_2$ →

CH$_3$CH$_2$CH$_2$CH$_2$COOH + SOCl$_2$ →

(CH$_3$CO)$_2$O + PhOH →

CH$_3$COOCH$_2$CH$_3$ + H$_2$NCH$_3$ →

cyclohexyl-COO$^-$ + CH$_3$I →

CH$_3$COOPh + HO$^-$ →

PhCOOH + PBr$_3$ →

PhCN + H$^+$/H$_2$O →

PhCONH$_2$ + SOCl$_2$ →

CH$_3$COCl + CH$_3$COOH →

cyclohexyl-COOH + CH$_2$N$_2$ →

Carboxylic Acid Derivatives: *Reactions of Acyl Compounds*

For each of the reactions shown below, predict the major organic product.

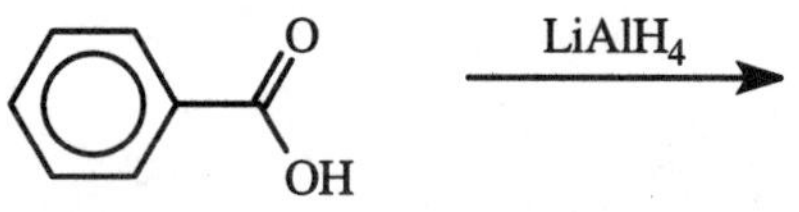

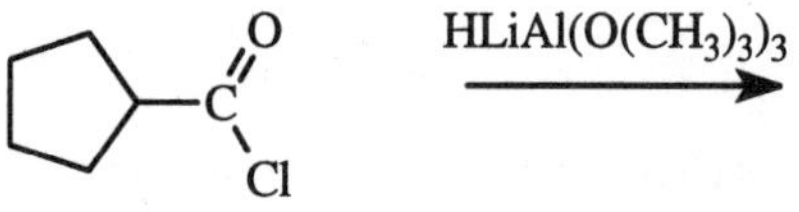

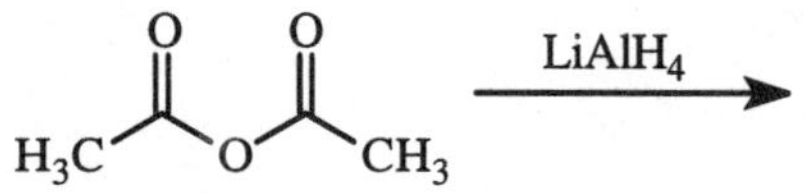

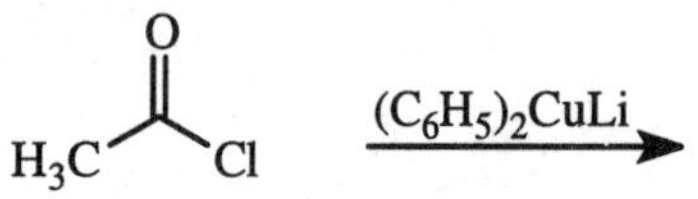

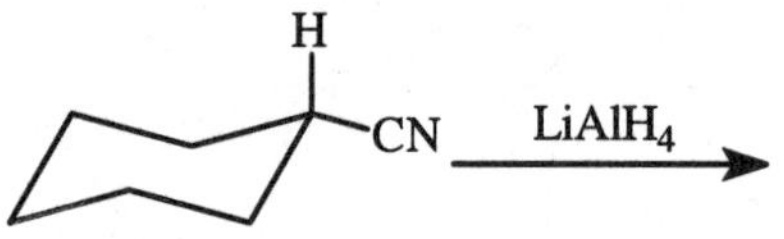

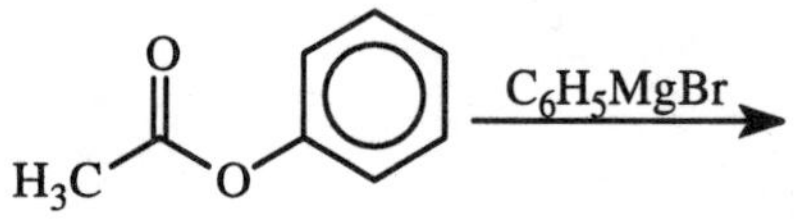

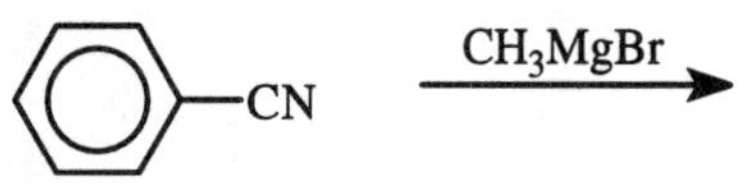

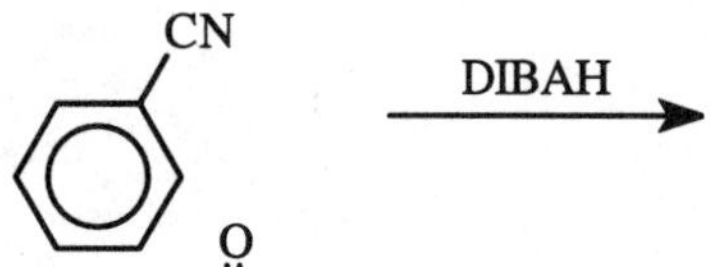

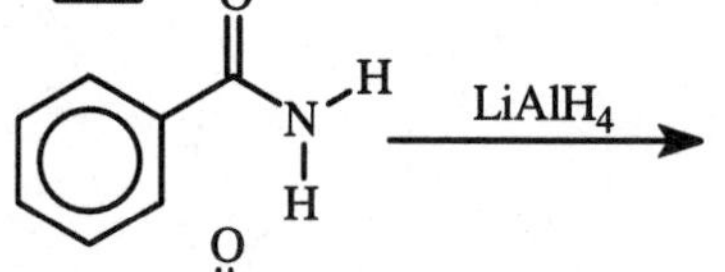

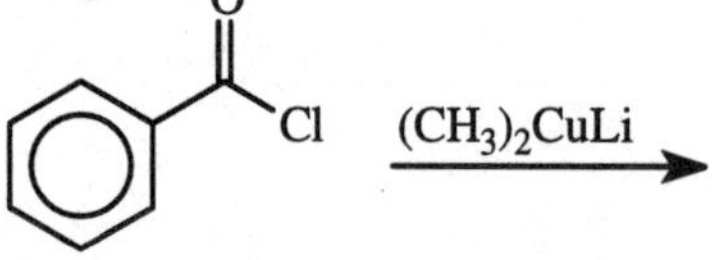

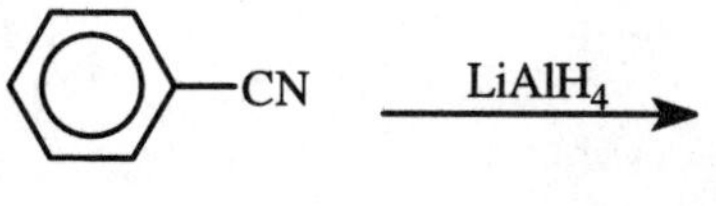

PhCOOH $\xrightarrow{\text{LiAlH}_4}$

cyclopentyl-C(=O)Cl $\xrightarrow{\text{HLiAl(O(CH}_3)_3)_3}$

CH$_3$CH$_2$CH$_2$CH$_2$–CN $\xrightarrow{\text{CH}_3\text{MgBr}}$

(CH$_3$CO)$_2$O $\xrightarrow{\text{LiAlH}_4}$

CH$_3$COCl $\xrightarrow{(\text{C}_6\text{H}_5)_2\text{CuLi}}$

cyclohexyl-CN $\xrightarrow{\text{LiAlH}_4}$

CH$_3$C(=O)O–C$_6$H$_5$ $\xrightarrow{\text{C}_6\text{H}_5\text{MgBr}}$

C$_6$H$_5$–CN $\xrightarrow{\text{CH}_3\text{MgBr}}$

C$_6$H$_5$–CN $\xrightarrow{\text{DIBAH}}$

C$_6$H$_5$C(=O)NH$_2$ $\xrightarrow{\text{LiAlH}_4}$

C$_6$H$_5$C(=O)Cl $\xrightarrow{(\text{CH}_3)_2\text{CuLi}}$

C$_6$H$_5$–CN $\xrightarrow{\text{LiAlH}_4}$

Carboxylic Acid Derivatives: *Synthesis*

Suggest a synthesis for each of the compounds shown below, beginning with the starting material shown on the left. Clearly show all intermediates and reaction conditions.

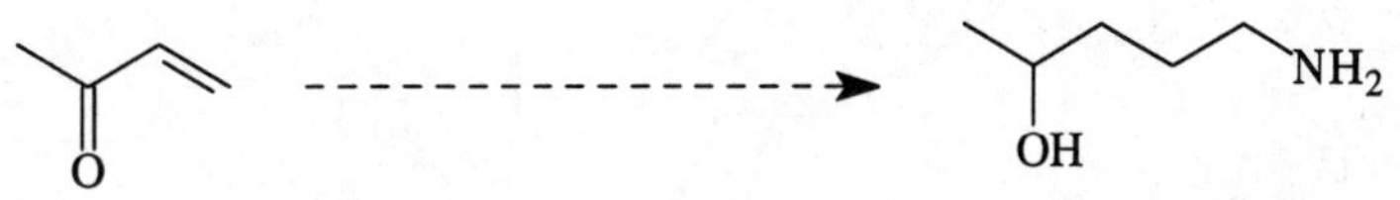

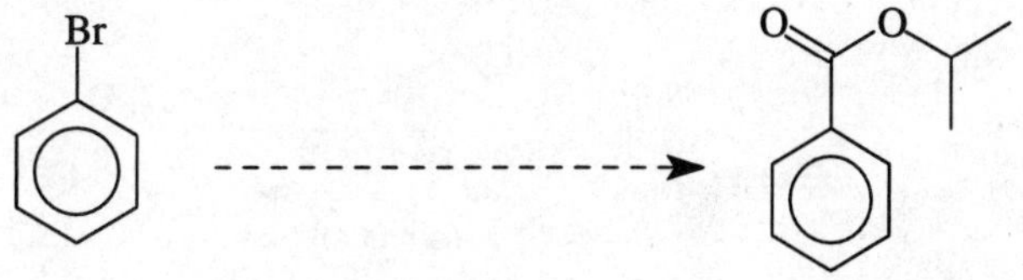

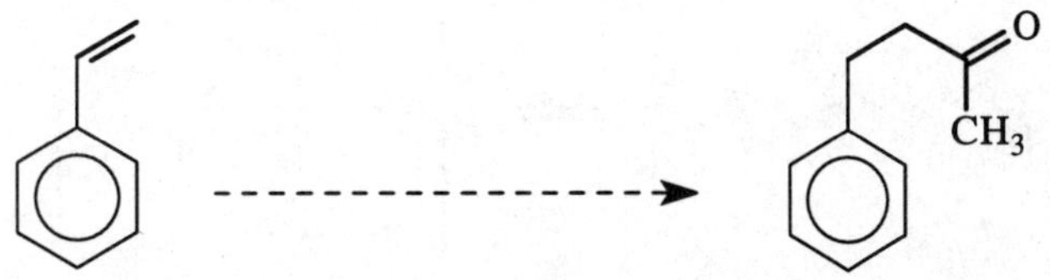

Carbonyl α-Substitution Reactions

Carbonyl compounds exist in equilibrium with a very small amount of a structural isomer, termed an enol. An enol is formed by abstraction of a proton from the α-carbon, delocalization of the electrons onto the carbonyl oxygen, and finally, protonation of the oxygen to give an alkene bonded to an alcohol (an enol!). It is important to note that this is a true equilibrium and the carbonyl compound and its enol are distinct different chemical species, not resonance forms. Since both proton abstraction and donation are required in the isomerization, keto-enol isomerization is catalyzed by both acids and bases.

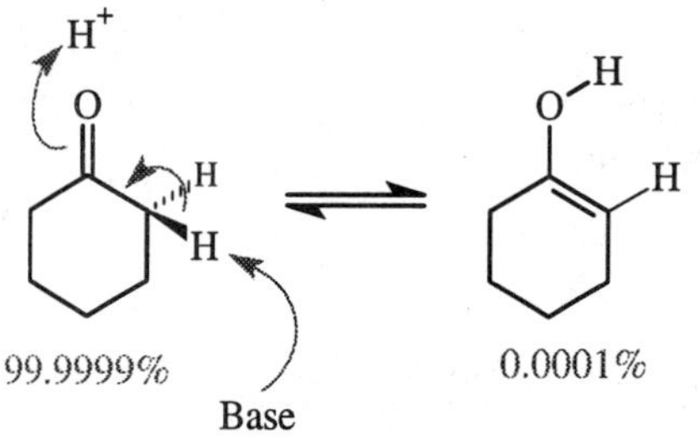

In spite of their low equilibrium concentrations, enols and enolate anions are useful in organic chemistry because they can be used as nucleophiles to attack electrophilic centers such as Br_2, alkyl halides, and other carbonyl carbons.

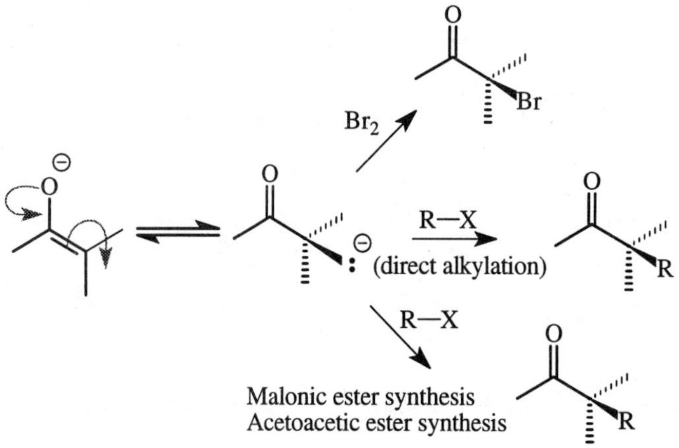

Bromination of an α-carbon is accomplished by reacting the carbonyl compound with bromine in an acidic solution (or in acetic acid as solvent). Under these conditions, the α-carbanion character of the enol attacks Br_2 to form the α-bromo carbonyl compound, as shown below.

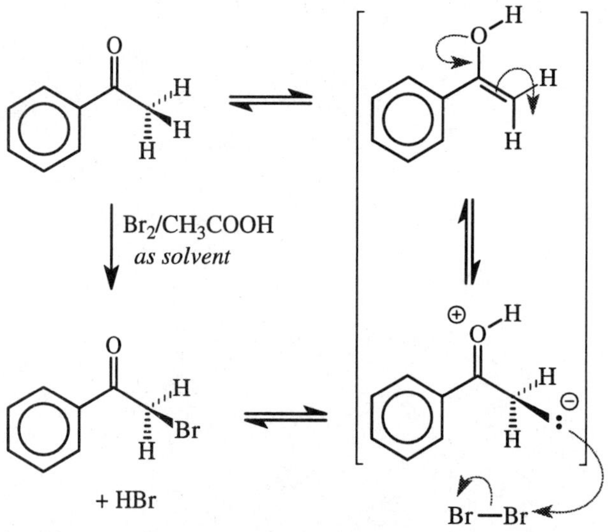

Chlorine can be introduced into the α-position conveniently using Cl_2 in aqueous HCl.

The same general rules apply regarding enol stability as for alkenes, that is, the more highly substituted enol is favored, unless it is crowded sterically.

most highly substituted enol → Br$_2$/CH$_3$COOH → **major product**

The α-halo ketones and aldehydes which are formed can undergo an E2 elimination reaction in the presence of base to give the α,β-unsaturated ketone or aldehyde; pyridine is commonly used as a base for this purpose.

...the mechanism is a simple E2 elimination

α,β-Unsaturated ketones can also be prepared using an intermediate organo-selenium compound. Reaction of an enolizable carbon with LDA (lithium diisopropylamide; a very strong base prepared from diisopropylamine and butyllithium) generates the stable enolate anion as the lithium salt. This reacts with benzeneselenyl bromide to give the α-selenium intermediate, which is not isolated, but is oxidized with H$_2$O$_2$ to a powerful leaving group which eliminates to form the α,β-unsaturated ketone. The real beauty of this reaction is that it is useful for nitriles and esters, as well as ketones.

Direct bromination in acetic acid is limited to ketones and aldehydes, but α-bromo acids can be prepared using the **Hell-Volhard-Zelinskii** reaction. This reaction involves the conversion of the acid to the intermediate acid bromide, enolization, bromination to give the α-bromo acid bromide. The final step is work-up with water, which hydrolyzes the acid bromide to the acid, yielding the α-bromo acid as the final product. An interesting twist on this reaction is to work up the product in alcohol; this variation yields the corresponding α-bromo ester.

In another variation on the halogenation reaction, reaction of a methyl ketone or acetaldehyde (the only methyl aldehyde) with I_2 in the presence of hydroxide anion generates the triiodo-ketone. Attack by hydroxide anion on this forms the corresponding carboxylic acid and iodoform, a yellow compound which is insoluble in the aqueous base. A precipitate of iodoform is a standard qualitative test for the presence of a methyl ketone.

The Malonic Ester Synthesis

One of the most useful reactions involving enolates is their alkylation, using an alkyl halide as electrophile. These reactions are best performed using compounds with acidic α-hydrogens, typically those adjacent to two carbonyls or nitrile groups. Esters of malonic acid (propanedioic acid) are commonly used for these reactions and the reactions are generally referred to as the malonic ester synthesis. The α-hydrogens of diethyl malonate have a pK_a of about 10, and in the presence of ethoxide in ethanol, it is completely converted to the enolate anion.

If an alkyl halide is present, the enolate anion can undergo an S$_N$2 reaction to give a mono-alkylmalonic ester.

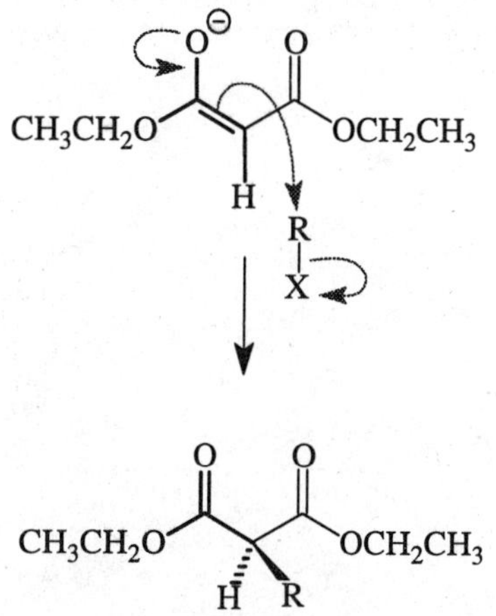

Acid hydrolysis of the diester generates the diacid, which is a β-carbonyl carboxylic acid. These are unstable, and undergo decarboxylation in the acidic solution to give the acid-enol, which isomerizes to give the final product, a simple carboxylic acid. You should note, however, that two carbons of the malonic ester have been added to the carbon skeleton of the alkyl halide which was used in the first step. Thus, using the malonic ester synthesis, two carbons (a CH$_2$ and a COOH) can be appended to virtually any primary or secondary alkyl halide.

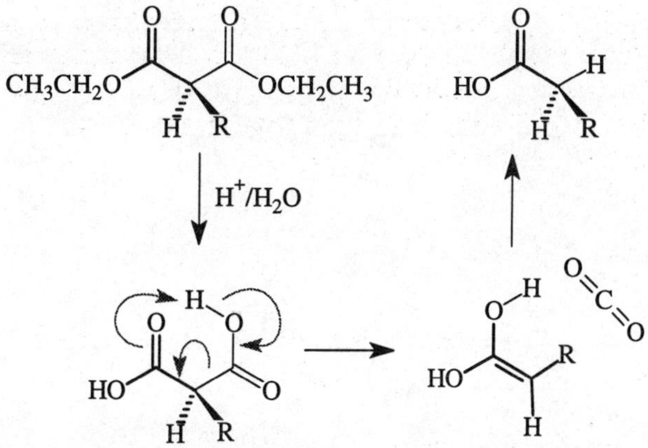

An example of a malonic ester synthesis is shown below. Reaction of benzyl bromide with diethylmalonate in ethoxide/ethanol yields the alkylated product. Acidification hydrolyzes the esters and the intermediate diacid decarboxylates to form the final product, 3-phenylpropanoic acid. The line in the figure shows the bond which was formed in the reaction and the benzyl unit and the -CH$_2$COOH (from the malonic ester) can be clearly identified.

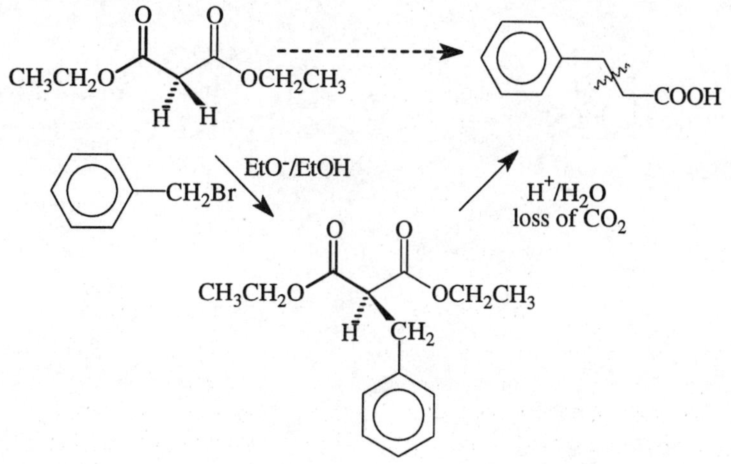

Identifying the alkyl halide necessary for a malonic ester synthesis simply requires you to remember that the carboxyl group and the α-carbon will come from the malonic ester, and the remainder of the molecule must come from the alkyl halide. For example, 3-methyl-3-phenylpropanoic acid can be prepared using (1-bromoethyl)benzene, as shown below:

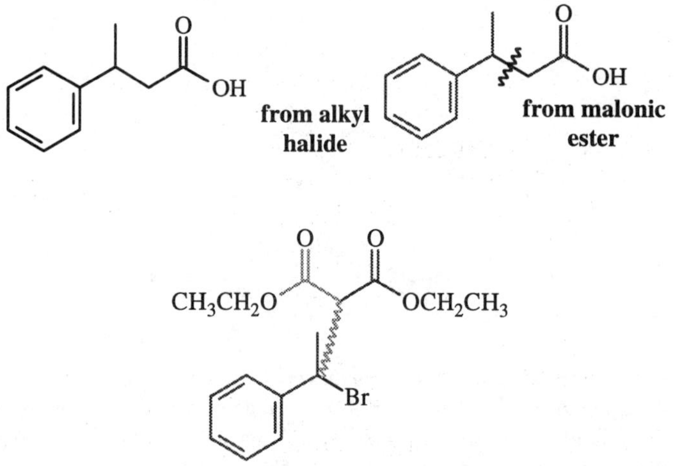

2-Methylpropanoic acid can be prepared by alkylating diethylmalonate twice with bromomethane,

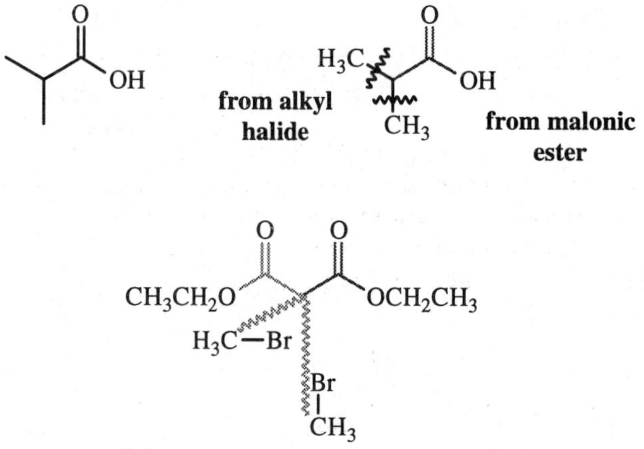

and 4,4-diphenylbutanoic acid can be prepared by alkylation with (2-bromo-1-phenylethyl)benzene.

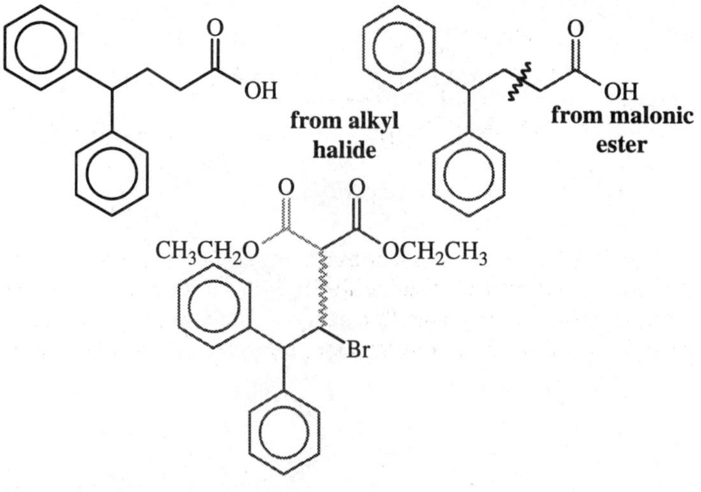

The Acetoacetic Ester Synthesis

In addition to malonic esters, esters of acetoacetic acid (3-oxobutanoic acid) are also commonly used for these reactions and the reactions are generally referred to as the acetoacetic ester synthesis. The α-hydrogens of ethyl acetoacetate have a pK_a of about 10, and in the presence of ethoxide in ethanol, it is completely converted to the enolate anion. Again, if an alkyl halide is present, this enolate anion can undergo an S_N2 reaction to give a mono-alkyl acetoacetic ester.

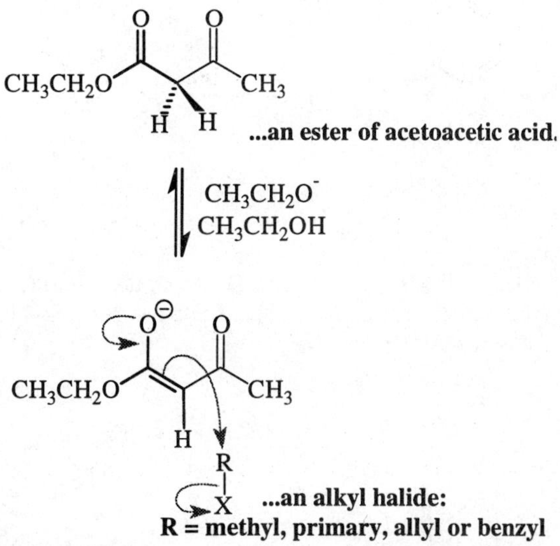

Acid hydrolysis generates a β-keto carboxylic acid. These are unstable, and undergo decarboxylation in the acidic solution to give the enol, which isomerizes to give the final product, a methyl ketone. You should note that three carbons of the acetoacetic ester have been added to the carbon skeleton of the alkyl halide which was used in the first step. Thus, using the acetoacetic ester synthesis, three carbons (a CH_2, a carbonyl and a CH_3) can be appended to virtually any primary, allyl or benzyl halide.

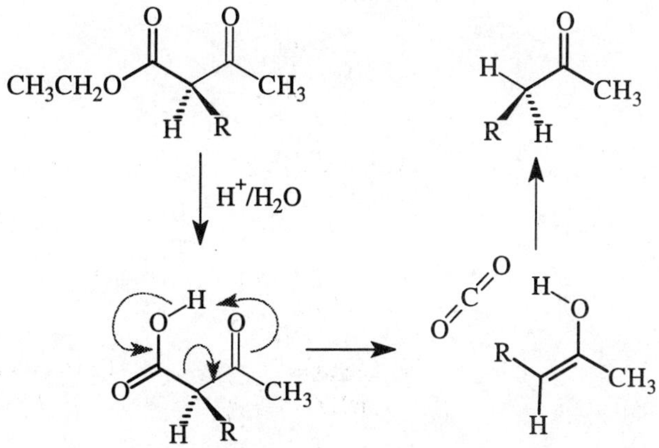

An example of an acetoacetic ester synthesis is shown below. Reaction of 1-bromopropane with ethylacetoacetate in ethoxide/ethanol yields the alkylated product. Acidification hydrolyzes the ester and the intermediate β-keto acid decarboxylates to form the final product, 2-hexanone. The red line in the figure shows the bond which was formed in the reaction and the propyl unit and the three carbons from the acetoacetic ester can be clearly identified.

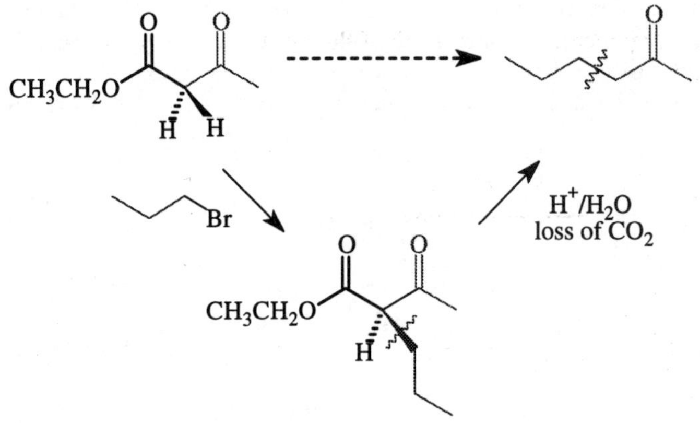

The alkylation of less acidic α-hydrogens can be accomplished using LDA to completely convert the carbonyl compound into the desired enolate anion. This is then reacted with an alkyl halide to give the alkylated product.

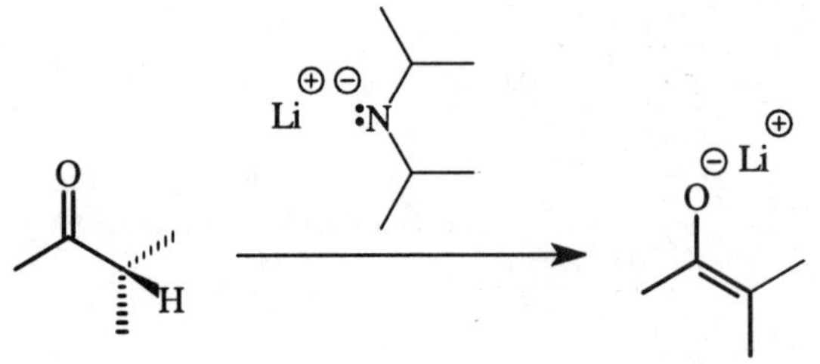

This reaction is useful with ketones, aldehydes, esters and nitriles, as shown below:

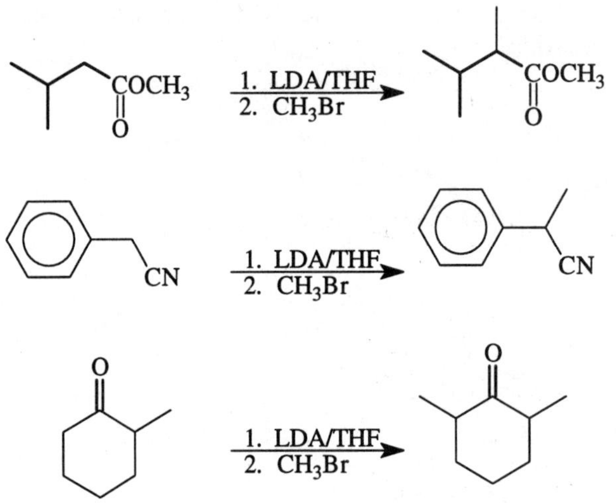

Carbonyl α–Substitution: *Reactions-I*

Predict the major organic product for each of the reactions shown below. Clearly show stereochemistry, if appropriate.

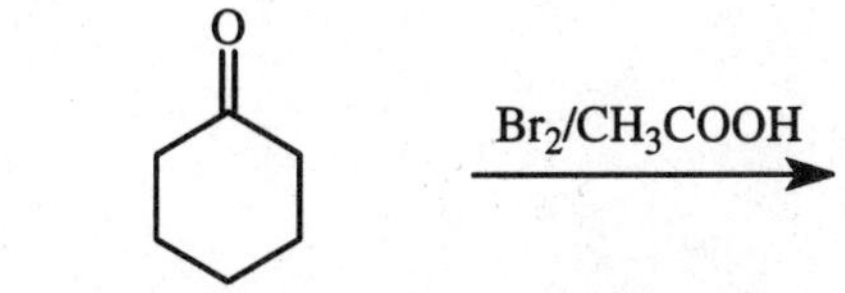

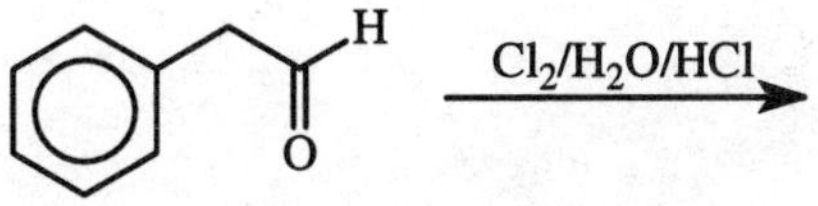

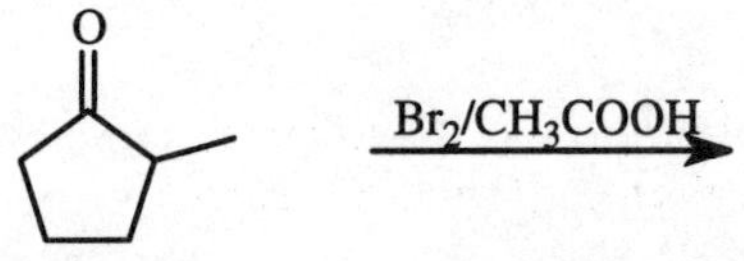

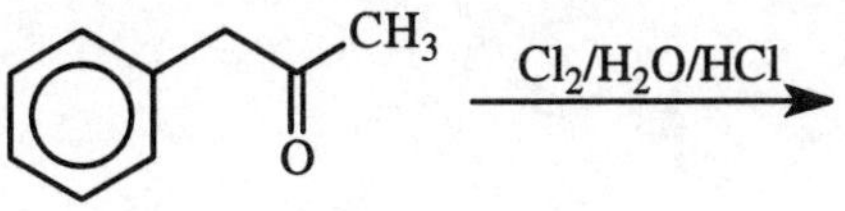

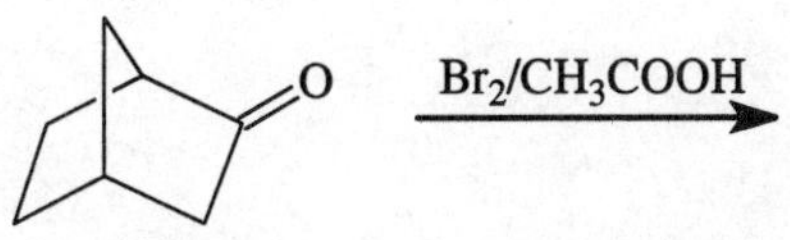

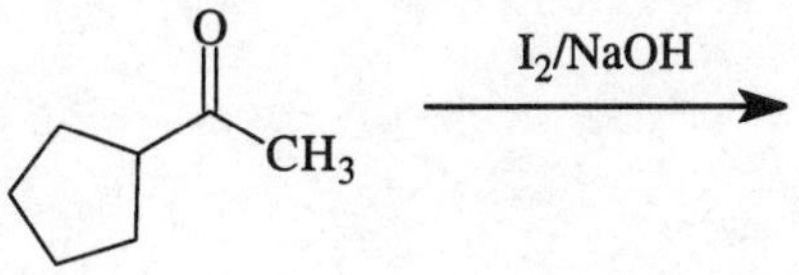

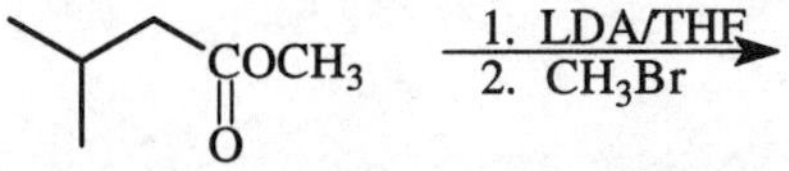

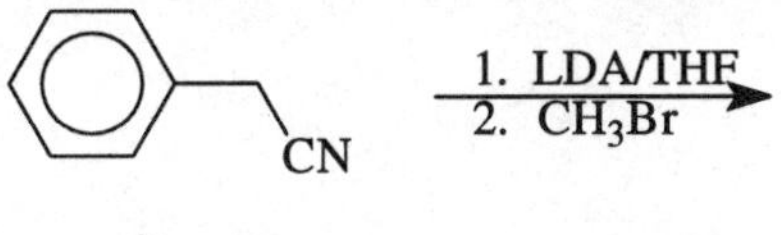

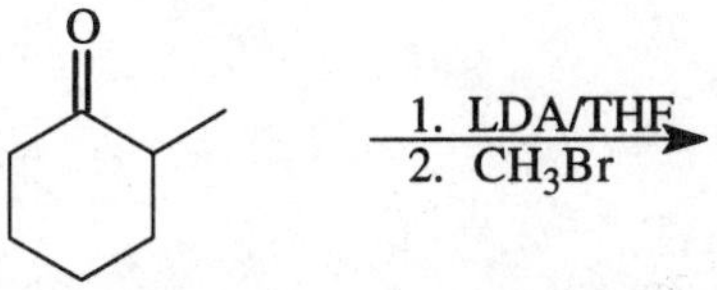

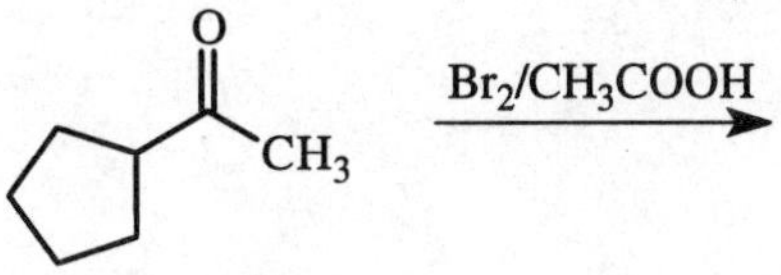

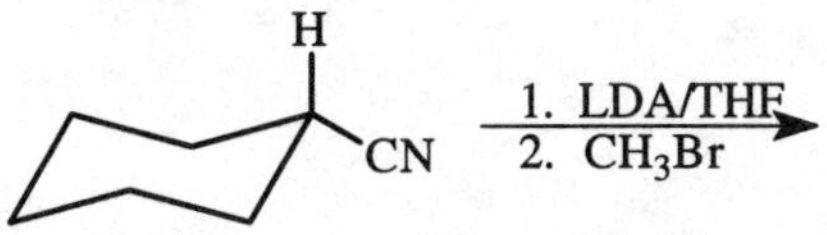

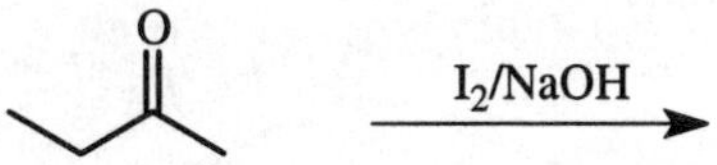

cyclohexanone + Br₂/CH₃COOH →

phenylacetaldehyde (PhCH₂CHO) + Cl₂/H₂O/HCl →

2-methylcyclopentanone + Br₂/CH₃COOH →

phenylacetone (PhCH₂COCH₃) + Cl₂/H₂O/HCl →

norcamphor (bicyclic ketone) + Br₂/CH₃COOH →

1-cyclopentyl-1-ethanone (cyclopentyl methyl ketone) + I₂/NaOH →

methyl isovalerate ((CH₃)₂CHCH₂COOCH₃) + 1. LDA/THF 2. CH₃Br →

phenylacetonitrile (PhCH₂CN) + 1. LDA/THF 2. CH₃Br →

2-methylcyclohexanone + 1. LDA/THF 2. CH₃Br →

1-cyclopentyl-1-ethanone + Br₂/CH₃COOH →

cyclohexanecarbonitrile (with H shown) + 1. LDA/THF 2. CH₃Br →

2-butanone + I₂/NaOH →

Carbonyl α–Substitution: *Reactions-II*

Predict the major organic product for each of the reactions shown below. Clearly show stereochemistry, if appropriate.

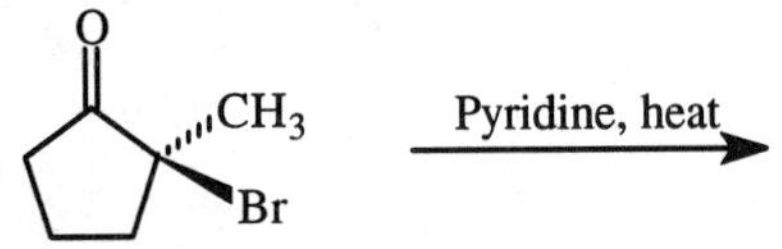

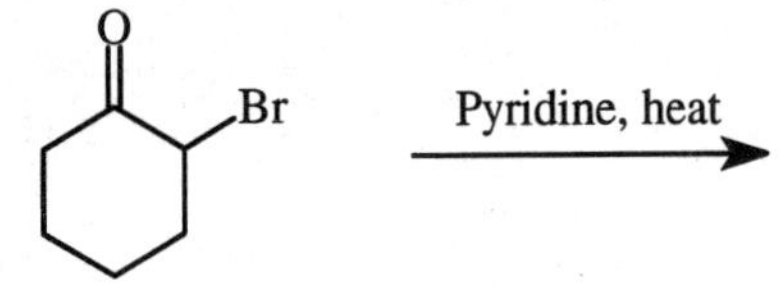

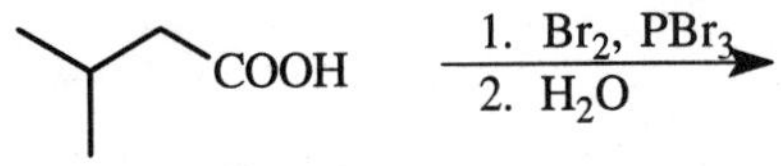

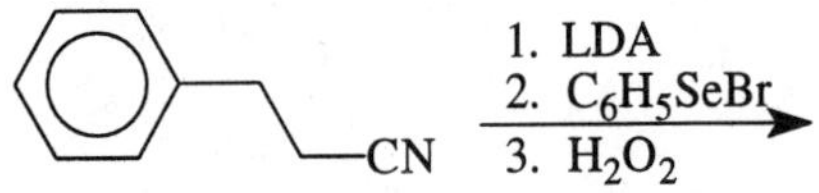

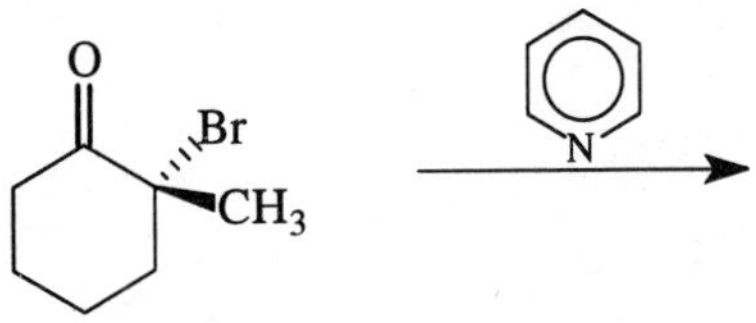

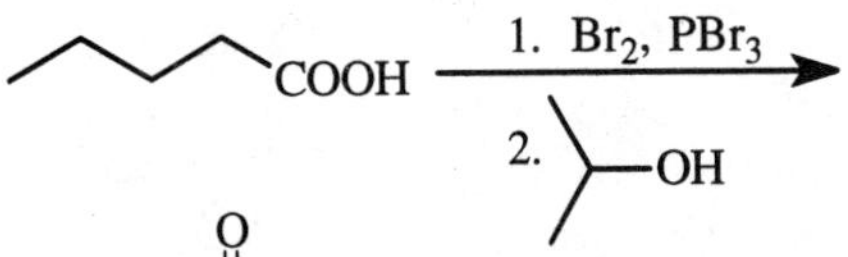

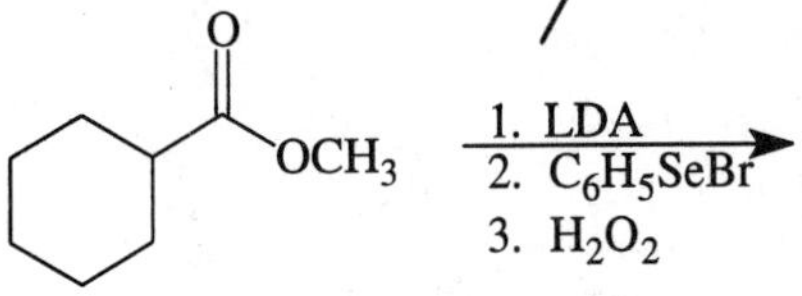

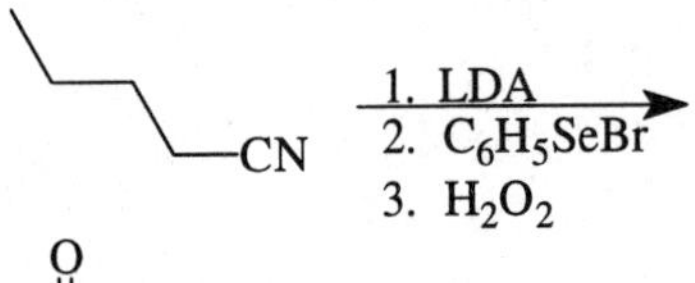

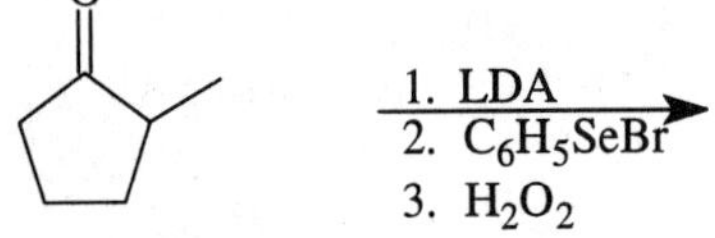

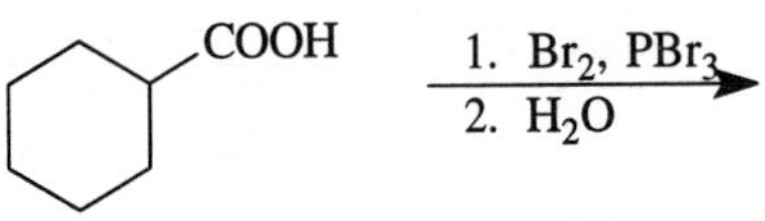

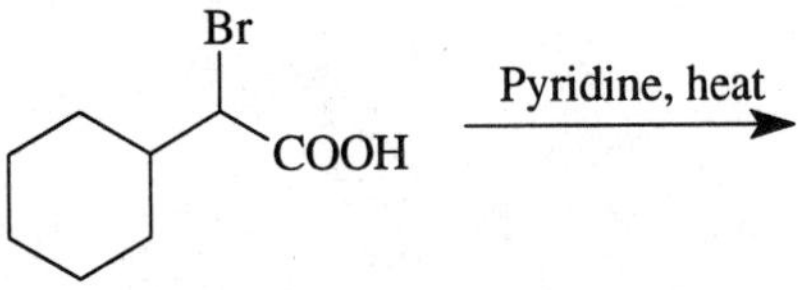

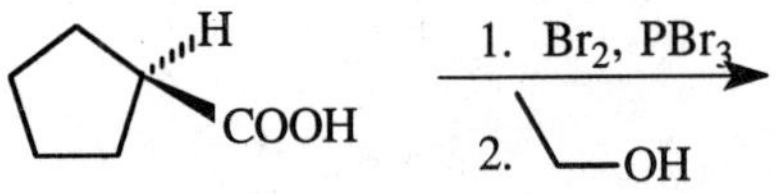

2-bromo-2-methylcyclopentanone → Pyridine, heat

2-bromocyclohexanone → Pyridine, heat

(CH₃)₂CHCH₂COOH → 1. Br₂, PBr₃ 2. H₂O

PhCH₂CH₂CN → 1. LDA 2. C₆H₅SeBr 3. H₂O₂

2-bromo-2-methylcyclohexanone → pyridine

CH₃CH₂CH₂CH₂COOH → 1. Br₂, PBr₃ 2. (CH₃)₂CHOH

methyl cyclohexanecarboxylate → 1. LDA 2. C₆H₅SeBr 3. H₂O₂

CH₃CH₂CH₂CH₂CN → 1. LDA 2. C₆H₅SeBr 3. H₂O₂

2-methylcyclopentanone → 1. LDA 2. C₆H₅SeBr 3. H₂O₂

cyclohexanecarboxylic acid → 1. Br₂, PBr₃ 2. H₂O

α-bromo-α-cyclohexylacetic acid → Pyridine, heat

(1-H)cyclopentanecarboxylic acid → 1. Br₂, PBr₃ 2. (CH₃)₂CHOH

Carbonyl α–Substitution Reactions: *The Malonic Ester Synthesis*

Suggest a synthesis for each of the compounds shown below, beginning with diethyl malonate and any other required materials. Clearly show all intermediates and reaction conditions.

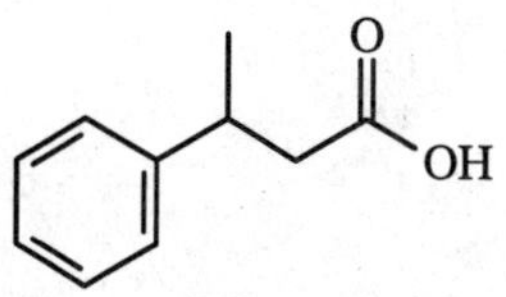

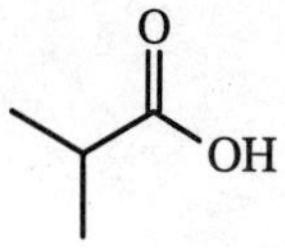

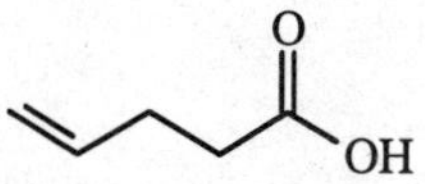

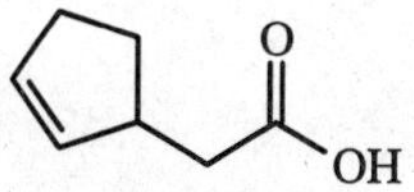

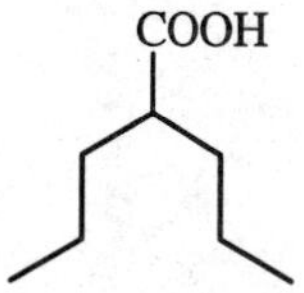

Carbonyl α–Substitution Reactions: *The Acetoacetate Ester Synthesis*

Suggest a synthesis for each of the compounds shown below, beginning with ethyl acetoacetate and any other required materials. Clearly show all intermediates and reaction conditions.

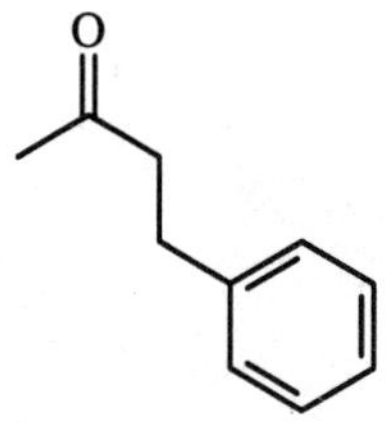

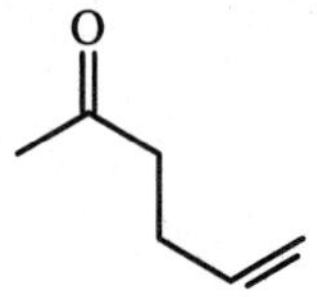

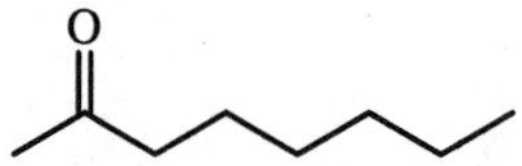

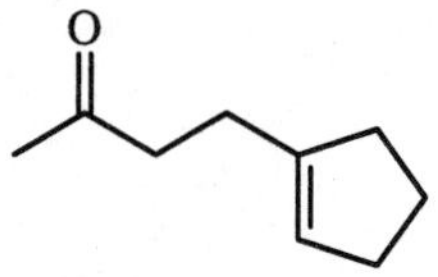

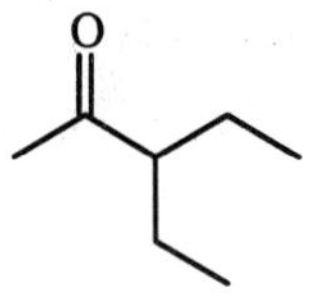

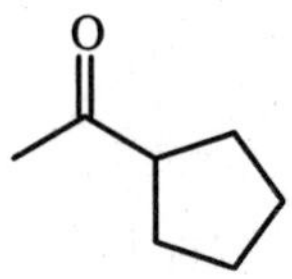

Carbonyl α–Substitution Reactions: *Synthesis*

Suggest a synthesis for each of the compounds shown below, beginning with the starting material shown on the left. Clearly show all intermediates and reaction conditions.

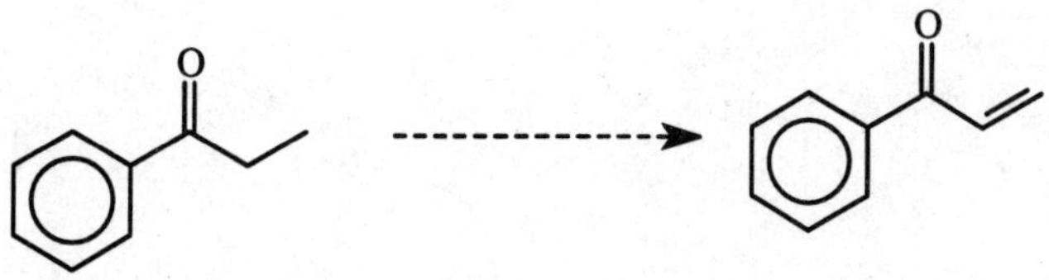

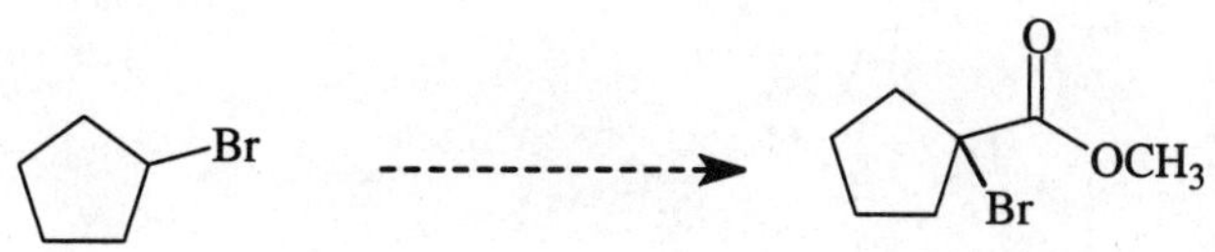

Carbonyl Condensation Reactions

The Aldol Condensation

As described previously, aldehydes and ketones enolize in base to produce a small equilibrium concentration of the corresponding enolate anion. If the concentration of carbonyl compound is sufficiently high, this enolate anion can function as a nucleophile towards the carbonyl carbon of other aldehydes in ketones in the solution. The result is the formation of a bond between the α–carbon of one mole of carbonyl compound and the carbonyl carbon of a second, to give a β–hydroxy aldehyde or ketone. This condensation reaction between two moles of an aldehyde or ketone is called the aldol condensation.

The reaction is general for both aldehydes and ketones, bearing at least one α–hydrogen.

The β–hydroxy carbonyl compounds which are formed are stable in base, but readily dehydrate in acid solution to give α,β–unsaturated carbonyl compounds.

As an example of the use of an aldol reaction in synthesis, consider the preparation of 2-ethyl-2-hexenal. This molecule is an α,β–unsaturated aldehyde. We can make these by the dehydration of β–hydroxy aldehydes in aqueous acid. The required β–hydroxy aldehyde, in turn can be prepared by the aldol condensation of two moles of butanal in aqueous base. To determine the structure of the required aldehyde, simply split the compound at the bond between the hydroxyl group and the α–carbon of the carbonyl, as shown on the right. As you can see, splitting like this breaks the molecule into two, four-carbon parts, both with an oxygen bonded to carbon #1.

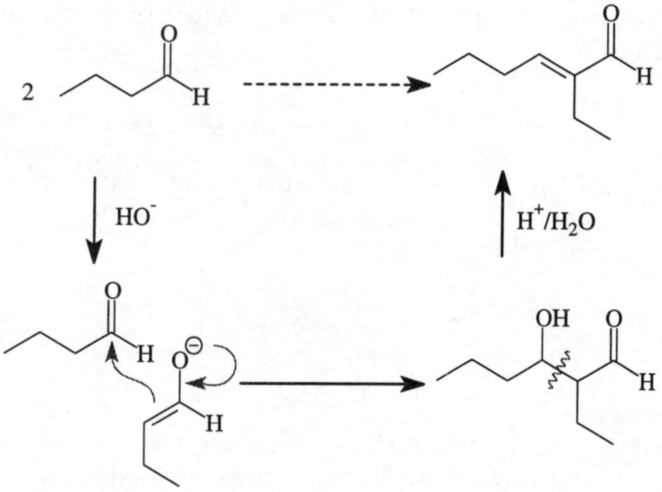

Drawing the structure of the reacting enol-aldehyde pair, as shown above, the mechanism of the condensation becomes clear.

It is also possible to utilize two different aldehydes or ketones in an aldol-type condensation reaction. In order to minimize self-condensations, generally one reactant is chosen which has no α–hydrogens, and that reactant is maintained in large excess over the second reactant (which will have α–hydrogens and can form the enolate anion). An example of this is the synthesis of 4-phenyl-3-buten-2-one, shown on the right. This α,β–unsaturated ketone can be prepared by dehydration of the β–hydroxy ketone, as shown. Splitting this between the α–carbon and the hydroxyl group, it is evident that it can be prepared by the condensation of the enol of 2-propanone (acetone) and benzaldehyde, as shown below.

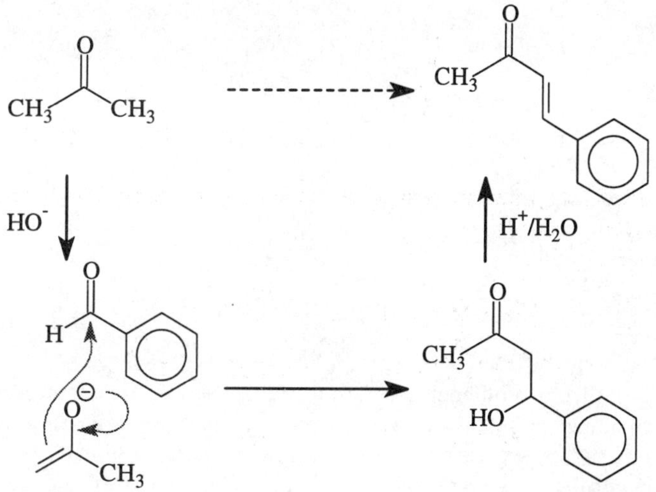

This molecule can also be prepared by the condensation of ethyl acetoacetate with benzaldehyde, followed by dehydration and decarboxylation, as shown below. This type of mixed aldol does not generate side-products due to self-condensation of the acetoacetate since the α–hydrogen of this molecule is acidic and it is converted essentially entirely to the enolate anion in the basic solution.

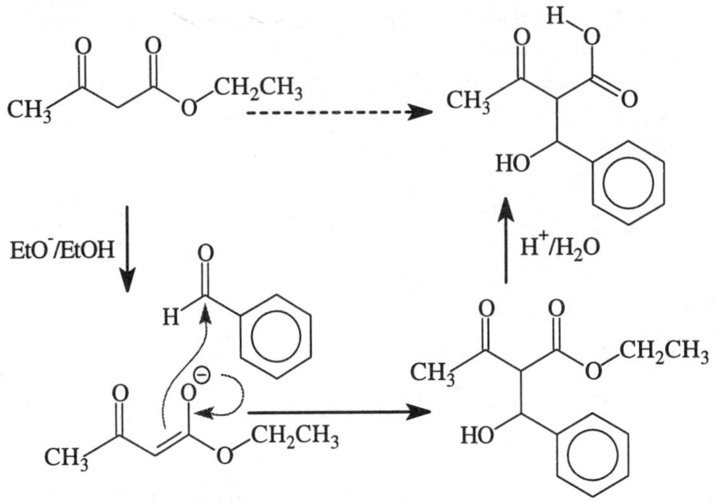

The Claisen Condensation

Esters, like aldehydes and ketones, enolize in base to produce a small equilibrium concentration of the corresponding enolate anion. As in the aldol condensation, this enolate anion can function as a nucleophile towards the carbonyl carbon of another mole of ester, and the result is the formation of a bond between the α–carbon of one mole of ester and the carbonyl of a second, to give a tetrahedral intermediate, as shown below for ethyl acetate. Expulsion of ethoxide from this intermediate re-forms the carbonyl and generates the condensation product, ethyl acetoacetate. This condensation reaction between two moles of an ester is called the **Claisen condensation**.

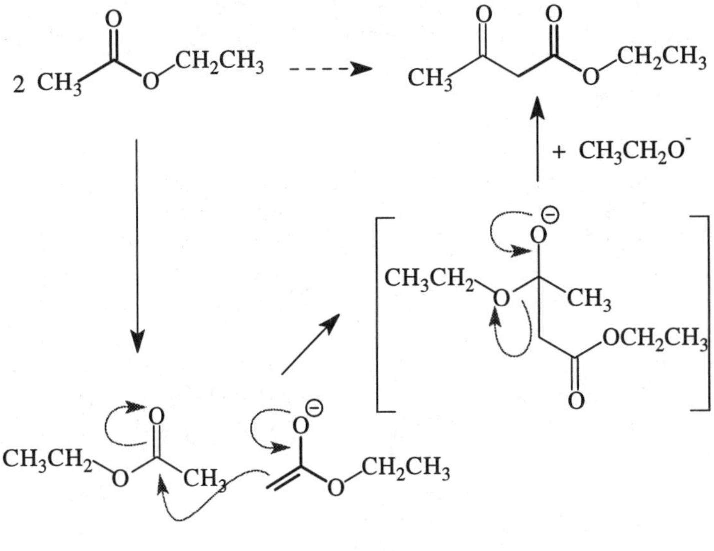

It is also possible to utilize two different esters in a Claisen-type condensation reaction (a mixed Claisen condensation). In order to minimize self-condensations, generally one reactant is chosen which has no α–hydrogens, and that reactant is maintained in large excess over the second reactant (which will have α–hydrogens and can form the enolate anion). An example of this is the synthesis shown below. This β–keto ester can be prepared by elimination of ethoxide from the tetrahedral intermediate formed by the addition of ethoxide anion to the ketone carbonyl. Splitting this tetrahedral intermediate between the α–carbon of the ester portion of the molecule and the tetrahedral center, it is evident that it can be prepared by the condensation of the enol of ethyl acetate and ethyl benzoate, as shown below.

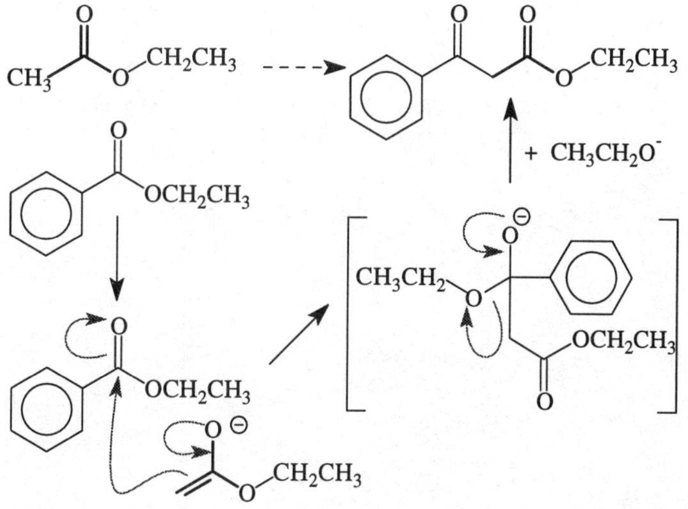

The β–keto esters which are formed in a Claisen condensation are stable in base, but readily decarboxylate in acid solution to give simple ketones.

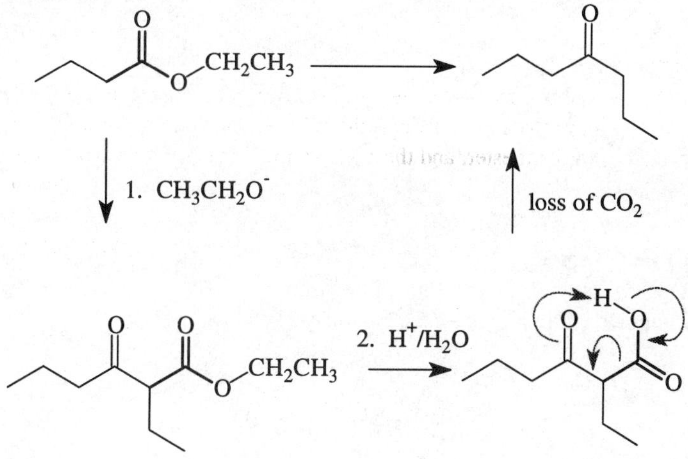

The decarboxylation reaction involves the hydrolysis of the ester to generate a β–keto carboxylic acid. These can undergo acid-catalyzed decarboxylation by the mechanism shown above to give the intermediate enol, which rapidly converts to the corresponding ketone. If the Claisen condensation involved the reaction of two moles of the same ester, the product formed will be a symmetrical ketone and the Claisen condensation is an excellent method for the preparation of symmetrical ketones. If a mixed Claisen condensation was utilized, more complex, unsymmetrical ketones can be prepared.

As an example of the use of a Claisen condensation in synthesis, consider the preparation of 4-heptanone. This molecule is a symmetrical ketone. We can make these by the decarboxylation of β–keto acid in aqueous acid. The required β–keto acid, in turn can be prepared by the Claisen condensation of two moles of ethyl butanoate in the presence of ethoxide anion.

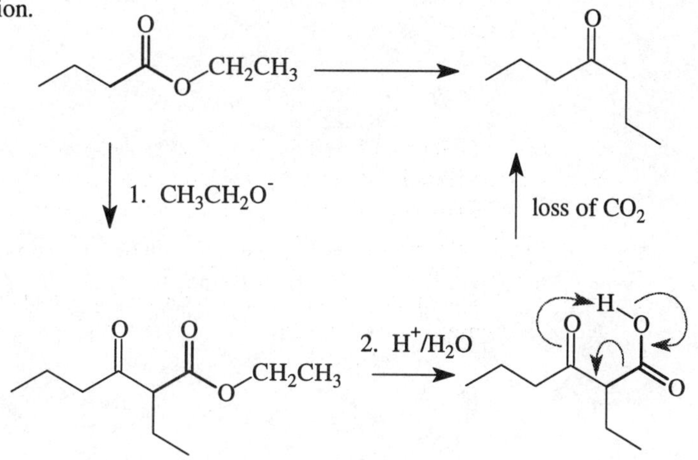

Remember, to determine the structure of the required ester, simply add ethoxide to the ketone carbonyl to form the tetrahedral intermediate, shown on the right, and split the compound at the bond between the tetrahedral center and the α–carbon of the ester portion of the molecule.

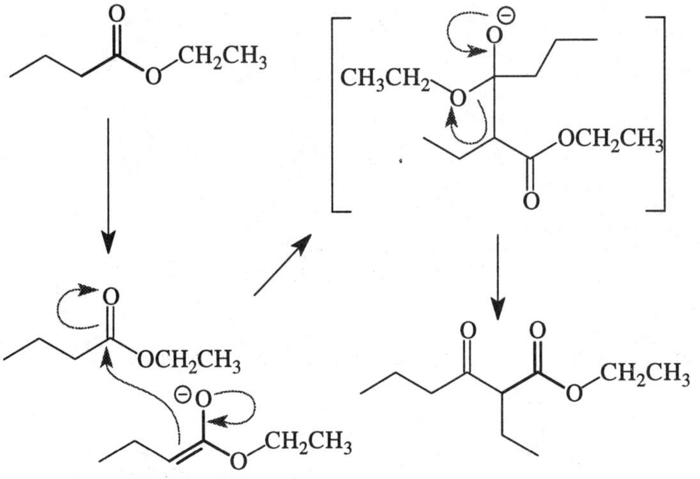

Conjugate Addition Reactions

In previous chapters, we have examined conjugate addition reactions of α,β–unsaturated ketones and aldehydes, and these reactions are summarized in the first three examples shown below.

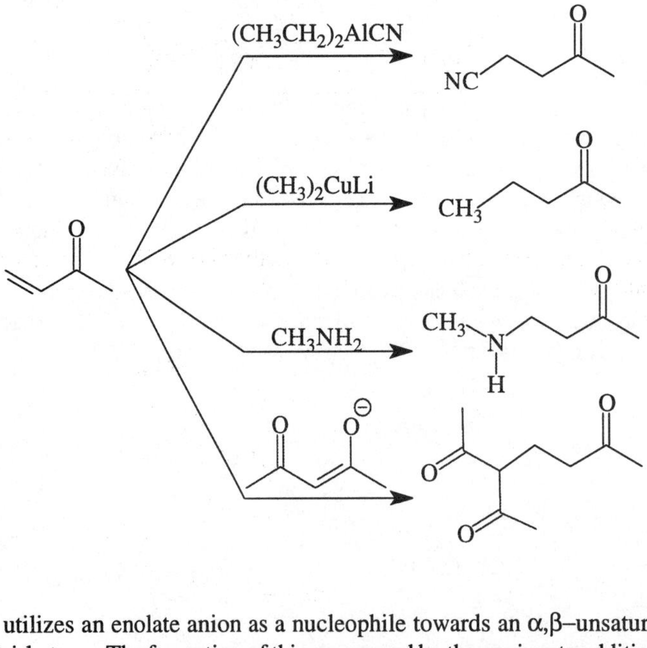

The last example utilizes an enolate anion as a nucleophile towards an α,β–unsaturated carbonyl to give, in this example, a branched tri-ketone. The formation of this compound by the conjugate addition follows the mechanism shown below. The enolate simply adds to the β–carbon, and then the intermediate enolate collapses, picking up a proton on the α–carbon to give the final product.

A variety of compounds with acidic hydrogens are suitable nucleophiles for conjugate addition, as shown in the examples below.

Nitroalkanes are also suitable nucleophiles, since the α–carbon is acidified by the adjacent nitrogen cation and the anion is resonance stabilized.

Enamines, likewise, can be utilized as nucleophiles in these reactions, yielding δ-di-ketones.

Carbonyl Condensations: *Aldol-Type Condensation Reactions*

Predict the major organic product for each of the reactions shown below. Clearly show stereochemistry, if appropriate.

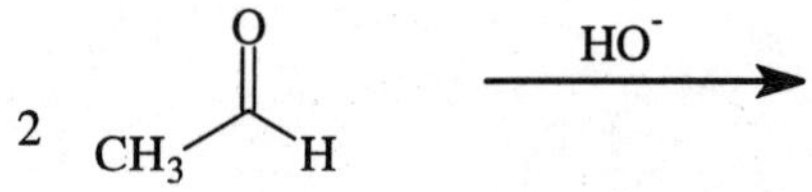

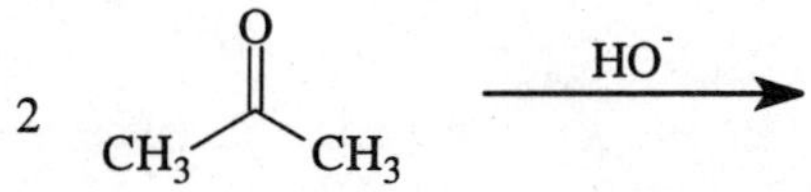

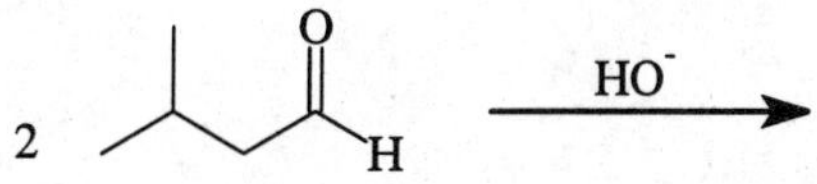

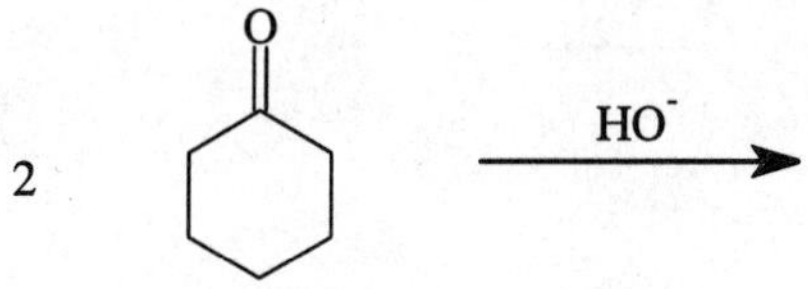

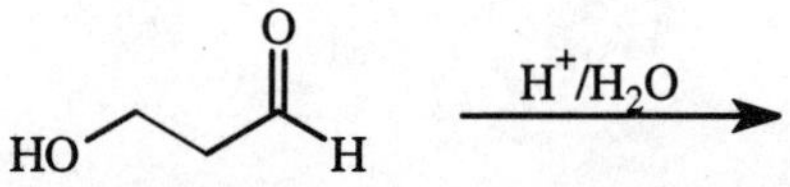

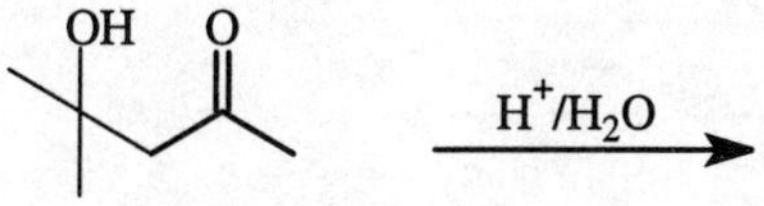

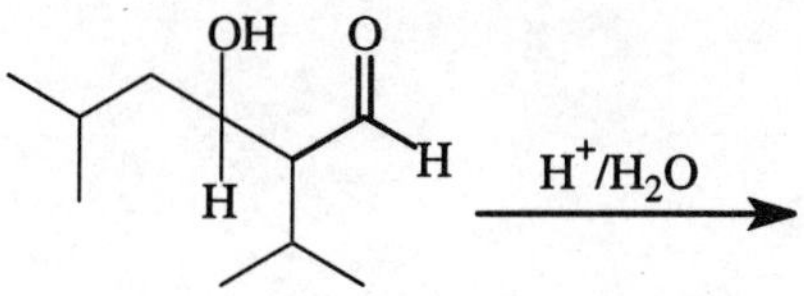

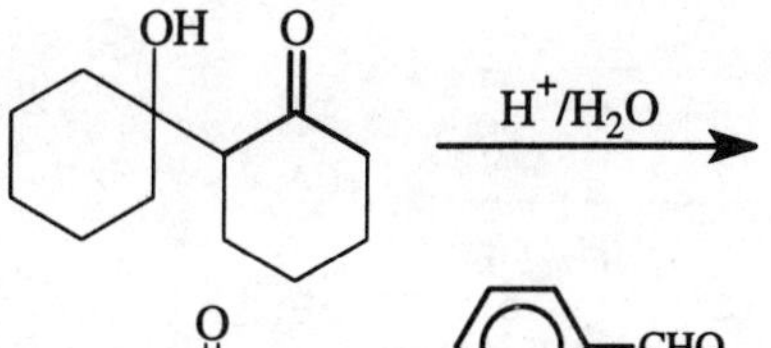

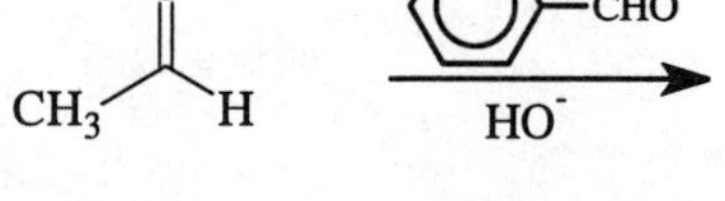

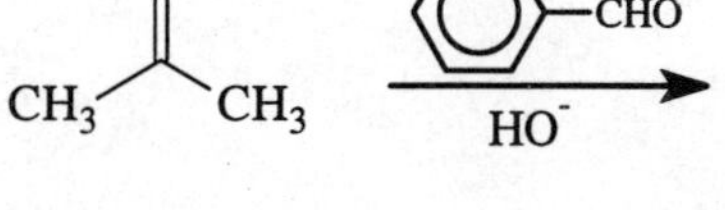

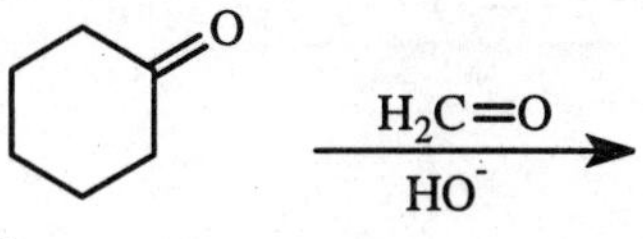

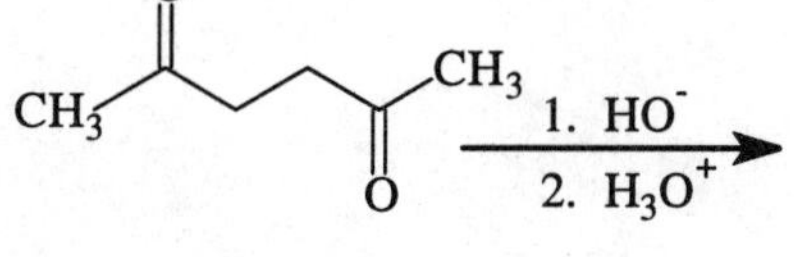

2 CH₃CHO $\xrightarrow{HO^-}$

2 CH₃COCH₃ $\xrightarrow{HO^-}$

2 (CH₃)₂CHCH₂CHO $\xrightarrow{HO^-}$

2 cyclohexanone $\xrightarrow{HO^-}$

HOCH₂CH₂CHO $\xrightarrow{H^+/H_2O}$

(CH₃)₂C(OH)CH₂COCH₃ $\xrightarrow{H^+/H_2O}$

(CH₃)₂CHCH₂CH(OH)CH(iPr)CHO $\xrightarrow{H^+/H_2O}$

1-(1-hydroxycyclohexyl)cyclohexan-2-one $\xrightarrow{H^+/H_2O}$

CH₃CHO + PhCHO $\xrightarrow{HO^-}$

CH₃COCH₃ + PhCHO $\xrightarrow{HO^-}$

cyclohexanone $\xrightarrow[HO^-]{H_2C=O}$

CH₃COCH₂CH₂COCH₃ $\xrightarrow[\text{2. } H_3O^+]{\text{1. } HO^-}$

Carbonyl Condensations: *Claisen-Type Reactions*

Predict the major organic product for each of the Claisen-type condensation reactions shown below. Clearly show stereochemistry, if appropriate.

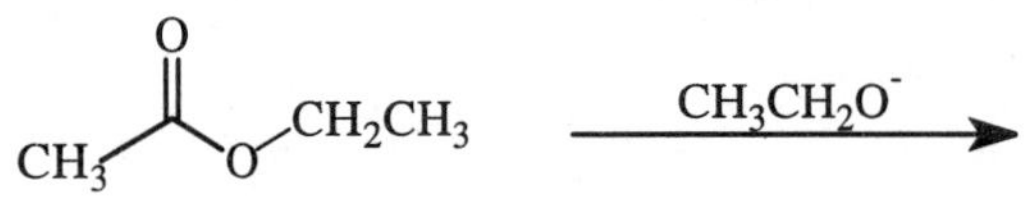

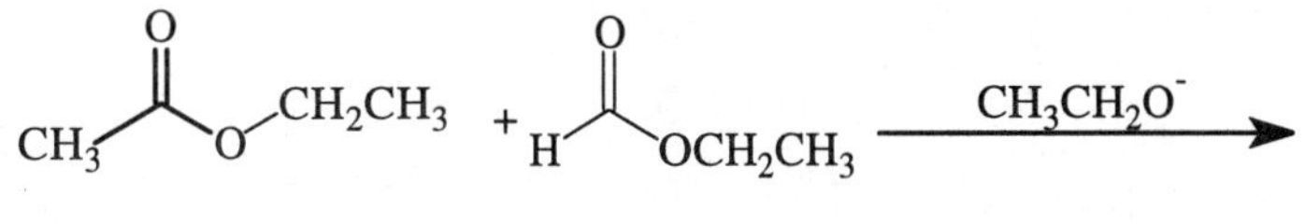

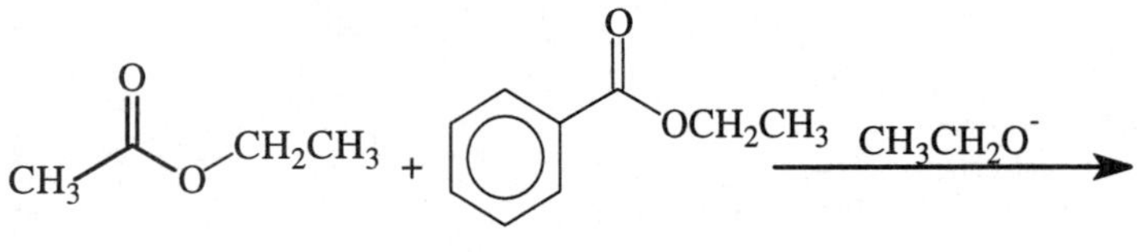

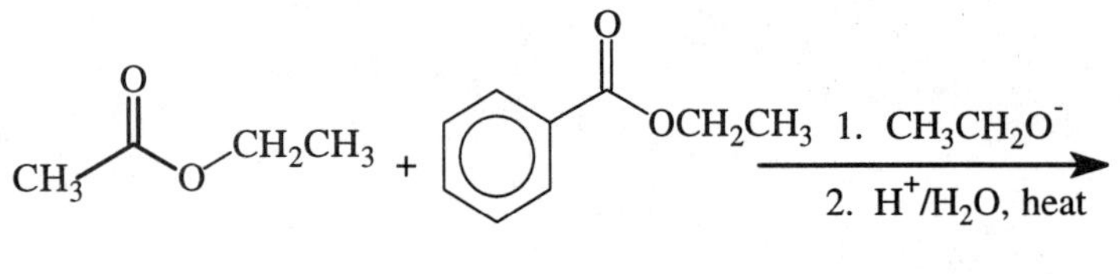

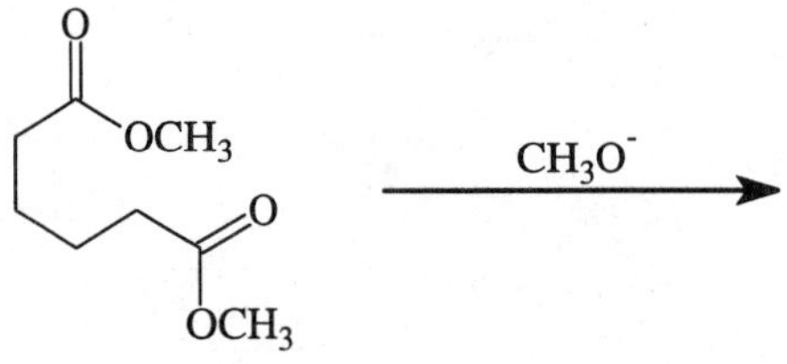

$$CH_3-C(=O)-O-CH_2CH_3 \xrightarrow{CH_3CH_2O^-}$$

$$CH_3-C(=O)-O-CH_2CH_3 + H-C(=O)-OCH_2CH_3 \xrightarrow{CH_3CH_2O^-}$$

$$CH_3-C(=O)-O-CH_2CH_3 + Ph-C(=O)-OCH_2CH_3 \xrightarrow{CH_3CH_2O^-}$$

$$CH_3-C(=O)-O-CH_2CH_3 + Ph-C(=O)-OCH_2CH_3 \xrightarrow[\text{2. } H^+/H_2O, \text{ heat}]{\text{1. } CH_3CH_2O^-}$$

$$CH_3O-C(=O)-CH_2CH_2CH_2-C(=O)-OCH_3 \xrightarrow{CH_3O^-}$$

Conjugate Additions: *Reactions*

Predict the major organic product for each of the addition reactions shown below. Clearly show stereochemistry, if appropriate.

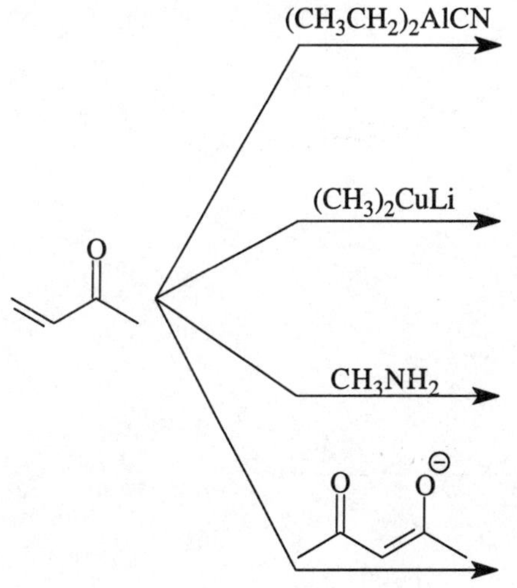

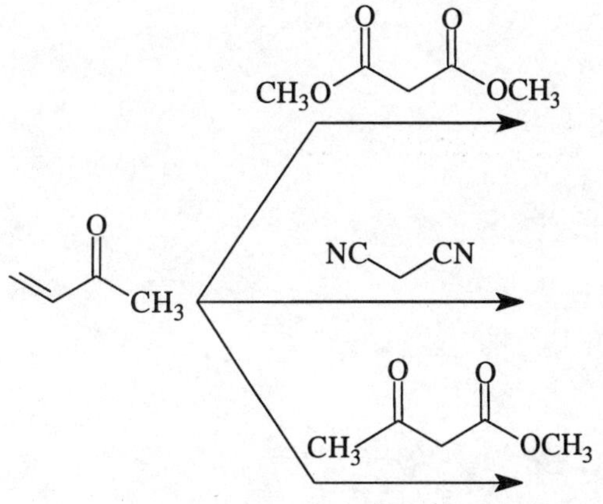

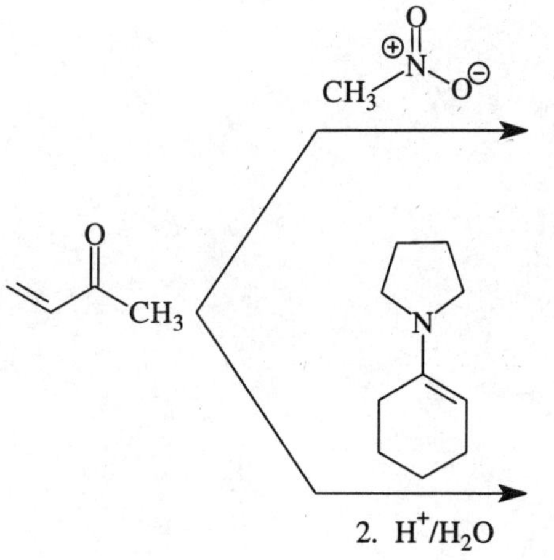

Carbonyl Compounds: *Reactions in Sequence*

Predict the major organic product for each of the reaction sequences shown below. Clearly show stereochemistry, if appropriate.

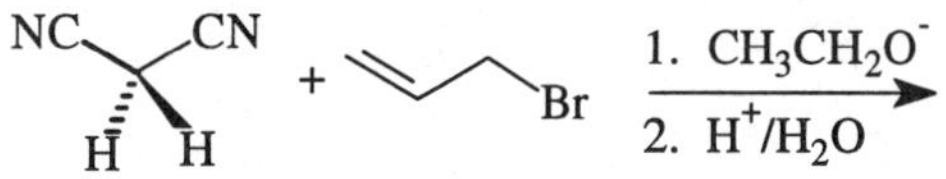

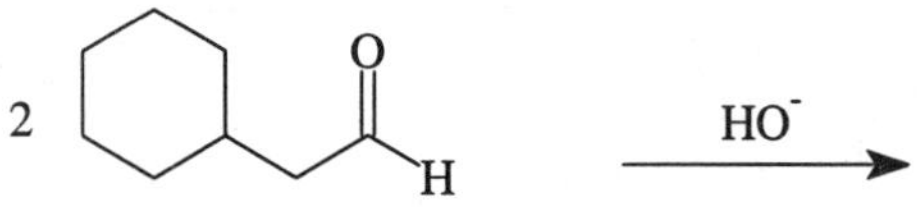

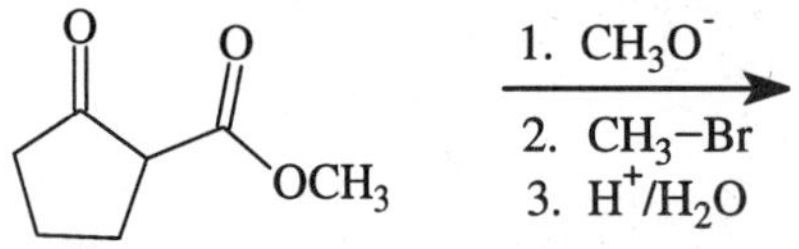

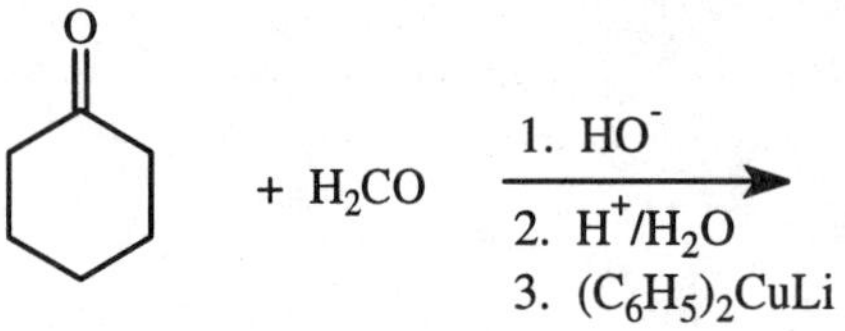

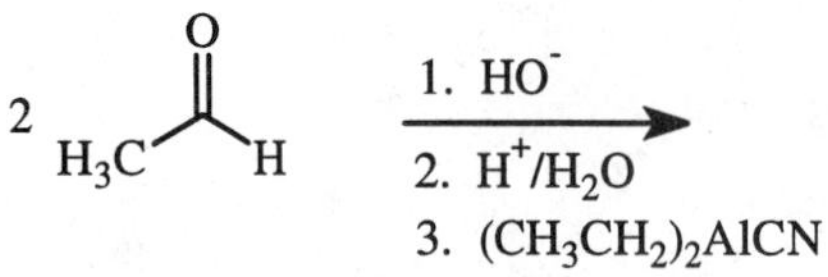

[Structure: methyl butanoate] 1. CH$_3$O$^-$ 2. H$^+$/H$_2$O

NC−CH(H)−CN (with stereochemistry H,H) + CH$_2$=CH−CH$_2$−Br   1. CH$_3$CH$_2$O$^-$ 2. H$^+$/H$_2$O

2 [cyclohexylacetaldehyde]   HO$^-$

[methyl 2-oxocyclopentanecarboxylate]   1. CH$_3$O$^-$ 2. CH$_3$−Br   3. H$^+$/H$_2$O

[cyclohexanone] + H$_2$CO   1. HO$^-$ 2. H$^+$/H$_2$O   3. (C$_6$H$_5$)$_2$CuLi

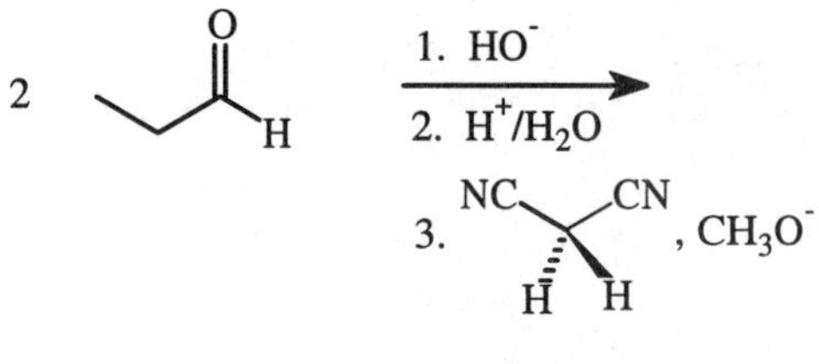

2 [propanal]   1. HO$^-$ 2. H$^+$/H$_2$O   3. NC−CH(H)−CN , CH$_3$O$^-$

2 [acetaldehyde, H$_3$C−CHO]   1. HO$^-$ 2. H$^+$/H$_2$O   3. (CH$_3$CH$_2$)$_2$AlCN

Aliphatic and Arylamines

Nomenclature

Simple amines are named as derivatives of the parent alkane, using the suffix **-amine**, or by using **-amino** to name a numbered substituent, using the following rules:

1. Select the longest continuous carbon chain, containing the amino group, and derive the parent name by replacing the -e ending with **-amine**, or by naming the nitrogen as an amino substituent.

2. Number the carbon chain, beginning at the end nearest to the amino group, or, to give the lowest number at the first point of difference.

3. Number the substituents and write the name, listing substituents alphabetically.

Thus for the following example, you would number from the end closest to the nitrogen, generating the names, **3-methylpentanamine** (or 1-amino-3-methylpentane) and **5-methyl-2-hexanamine** (or 2-amino-5-methyl-2-hexane), respectively.

3-methylpentanamine
or **1-amino-3-methylpentane**

5-methyl-2-hexanamine
or **2-amino-5-methyl-2-hexane**

In this example, however, you number to give the lowest number at the first point of difference, generating the name, 5-amino-2,3-dimethylhexane (not 2-amino-4,5-dimethylhexane).

Reactions of Aliphatic Amines

Simple amines and ammonia are strong nucleophiles and will undergo an $S_N 2$ reaction with alkyl halides (or alkyl groups with "good leaving groups") to give further substitution on the nitrogen, as described previously. They will also react with activated carbonyl compounds to undergo acyl transfer reactions; thus amides are readily formed by the reaction of amines with acid halides, acid anhydrides or carboxylate esters.

The reaction of an amine with a sulfonyl halide forms the **sulfonamide**. A common reagent utilized in this reaction is *p*-toluenesulfonyl chloride, producing the corresponding ***p*-toluenesulfonamide**.

[Reaction scheme: 3,4-dihydroxyphenethylamine + p-toluenesulfonyl chloride (CH₃-C₆H₄-SO₂Cl) → sulfonamide product (HO,OH-C₆H₃-CH₂CH₂NH-SO₂-C₆H₄-CH₃)]

The base solubility of sulfonamides forms the basis of the **Hinsberg test**, for distinguishing primary, secondary and tertiary amines; primary p-toluenesulfonamides undergo ionization in strong base to give the base-soluble anion, secondary p-toluenesulfonamides lack the acidic hydrogen and do not form a soluble anion; tertiary amines would yield highly unstable cationic quarternary sulfonamides, and generally do not react at all. **Thus, the formation of a base-soluble sulfonamide indicates the presence of a primary amine.**

[Equilibrium scheme: sulfonamide ⇌ (KOH) potassium salt of sulfonamide anion]

Reaction of primary amines with an excess of iodomethane converts the primary amine into the quarternary ammonium salt. The cationic nitrogen which is now formed is a good leaving group, and will undergo E2 elimination on reaction with Ag₂O to give the alkene, in a reaction known as the **Hofmann Elimination**.

[Mechanism: HO⁻ abstracts β-hydrogen from quaternary ammonium salt (Ar-CH₂-CH₂-N⁺(CH₃)₃), Ag₂O, H₂O → alkene product (Ar-CH=CH₂ + N(CH₃)₃), where Ar = 3,4-dihydroxyphenyl]

The Hofmann Elimination is unusual for an E2 elimination because the **least substituted alkene** is typically formed.

$$\text{sec-butylamine} \xrightarrow{\substack{1.\ CH_3Br\ (excess) \\ 2.\ Ag_2O,\ H_2O}} \text{1-butene} \quad \xcancel{\longrightarrow} \quad \text{2-butene}$$

Amides and acid azides can also be converted to amines using the **Hofmann** and **Curtius** rearrangements, respectively. Both the amide and acyl azide can be prepared from an intermediate acid halide, and the reaction results in the shortening of the alkyl chain by one carbon (the carbonyl is lost as CO_2).

cyclopentanecarboxamide $\xrightarrow{NaOH,\ Br_2/H_2O}$ cyclopentylamine + CO_2

The Hofmann rearrangement...

cyclopentanecarbonyl azide ($N=N^+=N^-$) $\xrightarrow{H_2O,\ heat}$ cyclopentylamine + CO_2 + $N\equiv N$

The Curtius rearrangement...

Reactions that Yield Aliphatic Amines

Simple amines and ammonia are strong nucleophiles and will undergo an S_N2 reaction with alkyl halides (or alkyl groups with "good leaving groups") to give further substitution on the nitrogen. Ammonia reacts with alkyl halides to give a mixture, consisting largely of the corresponding primary and secondary amines, with a trace of tertiary and quarternary amines. Since a mixture is obtained, this is generally a poor method for the preparation of amines.

$CH_3(CH_2)_6CH_2Br$ + NH_3 →

- $CH_3(CH_2)_6CH_2NH_2$ — 45%
- $[CH_3(CH_2)_6CH_2]_2NH$ — 43%
- $[CH_3(CH_2)_6CH_2]_3N$ — Trace
- $[CH_3(CH_2)_6CH_2]_4N^+\ Br^-$ — Smaller Trace

Better methods for the preparation of primary amines involve the **reduction of a intermediate alkyl azide**, or the **Gabriel Synthesis**, involving an intermediate phtalimide. In the first method, the alkyl halide reacts with azide anion in a simple S_N2 reaction to give the intermediate alkyl azide. This is generally not isolated, but is reduced immediately with $LiAlH_4$ to give the corresponding primary amine.

$$CH_3(CH_2)_6CH_2Br + NaN_3 \longrightarrow CH_3(CH_2)_6CH_2N\overset{\oplus}{=}N\overset{\ominus}{=}N$$

$$\downarrow \text{2. LiAlH}_4$$

$$CH_3(CH_2)_6CH_2NH_2$$

In the **Gabriel Synthesis**, pthalimide anion is reacted with the alkyl halide to give the intermediate *N*-substituted pthalimide. On acid hydrolysis, this gives pthalic acid and the corresponding primary amine.

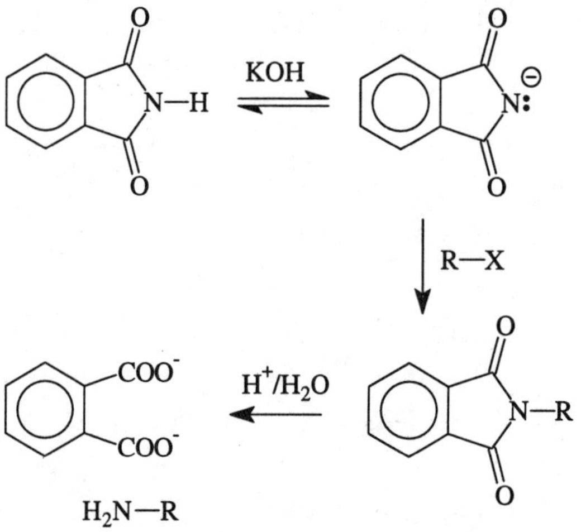

Primary amines can also be prepared by the reduction of two other functional groups; **nitriles** and **amides**. The reduction of a nitrile by LiAlH$_4$ gives the primary amine, as shown below. Reduction of a an amide can yield a primary, secondary or tertiary amine, depending on the amide.

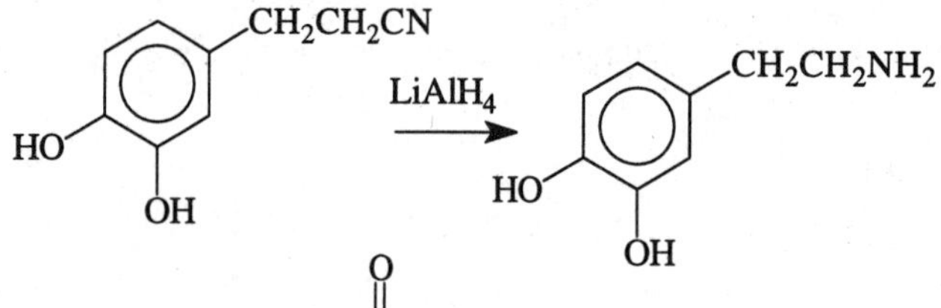

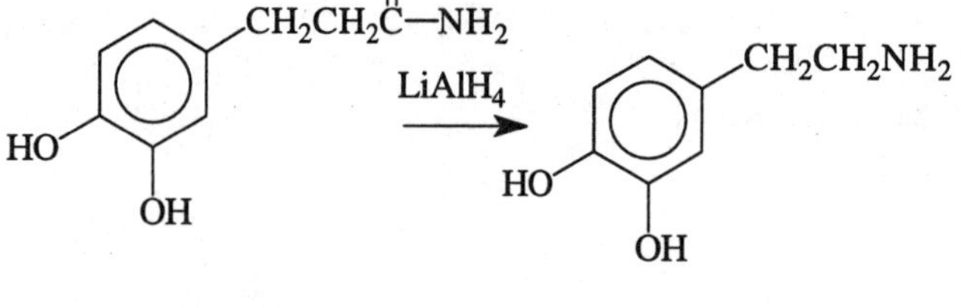

Amines can also be prepared by the process of **reductive amination of aldehydes and ketones**. An aldehyde or ketone will react with ammonia to give an intermediate imine. Reduction of this imine with Raney Ni/H$_2$ gives the corresponding amine.

[Reaction scheme: 3,4-dihydroxyphenylacetaldehyde + NH₃, H₂/Raney Ni → 3,4-dihydroxyphenethylamine, via iminium intermediate]

Primary and secondary amines can also be utilized in reductie amination reactions, typically using **cyanoborohydride** (NaBH₃CN) to trap the intermediate immonium ion.

[Reaction scheme: 3,4-dihydroxyphenylacetaldehyde + CH₃—NH₂, NaBH₃CN → N-methyl-3,4-dihydroxyphenethylamine, via iminium intermediate]

Reactions of Aryl Amines

Aryl amines, like aliphatic amines and ammonia, are strong nucleophiles and will undergo an S$_N$2 reaction with alkyl halides (or alkyl groups with "good leaving groups") to give further substitution on the nitrogen, as described previously. They will also react with activated carbonyl compounds to undergo acyl transfer reactions; thus amides are readily formed by the reaction of amines with acid halides, acid anhydrides or carboxylate esters. The conversion of an aryl amine into an amide is a convenient method for limiting ring bromination to the *para-* position on the ring, the intermediate amide being simply hydrolyzed in a second step.

[Reaction scheme: aniline + CH₃COCl → acetanilide; acetanilide + 1. Br₂, 2. NaOH, H₂O → 4-bromoaniline]

This intermediate step is necessary, since the free amine highly activates the ring to substitution, yielding tri-substitution.

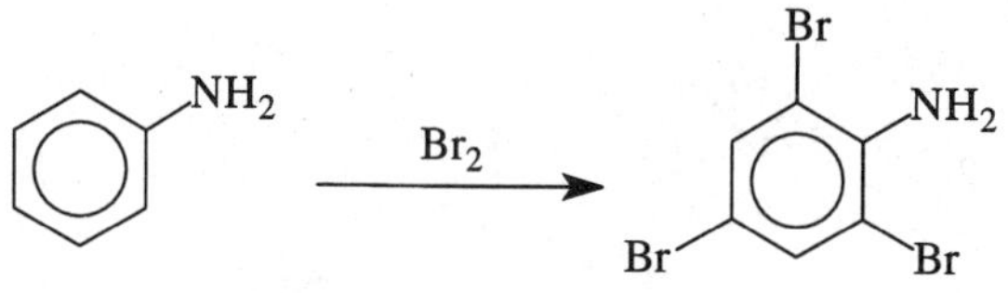

[Reaction: aniline + Br₂ → 2,4,6-tribromoaniline]

Perhaps the most useful reactions of aryl amines involve the intermediate conversion into the corresponding **diazonium salt** by reaction with nitrous acid.

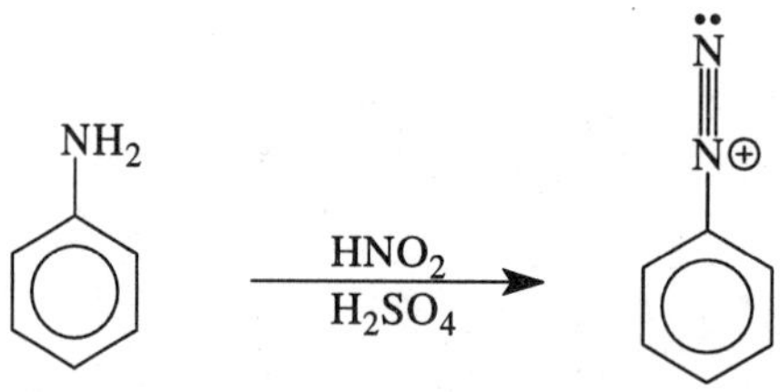

[Reaction: aniline + HNO₂/H₂SO₄ → benzenediazonium cation]

These diazonium salts undergo a series of replacement reactions, collectively known as the **Sandelmyer Reaction** to give aryl nitriles or aryl halides. A copper salt is generally required to catalyze the reaction, with the exception of iodination, which occurs spontaneously.

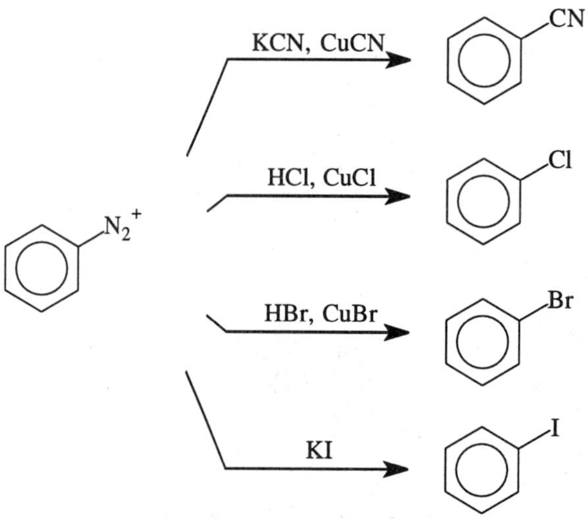

[Reactions of PhN₂⁺:
- KCN, CuCN → PhCN
- HCl, CuCl → PhCl
- HBr, CuBr → PhBr
- KI → PhI]

Aryl diazonium salts also undergo reduction to yield the arene on reaction with **phosphouous acid** (not phosphoric), and hydrolysis in the presence of aqueous acid to give the corresponding phenol. The cationic nitrogen of diazonium salts also adds to the para- position of highly activated aryl rings (generally aryl amines and phenols) to give **coupling products,** as shown below.

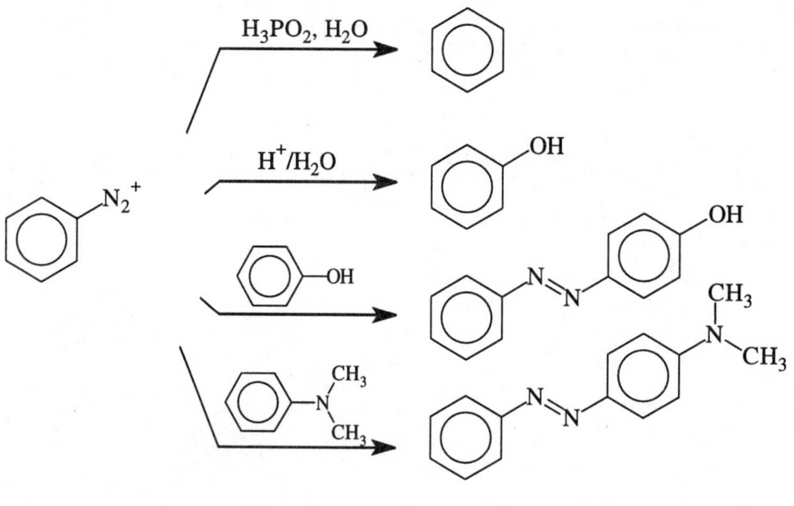

[Reactions of PhN₂⁺:
- H₃PO₂, H₂O → benzene
- H⁺/H₂O → phenol
- PhOH → Ph-N=N-C₆H₄-OH
- PhN(CH₃)₂ → Ph-N=N-C₆H₄-N(CH₃)₂]

Reactions of Amines

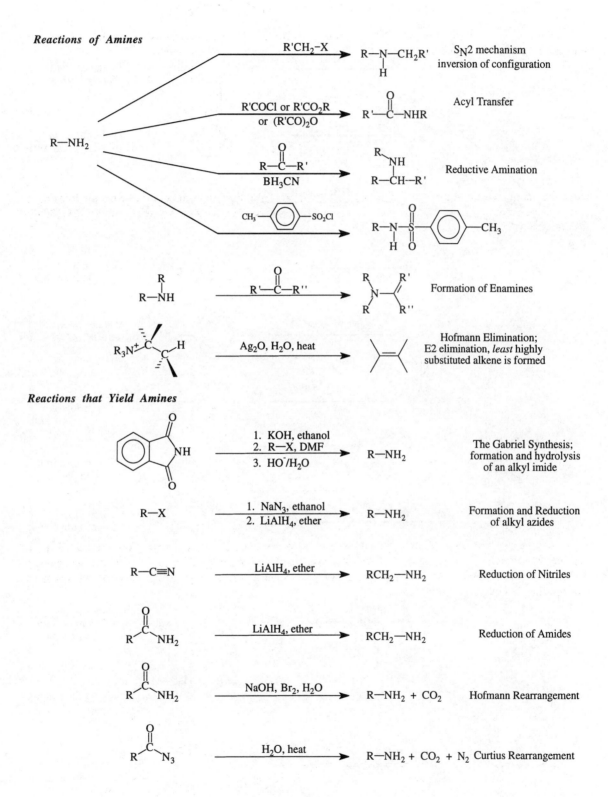

Reactions of Arylamines & Diazonium Salts

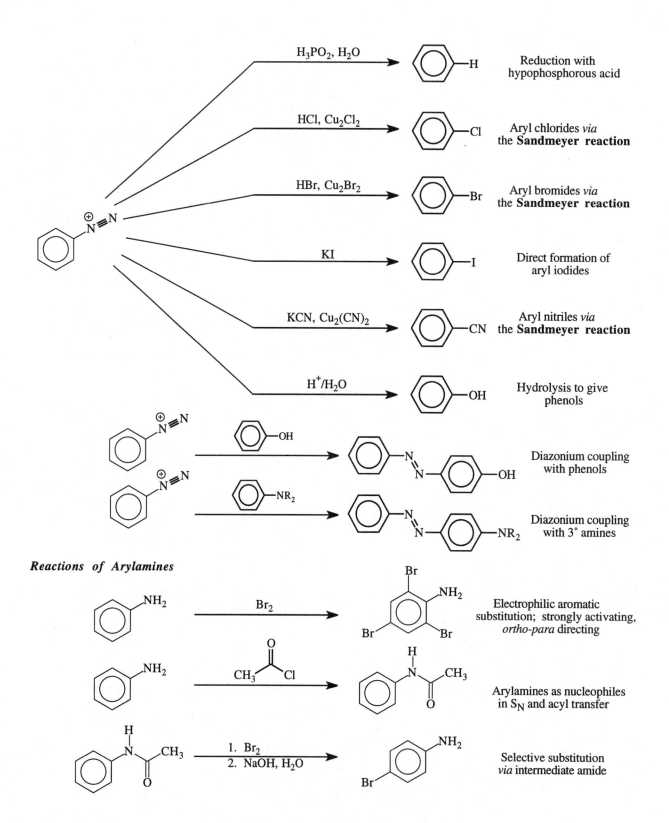

Reactions of Phenols

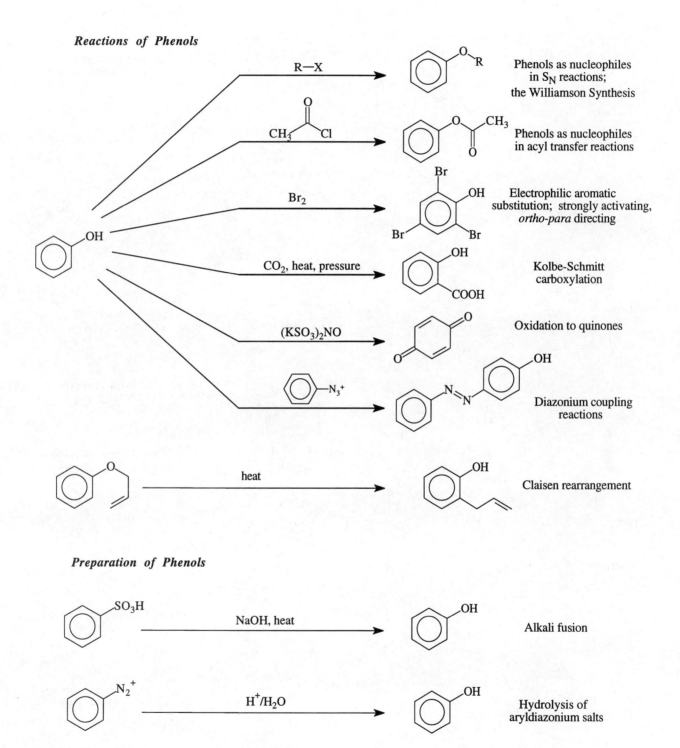

Amines & Arylamines: *Nomenclature*

Provide an acceptable IUPAC name for each of the following compounds.

Amines: *Reactions*

Predict the major organic product for each of the following reactions. Clearly show stereochemistry if appropriate.

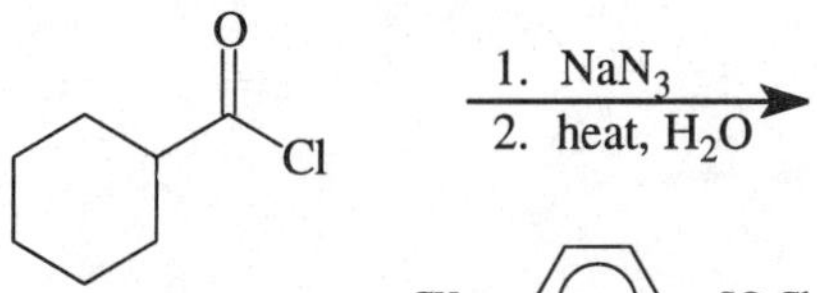

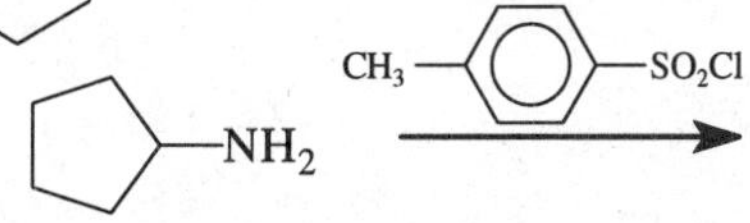

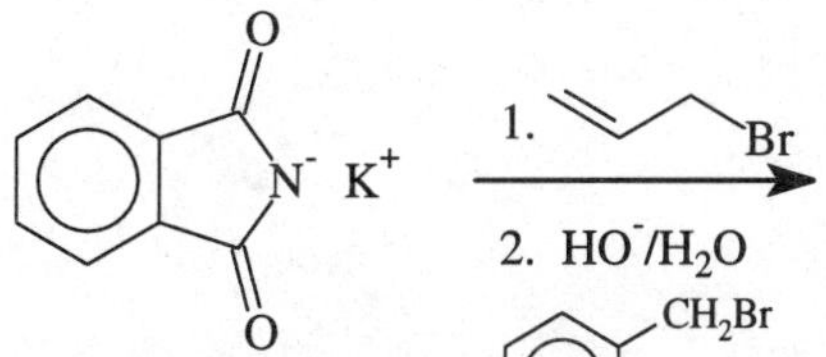

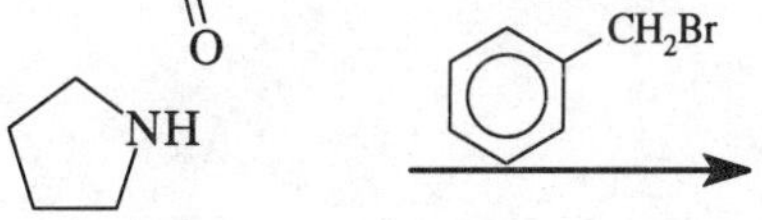

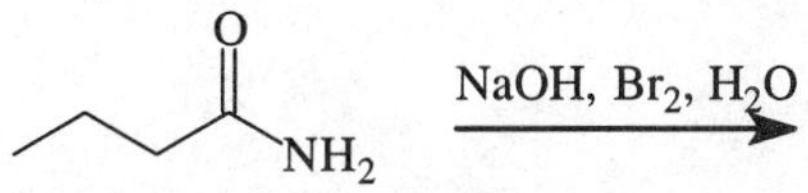

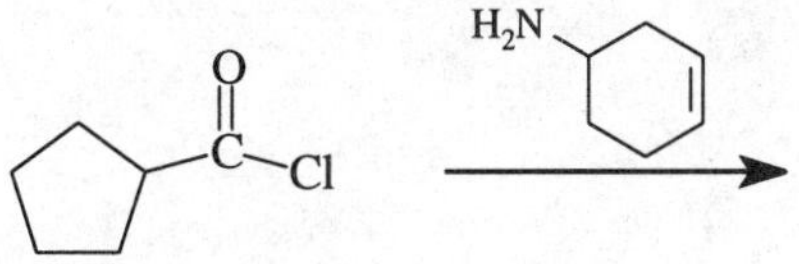

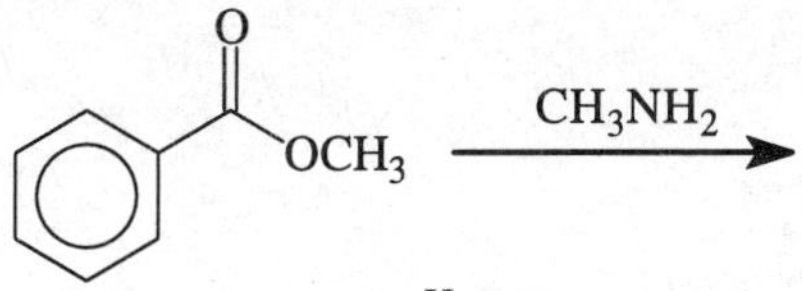

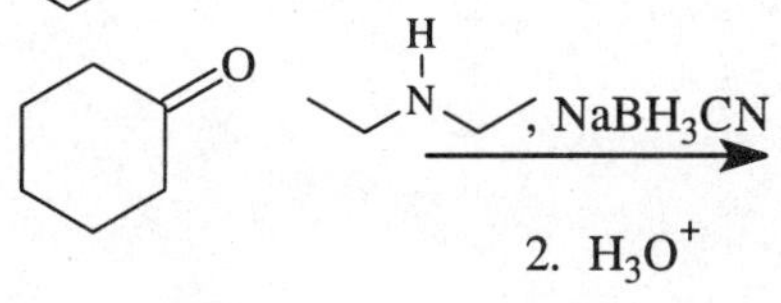

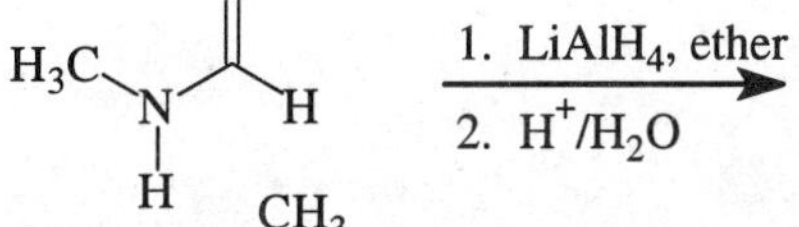

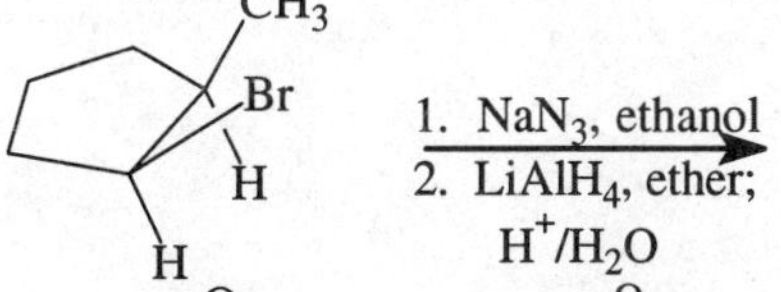

cyclohexanecarbonyl chloride $\xrightarrow{\text{1. NaN}_3 \quad \text{2. heat, H}_2\text{O}}$

cyclopentyl-NH$_2$ + CH$_3$-C$_6$H$_4$-SO$_2$Cl $\longrightarrow$

phthalimide N$^-$ K$^+$ $\xrightarrow{\text{1. allyl-Br} \quad \text{2. HO}^-/\text{H}_2\text{O}}$

pyrrolidine (NH) + PhCH$_2$Br $\longrightarrow$

butanamide $\xrightarrow{\text{NaOH, Br}_2, \text{H}_2\text{O}}$

cyclopentanecarbonyl chloride + 3-aminocyclohexene (H$_2$N-) $\longrightarrow$

methyl benzoate (OCH$_3$) + CH$_3$NH$_2$ $\longrightarrow$

cyclohexanone + CH$_3$CH$_2$NHCH$_2$CH$_3$, NaBH$_3$CN ; 2. H$_3$O$^+$ $\longrightarrow$

H$_3$C-N(H)-CHO with CH$_3$ $\xrightarrow{\text{1. LiAlH}_4, \text{ether} \quad \text{2. H}^+/\text{H}_2\text{O}}$

1-bromo-1-methylcyclopentane (stereochem shown) $\xrightarrow{\text{1. NaN}_3, \text{ethanol} \quad \text{2. LiAlH}_4, \text{ether; H}^+/\text{H}_2\text{O}}$

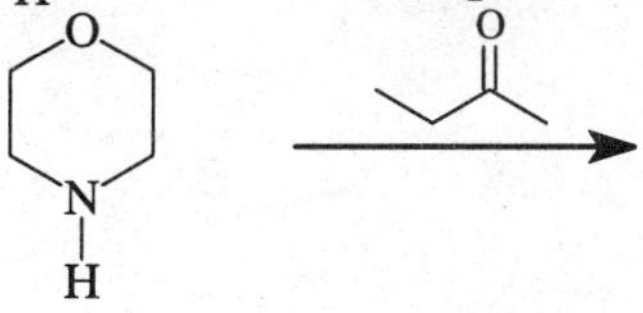

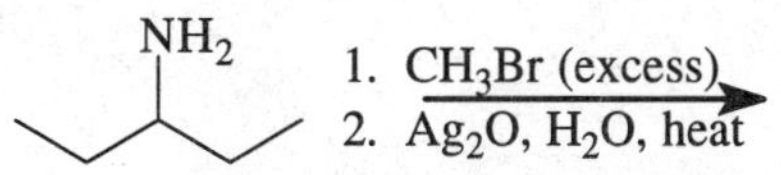

morpholine (N–H) + 2-butanone $\longrightarrow$

2-aminobutane (sec-butylamine, NH$_2$) $\xrightarrow{\text{1. CH}_3\text{Br (excess)} \quad \text{2. Ag}_2\text{O, H}_2\text{O, heat}}$

Aryl Amines: *Reactions of Diazonium Salts*

Predict the major organic product for each of the following reactions. Clearly show stereochemistry if appropriate.

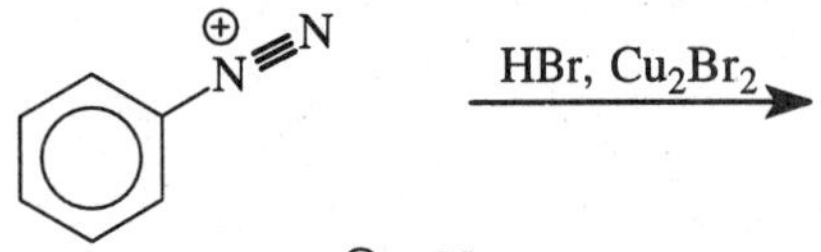

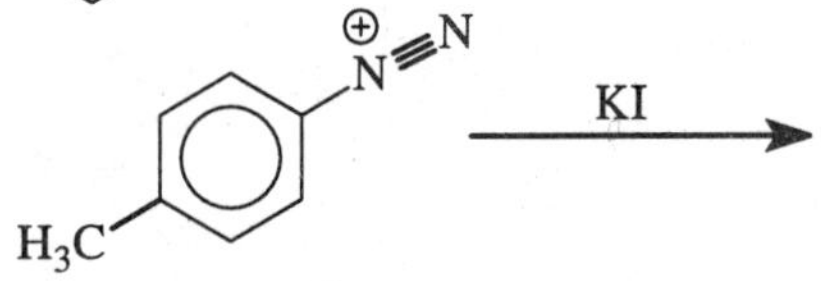

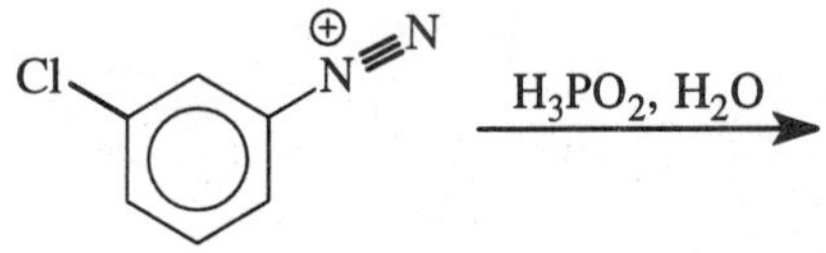

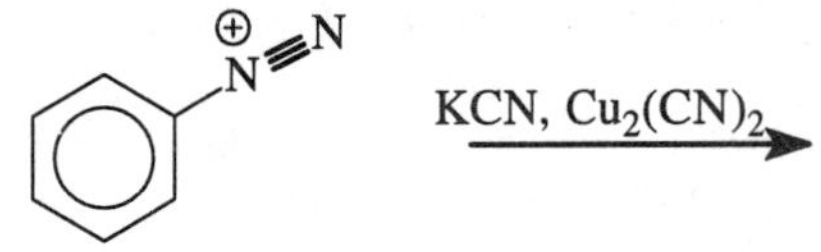

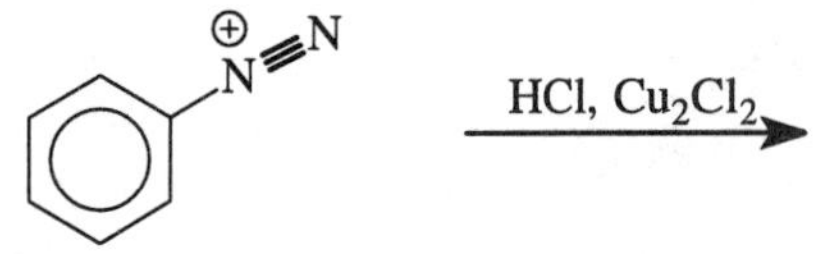

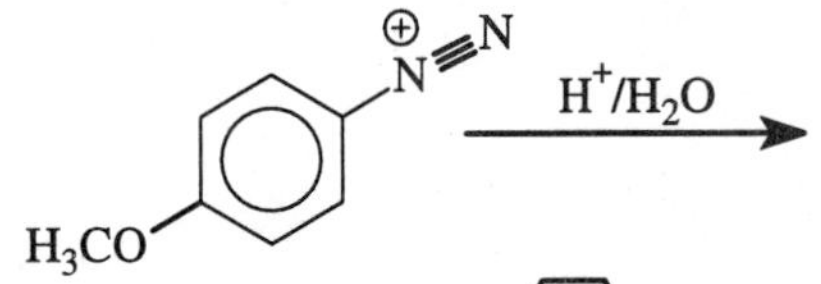

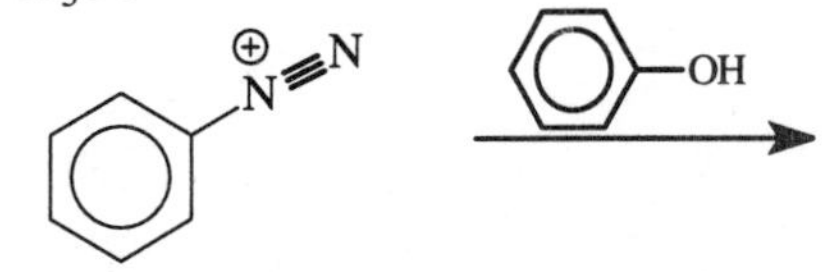

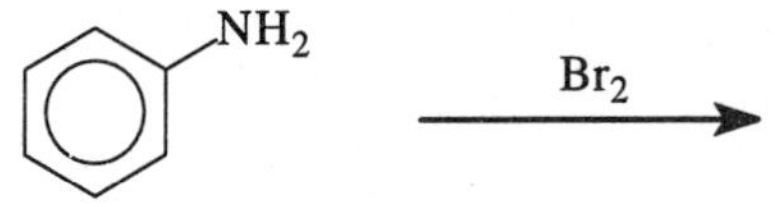

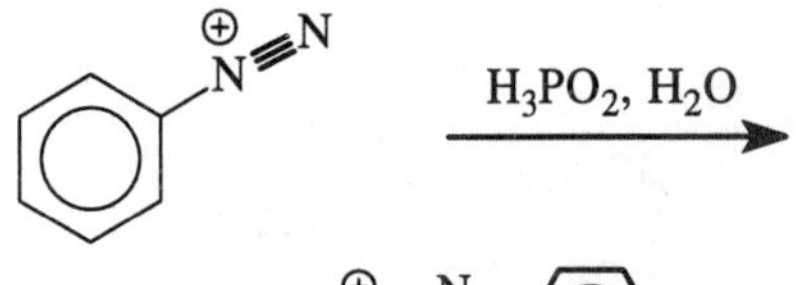

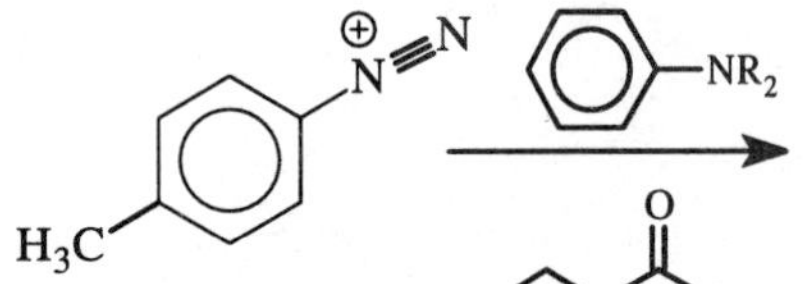

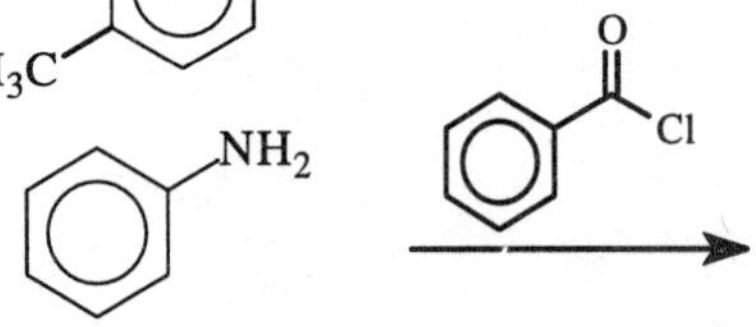

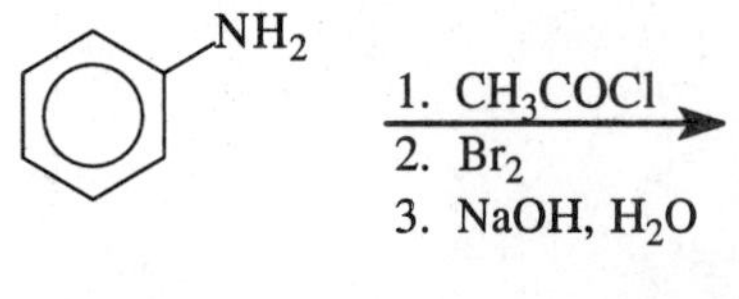

PhN$_2^+$ $\xrightarrow{\text{HBr, Cu}_2\text{Br}_2}$

4-CH$_3$-C$_6$H$_4$-N$_2^+$ $\xrightarrow{\text{KI}}$

3-Cl-C$_6$H$_4$-N$_2^+$ $\xrightarrow{\text{H}_3\text{PO}_2, \text{H}_2\text{O}}$

PhN$_2^+$ $\xrightarrow{\text{KCN, Cu}_2(\text{CN})_2}$

PhN$_2^+$ $\xrightarrow{\text{HCl, Cu}_2\text{Cl}_2}$

4-CH$_3$O-C$_6$H$_4$-N$_2^+$ $\xrightarrow{\text{H}^+/\text{H}_2\text{O}}$

PhN$_2^+$ + C$_6$H$_5$-OH $\longrightarrow$

C$_6$H$_5$-NH$_2$ $\xrightarrow{\text{Br}_2}$

PhN$_2^+$ $\xrightarrow{\text{H}_3\text{PO}_2, \text{H}_2\text{O}}$

4-CH$_3$-C$_6$H$_4$-N$_2^+$ + C$_6$H$_5$-NR$_2$ $\longrightarrow$

C$_6$H$_5$-NH$_2$ + C$_6$H$_5$-C(O)Cl $\longrightarrow$

C$_6$H$_5$-NH$_2$ $\xrightarrow[\text{2. Br}_2]{\text{1. CH}_3\text{COCl}}$ $\xrightarrow{\text{3. NaOH, H}_2\text{O}}$

Aliphatic Aryl Amines: *Synthesis*

Suggest a synthesis for each of the compounds shown beolw on the right, beginning with the starting material shown on the left and any other required materials. Be sure to clearly show all intermediates and reaction conditions.

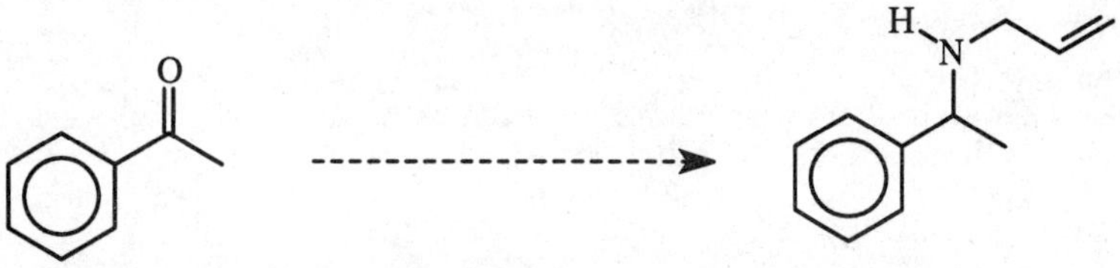

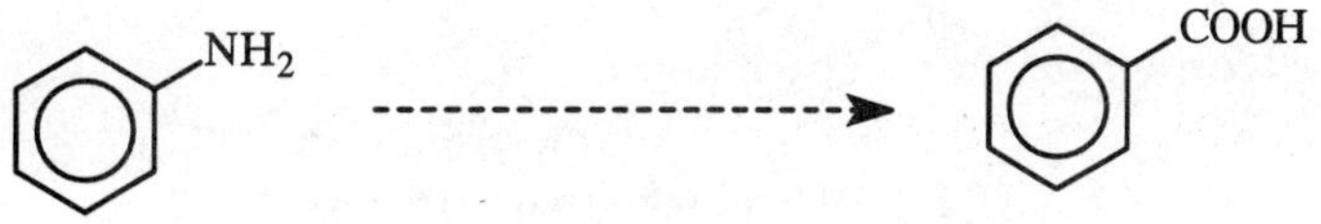

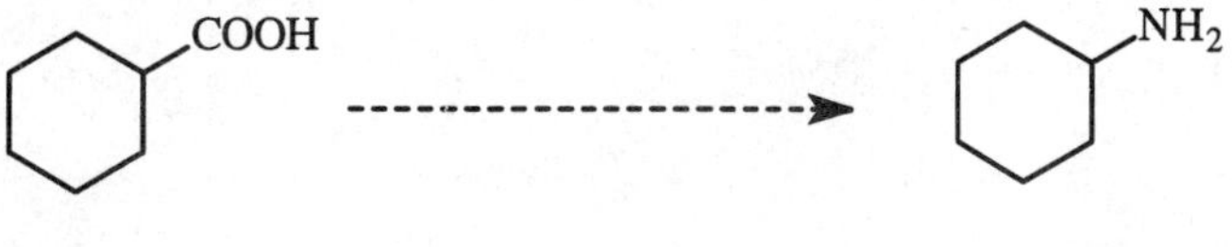

"Pushing Electrons": Representing Reaction Mechanisms

The **mechanism** of a reaction is a step-by-step description of how that reaction is thought to occur. In microscopic detail, chemical reactions involve multiple steps, including encounter equilibria, bond-making and bond-breaking, diffusion steps and re-hybridization ("heavy atom rearrangement"). In undergraduate organic chemistry, mechanisms are generally simplified to include only bond-making and breaking steps and "curved arrows" are generally utilized to indicate the flow of electrons within these steps. Although these simplified mechanisms are only approximations, they are useful since they allow a clearer understanding of how a given reaction proceeds, and allow reactions to be organized by mechanisms, making the study of organic chemistry simpler and more logical.

The process of *"pushing electrons"* using curved arrows requires a few simple conventions:

- Movement of a **pair** of electrons is shown using an arrow with a standard double-barbed arrowhead.
- Movement of a single electron (a **radical** reaction) is shown using an arrow with only a single barb.
- The arrow is drawn from the **middle of the electron pair** (or single electron) which is moving, **to the atom accepting the electron**(s).

Thus, **homolytic** cleavage of a carbon-bromine bond (a radical process) is shown below in (**a**) and the **heterolytic** cleavage is shown in (**b**).

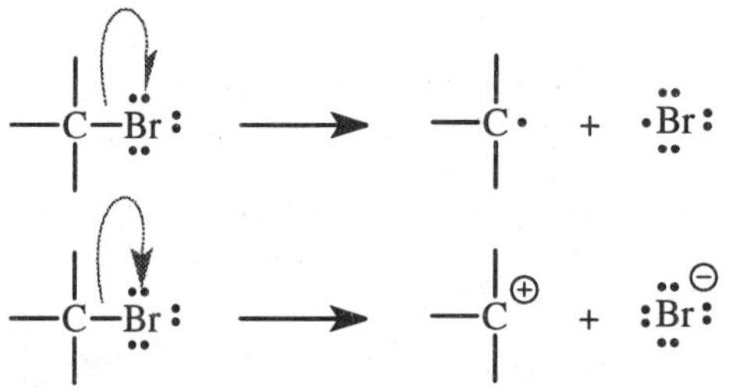

The heterolytic cleavage (**b**) leads to the formation of a pair of ions, which is indicated in the mechanism by the positive and negative charges. Two examples of this type of heterolytic cleavage reaction which are discussed in elementary organic chemistry are the **E1** and **S$_N$1** reaction mechanisms. Both of these reactions are unimolecular (the "1") meaning that only one of the reactant molecules is present in the rate-limiting transition state. The prefixes on these general mechanisms refer to elimination ("E") and substitution, nucleophilic ("S$_N$").

Treating the S$_N$1 reaction first, the rate of reaction of 2-methyl-2-propanol with HCl to give 2-chloro-2-methylpropane has been found to be independent of the concentration of chloride anion in the reaction mixture, suggesting that only the alcohol is present in the rate-limiting transition state and that reaction with chloride anion is fast. Thus, the rate-limiting step most likely involves acid-catalyzed loss of the hydroxyl group to give an intermediate **carbocation** (the rate-limiting step is the slowest step in a given reaction sequence). Reaction of the carbocation with chloride anion is then very fast, to give the final product. The major bond-making and bond-breaking steps in this process can be represented by the mechanism shown below.

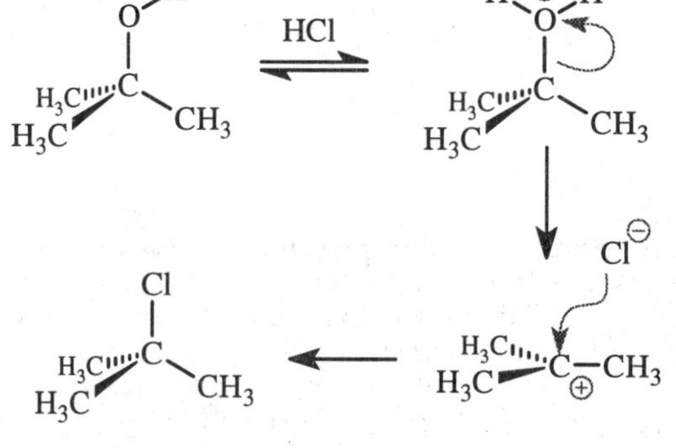

The protonation step is a rapid equilibrium process (described by the K_a for the protonated alcohol). The slow, or rate-limiting step is the breaking of the carbon-oxygen bond, and the fast step is the addition of chloride anion. Chloride anion concentration does not affect the rate of the reaction since the step involving the nucleophilic attack occurs *after* the rate-limiting step.

An example of an unimolecular elimination reaction (E1) is shown in the scheme below.

In this reaction, 2-chloro-2-methylpropane reacts with a base to lose HCl and form the alkene, 2-methylpropene. Again, the rate of this reaction is found to be independent of the concentration of the base that is used, requiring that the only reactant in the rate-limiting transition state is the alkyl halide and that loss of the proton from the carbocation is very fast.

In the mechanistic representation of this reaction, the chlorine is shown to depart with a pair of electrons to give the carbocation and chloride anion. To show the process of elimination, the electron pair from the base is shown removing the hydrogen from the carbon adjacent to the carbocation. Concurrent with this, the electron pair from the carbon-hydrogen bond is shown moving into the space between the carbocation and the carbon, to form a carbon-carbon double bond; again, this process must be fast relative to the spontaneous loss of chlorine from the starting material in order for the reaction to be independent of the concentration of the base.

Implied in the mechanism, although not generally discussed, is the re-hybridization of the CH_2 group and the overlap of the adjacent p-orbitals to form the alkene π-system. There are also major changes in the solvation of the (neutral) starting material as it forms ions, and further changes as the neutral alkene is produced. While these steps are generally ignored, examples exist where these types of processes are partially, or largely, rate-limiting.

Both of the examples described above are multiple step reactions involving a high-energy intermediate (the carbocation). These types of reactions are said to occur **stepwise**, and generally one step in the sequence is slow and is rate-limiting. In a **concerted** process, all of the reactant molecules are present together in the rate-limiting transition state and all of the bond-making and bond-breaking steps occur *simultaneously*. Two examples of these types of concerted processes are the S_N2 and the E2 reaction mechanisms. An example of an S_N2 process is shown below:

The rate of the reaction of 2-chloropropane with hydroxide anion is dependent on both the concentration of the alkyl halide and on the concentration of hydroxide anion (the reaction is bimolecular; the "2" in S_N2). This means that both of these must be present together in the rate-limiting transition state. The process of substitution within this transition state can be shown using curved arrows as shown above. The electron pair on hydroxide anion is shown to attack to carbon bearing the chlorine (the leaving group) at the same time that the carbon-chlorine bond is breaking.

In one simultaneous, coupled movement, the bond from the oxygen to the carbon has formed, the bond from the carbon to the chlorine has undergone heterolytic cleavage, and the central carbon has undergone stereochemical inversion (**S**-2-chlorobutane has formed **R**-2-butanol). A concerted reaction such as this does not have a "high energy" intermediate (such as the carbocation, described for the S_N1 process), but occurs in a **single step through a concerted transition state.**

The second common example of a concerted bimolecular process is the E2 elimination.

$$\text{Base:} \quad \underset{\underset{R}{\overset{R}{|}}}{\overset{H}{\underset{|}{C}}} - \underset{\underset{Cl}{|}}{\overset{R}{\underset{|}{C}}} - R \quad \longrightarrow \quad \underset{R}{\overset{R}{C}}=\underset{R}{\overset{R}{C}} \quad + \text{ HCl}$$

In an E2 elimination reaction, the rate is dependent on the concentration of the alkyl halide and on the concentration of the base. Again, both of these must be together in the rate-limiting transition step, and the electron flow in this transition state is typically drawn as shown above.

In a coupled, **concerted** process, the electron pair on the base removes the proton from the α–carbon **at the same time** as the electron pair is moving into the space between the carbons at the same time as the carbon-chlorine bond is breaking. Again, there is no intermediate, only the concerted transition state.

The reaction mechanisms shown in the animations on the CD accompanying this workbook include examples of concerted bimolecular processes, along with examples of highly complex multiple step reactions. For all of these examples, however, the mechanism can be simplified and shown as a discrete set of electron movements using the curved arrow approach. In those examples where the rate-limiting transition state is well-characterized, it is shown as part of the animation. For a more detailed discussion of each of the reactions in this section, please see the discussion in your text.

Organic Chemistry On-Line: *The Workbook* 185

Reaction Mechanisms Problem Set

In the space provided, draw a detailed mechanism for the reaction shown on the left. Use "curved arrows" to show electron flow, bond-making and bond-breaking and clearly show all intermediate charges.

Addition of HBr to an Alkene

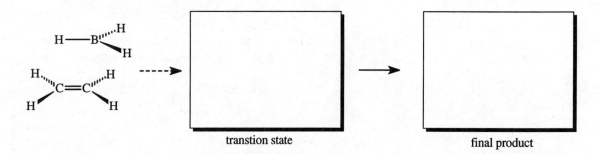

Addition of Diborane to an Alkene

Alkylation of an Acetylide Anion

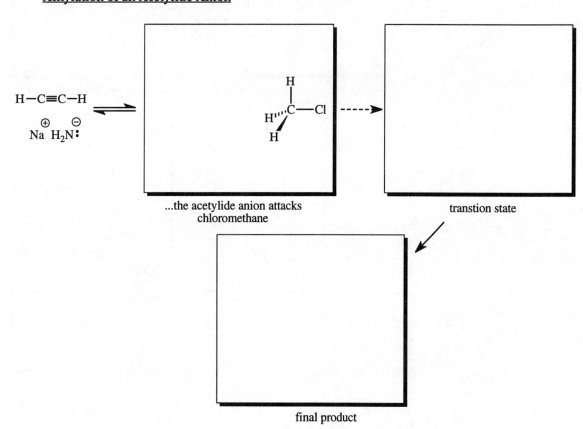

Reaction Mechanisms: *"Pushing Electrons"*

Electrophilic Aromatic Substitution: The Bromination of Benzene

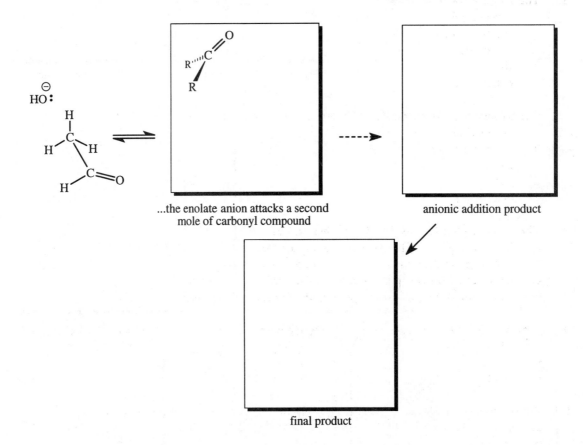

Carbonyl Condensation Reactions: The Addition of an Enolate Anion to a Carbonyl

Practice Examinations

The following pages contain **six 25-question practice examinations** covering sections in the first twenty-four chapters in John McMurry's *Organic Chemistry*, and **two fifty-question examinations** covering all twenty-four chapters. These examinations are multiple-choice and are written in the style that is often encountered on pre-professional entrance examinations (i.e., the MCAT or graduate school placement examinations). Because organic chemistry is a *structural science*, many classroom examinations feature questions in which students must draw structures and mechanisms, such as those seen in the bulk of this workbook. While this approach is very effective at teaching the fundamentals of organic chemistry, students often receive little practice with the multiple-choice format, and sometimes feel poorly prepared for this type of standardized testing.

In *most* of the examinations in this section, references are provided referring the student back to McMurry's fifth edition. These examinations also appear on the accompanying CD, and the answers to the questions, together with a short explanation are also presented. As always, it is recommended that you complete these exercises in your workbook *prior* to interactively checking your answers on the CD-tutorial.

Practice Exam #1 – *McMurry*, Chapters 1 - 6

(Use the following information for questions **1-2**.) Consider the molecule NH_2Cl, chloroamine. [see McMurry, **Section 1.6**]

1. The <u>shape</u> of the molecule is
    a. tetrahedral
    b. trigonal planar
    c. pyramidal
    d. flat
    e. octahedral

2. The arrangement of electron pairs about the central N atom is [see McMurry, **Section 1.6**]
    a. tetrahedral
    b. trigonal planar
    c. pyramidal
    d. flat
    e. octahedral

3. Chloroamine (NH_2Cl) [see McMurry, **Section 2.2**]
    a. has a strong bond dipoles and a net dipole moment.
    b. has no bond dipoles and a net overall dipole moment.
    c. has a strong bond dipoles and NO net overall dipole moment.
    d. has no bond dipoles bond dipoles and NO net dipole moment.

4. Which of the following substances is the <u>worst</u> Brønsted-Lowry acid? [see McMurry, **Section 2.7**]
    a. CH_4
    b. HCl
    c. $HClO_3$
    d. H_2SO_4
    e. H_3PO_4

5. Which element in CH_3OCH_2Li is the most electronegative? [see McMurry, **Section 2.1**]
    a. C
    b. H
    c. O
    d. Li
    e. H_3

6. Which of the following substances is the best Lewis acid? [see McMurry, **Section 2.11**]
    a. CH_3CH_2OH
    b. $NH(CH_3)_2$
    c. H_2
    d. H_2S
    e. BF_3

Choose the structure that contains the listed functional group: [see McMurry, **Section 3.1**]

(structures a–e shown: a = morphine-like structure with HO, O, OH; b = benzamide (PhC(O)NH₂); c = cyclic structure with COOH; d = cyclohexane with CH₃, CH₃CH₂–, CN; e = benzaldehyde Ph–CHO)

7. Alcohol

8. Carboxylic acid

9. Ethyl group

10. Aldehyde

11. Amide

12. How many of the constitutional isomers of C_6H_{14} can be correctly named as "butane" using the IUPAC naming system? [see McMurry, **Section 3.2,3**]

    a. 0   d. 3
    b. 1   e. 4
    c. 2

(Use the following information for questions 13-14) Consider the following molecule:

(Form A and Form B chair conformations of 1,3-dimethyl-1-ethylcyclohexane shown)

13. Which of the following is true? [see McMurry, **Section 4.11**]
    a. A is more stable than B due to reduced 1,3-diaxial destabilization.
    b. B is more stable than A due to reduced 1,3-diaxial destabilization
    c. A is more stable than B due to enhanced 1,3-diaxial stabilization
    d. B is more stable than A due to enhanced 1,3-diaxial stabilization
    e. A and B are exactly the SAME MOLECULE.

14. Which of the following statements is(are) correct? [see McMurry, **Section 4.11**]
    a. A can be converted into B by a ring flip.
    b. A can be converted into B by single rotation about a single bond.
    c. A and B are exactly the SAME MOLECULE.
    c. Both *a* and *b* are correct.
    d. None of these is correct.

15. Consider the conformers of pentane defined by rotation about the C2–C3 bond axis. Which of the following Newman projections represents the <u>lowest</u> energy minima? [see McMurry, **Section 4.3**]

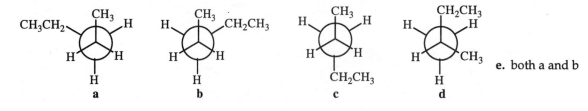

e. both a and b

16. Ring strain is very high for cyclopropane. The best explanation for this is [see McMurry, **Section 4.3,7**]
    a. Smaller bond angles produce increased eclipsing interactions.
    b. The molecules are deficient in H compared to non-cyclic alkanes.
    c. Orbital overlap is reduced by the strained geometry.
    d. The molecules are flat.
    e. More than one of the above.

17. Which of the following statements is(are) correct? [see McMurry, **Section 5.2**]
    a. Heterolytic cleavage leads to radical (unpaired-electron) reactions.
    b. Homolytic cleavage leads to polar (charged-species) reactions.
    c. Both *a* and *b* are correct.
    d. Neither *a* nor *b* are correct.

(Use the following information for questions 18-19.) Consider the polar reaction shown below.

18. The first step involves [see McMurry, **Section 5.5**]
    a. attack of the Cl on the double bond
    b. loss of a proton (H⁺) from the alkene
    c. attack on the proton of HCl by the sigma-bond electrons between the C's
    d. attack on the proton of HCl by the pi-bond electrons between the C's
    e. formation of a negatively charged carbon species, a <u>carbanion</u>

19. In the second step of this reaction
    a. Cl is a nucleophile and the carbocation is an electrophile
    b. the HCl proton is a nucleophile
    c. the sigma-bond electrons act as a electrophile
    d. the pi-bond electrons act as an electrophile
    e. homolytic cleavage occurs

Use the following reaction for Questions 20 – 23.

20. Which of the products above is preferred? [see McMurry, **Section 6.9**]
    a. A
    b. B
    c. Neither is preferred. There is a 50:50 mix of the two.
    d. The reaction does not occur under these conditions.

21. Which of the following intermediates would lead to product A if the reaction leading to that product did occur? [see McMurry, **Section 6.9,10**]

    a. H₃C−C⁺(CH₃)−CH₃   b. H₂C⁺−CH₂−CH₃

    c. Neither intermediate is possible.

22. Which of the following statements is <u>TRUE</u>? [see McMurry, **Section 6.9,10**]
    a. A is the preferred product because the reaction leading to A is slower than that leading to B.
    b. B is the preferred product because the reaction leading to B is slower than that leading to A.
    c. A is the preferred product because the intermediate leading to A is lower in energy.
    d. B is the preferred product because the intermediate leading to B is lower in energy.
    e. The reaction does not occur.

23. The rule governing the product distribution is [see McMurry, **Section 6.9**]
    a. Zaitsev's Rule
    b. Markovnikov's Rule
    c. Carbocation rearrangement
    d. The "nonspecific rule"
    e. The Rule of Thumb

24. Which of the following is the correct designation for the molecule shown? [see McMurry, **Section 6.6**]

    a. cis   d   Z
    b. trans   e. None of these.
    c. E

25. Compared to single bonds, double bonds are:
    a. longer and weaker   d. shorter and stronger
    b. shorter and weaker   e. orthogonal
    c. longer and stronger

Practice Exam #2 – McMurry, Chapters 7 - 12

1. What is the major product (X) when styrene (shown below) is reacted with H$_2$ over a Ni catalytic surface? [see *McMurry*, Section 7.7]

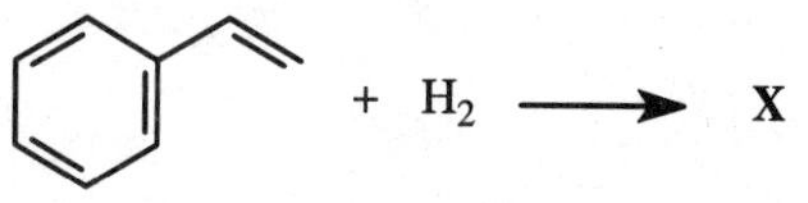

   styrene + H$_2$ $\longrightarrow$ X

   a. 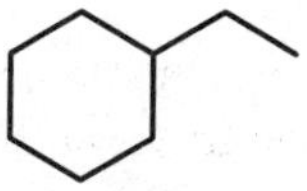 ethylcyclohexane
   b. 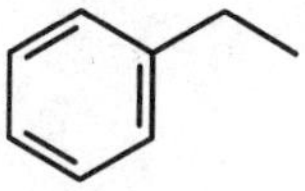 ethylcyclohexadiene
   c. 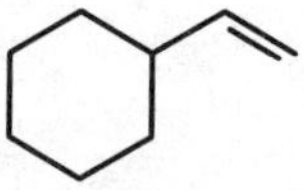 vinylcyclohexane
   d. 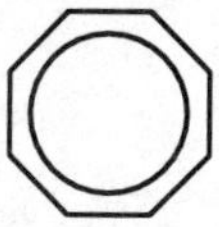 cyclooctatetraene
   e. No reaction occurs under these conditions.

2. Which product(s) would be produced by acid catalyzed dehydration of the following alcohol? [see *McMurry*, Section 7.1]

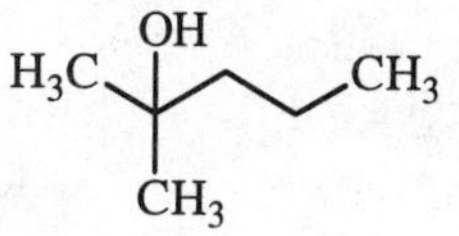

   a. 2-methylpentene
   b. 2-methyl-2-pentene
   c. 2-methylpentene <u>and</u> 2-methyl-2-pentene
   d. 4-methyl-1-pentene
   e. all of these

3. What product would result from the reaction of 1,2-dimethylcyclohexene with hydrogen gas in the presence of a Pt catalyst? [see *McMurry*, Section 7.7]

   a. 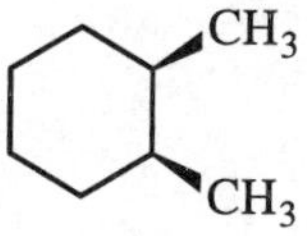 cis-1,2-dimethylcyclohexane (both CH$_3$ wedge)
   b. 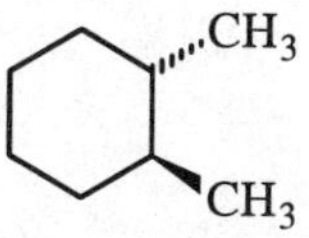 1,2-dimethylcyclohexane (both CH$_3$ dash)
   c. 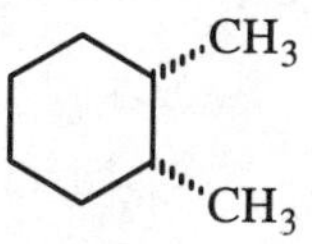 trans-1,2-dimethylcyclohexane (one dash, one wedge)
   d. a <u>and</u> b
   e. a <u>and</u> c

4. The role of peroxides in the polymerization of ethylene is [see *McMurry*, **Section 7.11**]
   a. radical producing chain initiator
   b. Lewis Acid producing chain initiator
   c. polar reaction catalyst
   d. radical producing chain propagator
   e. radical producing chain terminator

5. A triple bond is composed of: [see *McMurry*, **Section 8.1**]
   a. three sigma bonds
   b. two sigma bonds and one pi bond
   c. one sigma bond and two pi bonds
   d. sp^3 bonds
   e. sp bonds

6. Which of the following is required to form an acetylide anion? [see *McMurry*, **Section 8.8**]
   a. a VERY powerful base.
   b. a terminal alkyne
   c. peroxides
   d. a <u>and</u> b
   e. b <u>and</u> c

7. Which of the following is a good operational way to identify a chiral center? [see *McMurry*, **Section 9.1,2**]
   a. Determine if a plane of symmetry exists.
   b. Determine if there are four different attachments at a carbon center.
   c. Check for the existence of mirror images.
   d. See if the molecule is optically active (*i.e.* does it have color).

8. The reaction shown below can be said to be:

   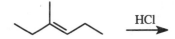

   a. stereoselective.
   b. regioselective.
   c. both a <u>and</u> b
   d. neither a <u>nor</u> b
   e. The answer choices make no sense in this context

9. Molecules that rotate plane polarized light to the left are: [see *McMurry*, **Section 9.3,4**]
   a. levorotary
   b. dextrorotary
   c. R
   d. S

(The following information and answer choices will be used for Questions **10 – 11**.) Classify the following pair of streoisomers as: [see *McMurry*, **Section 9.1**]

   a. enantiomers   b. diasteroemers   c. identical
   d. identical **and** meso compounds   e. different chemical compounds and **not** isomeric

10.

11.

12. Why is preparing a monohaloalkane by radical reaction with X₂ a poor reaction synthetically? [see McMurry, **Section 10.4**]
    a. The reaction is slow.
    b. The reaction is not regioselective.
    c. The reaction requires enormous light flux to begin.
    d. The reaction is massively exothermic and produces a huge amount of heat.

13. Which of the following is the most stable radical? [see McMurry, **Section 10.6**]

    a.

    b.

    c.

    d.

14. Why is SOCl₂ used instead of HCl to replace –OH groups on secondary alcohols? [see McMurry, **Section 10.7**]
    a. SOCl₂ is resonance stabilized, leader to faster reaction.
    b. There are no radical byproducts with SOCl₂.
    c. SOCl₂ has a steric advantage over the chloride anion.
    d. Reactions are run under mild conditions, leading to fewer rearrangements.

15. Which of the following reagents are used to produce a Grignard Reagent? [see McMurry, **Section 10.8**]
    a. Ge / H₃O⁺
    b. RX / Mg
    c. RH / NBS
    d. SOCl₂ / Mg
    e. THF / Mg

(Use the following for questions 16 – 17.) The monosubstitution reaction of methane with Cl_2 proceeds by a stepwise radical mechanism in the presence of UV light. Classify each of the following reactions as: [see McMurry, Section 10.4]

    a. chain doubling
    b. chain initiating
    c. chain propagating
    d. chain terminating
    e. none of these

16. $CH_4 + \dot{Cl} \longrightarrow \dot{CH_3} + HCl$

17. $2\dot{Cl} \longrightarrow Cl_2$

18. Consider nucleophilic substitution reactions of alkylhalides, where a moderately reactive nucleophile is used. Under which of the following sets of conditions is the S_N2 mechanism most likely predominant? [see McMurry, Section 11.4,5]

    a. primary substrate, acidic conditions
    b. secondary substrate, acidic conditions
    c. tertiary substrate, acidic conditions
    d. secondary substrate, basic conditions
    e. cannot differentiate

Use the following reaction for Questions 19 and 20. [see McMurry, Section 11.5]

    A. $CH_3CH_2CH_2F \xrightarrow[H_2O]{HO^-} CH_3CH_2CH_2OH + F^-$

    B. $CH_3CH_2CH_2I \xrightarrow[H_2O]{HO^-} CH_3CH_2CH_2OH + I^-$

19. What is the predominant mechanism type for both reactions?
    a. S_N1 b. S_N2
    c. E1   d. E2
    e. radical

20. Which reaction is faster and why?
    a. A, because F^- is a better leaving group that I^-
    b. B, because I^- is a better leaving group that F^-
    c. A, because F stabilizes the reactant better than I
    d. B, because I stabilizes the reactant better than F
    e. The reaction rates are equivalent.

Use the following answer choices for questions 21 - 24.

I.	substitution	VI.	unimolecular
II.	elimination	VII.	nucleophilic
III.	addition	VIII.	electrophilic
IV.	rearrangement	IX.	single step
V.	bimolecular	X.	multiple step

21. An E1 reaction is: [see McMurry, Section 11.14]
    a. VIII, IX
    b. VIII, IV, IIX
    c. VIII, VI
    d. II, IX
    e. II, VI

22. An S_N1 reaction is: [see McMurry, **Section 11.6**]
    a. I, VIII
    b. I, VI, IX
    c. I, X
    d. III, VI
    e. III, VII, X

23. An E2 reaction is: [see McMurry, **Section 11.11**]
    a. II, IX, V
    b. II, X, VI
    c. II, IX, VI
    d. IV, VIII, VI
    e. IV, VIII, IX

24. An S_N2 reaction is: [see McMurry, Section **11.4,5**]
    a. I, X
    b. I, IX, VII
    c. I, V, X
    d. III, V
    e. III, X

25. A synthetic chemist performed a series of reactions designed to produce bonds between C and N. On analysis of the IR spectra of the resulting compounds, there were characteristic bands resulting from various types of C---N connections. The observed bands fell into three ranges: [see McMurry, **Section 12.9**]

    1. 1020-1250 cm^{-1}
    2. 1471-1689 cm^{-1}
    3. 2000-2273 cm^{-1}

    The bands were assigned (in order 1, 2, 3) to functional groups as:

    a. amine, nitrile, imine
    b. amine, imine, nitrile
    c. nitrile, amine, imine
    d. nitrile, imine, amine
    e. imine, nitrile, amine

Practice Exam #3 – McMurry, Chapters 13 - 16

(Use the following five answer choices and the molecule shown below for questions 1 – 5 relating to NMR spectroscopy. Answers may be used more than once.)

$$\underset{1}{H_3C}-\underset{2}{CH_2}-O-\underset{4}{CH}\underset{3}{\overset{5}{\underset{CH_3}{\overset{CH_3}{|}}}}$$

A. downfield
B. upfield
C. quartet
D. triplet
E. equivalent

1. The protons labeled 2 are _____ relative to protons 1. [see *McMurry*, **Section 13.3**]

2. The protons labeled 3 are _____ relative to protons 5. [see *McMurry*, **Section 13.3, 8**]

3. The NMR signal from protons labeled 2 are split into a _____. [see *McMurry*, **Section 13.11**]

4. The NMR signal from protons labeled 1 are split into a _____. [see *McMurry*, **Section 13.11**]

5. A <u>less</u> shielded proton falls _____ on an NMR spectrum in relation to a <u>more</u> shielded proton. [see *McMurry*, **Table 13.3**]

6. A compound of the formula $C_5H_{10}O$ gave the following spectral information:

   <u>^{1}H Spectrum:</u>   <u>IR Spectrum:</u>
   6 H doublet @ δ 1.10   strong peak at 1720 cm^{-1}
   3 H singlet @ δ 2.10
   1 H septet @ δ 2.50

   Which of the following is a reasonable structure for the compound? [see *McMurry*, **Section 13.3 & Chapter 12**]

   a. $H_3C\diagdown\diagup\diagdown\underset{H}{\overset{O}{\diagup\!\!\!\!\diagdown}}$

   b. $H_3C\diagdown\diagup\underset{}{\overset{O}{\diagup\!\!\!\!\diagdown}}CH_3$

   c. $H_3C-\underset{CH_3}{\overset{CH_3}{|}}-\underset{H}{\overset{O}{\diagup\!\!\!\!\diagdown}}$

   d. $H_3C\diagdown\underset{}{\overset{O}{\diagup\!\!\!\!\diagdown}}CH_3$

   e. $H_3C\diagdown\underset{CH_3}{\overset{O}{\diagup\!\!\!\!\diagdown}}CH_3$

7. The 1,2 monoaddition product(s) for HCl reacting with 2,4-hexadiene is(are) [see McMurry, **Section 14.5**]
   a. No reaction.
   b. 5-chloro-2-hexene
   c. 4-chloro-2-hexene
   d. 2-chloro-3-hexene
   e. more than one of the above

(Use the following information for questions 8 – 11.) [see McMurry, **Section 14.12 & Chapters 12 and 13**]

   ¹H NMR: 3 H s @ 1.96 δ   **Molecular Formula:** $C_9H_{10}O_2$
   2 H s @ 5.00 δ
   5 H s @ 7.22 δ   **UV-Visible:** Intense peak at 255 nm
   Major IR peaks: 1610, 1495, 1735 cm⁻¹

8. Which of the following is ruled out by the IR information?
   a. carboxylic acid
   b. extended conjugated system.
   c. ester
   d. acid anhydride
   e. aromatic systems

9. The NMR peak at 7.22δ is indicative of the presence of a(n) [see McMurry, **Section 13.9**]
   a. carboxylic acid.
   b. extended conjugated system.
   c. ester.
   d. acid anhydride.
   e. aromatic system.

10. The fact that all of the peaks in the NMR are singlets means that [see McMurry, **Section 13.9**]
    a. a decoupling signal has been added to the spectrum to reduce clutter.
    b. the three sets of hydrogens are interchangeable with one another on the NMR timescale.
    c. equivalent protons on adjacent carbons do not spilt one another.
    d. the three sets of hydrogens are isolated from one another.
    e. no information can be gained.

11. Which of the following is the best structure given the information above? [see McMurry, **Section 13.9, 14.12**]

    a. HO~~~~~~~~OH
    b.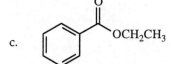
    c. (benzoyl ethyl ester: Ph-C(=O)-OCH₂CH₃)
    d. (bicyclic diketone structure)
    e. (methyl phenylacetate: Ph-CH₂-C(=O)-OCH₃)

12. Which of the following is (are) true? [see *McMurry*, **Section 15.7**]
    I. Pyrrole is aromatic.   IV. Pyridine is aromatic.
    II. Pyrrole is basic.   V. Pyridine is basic.
    III. Pyrrole is neutral.   VI. Pyridine is neutral.
    a. I and IV only
    b. II and V only
    c. I and II, IV and V
    d. I and III, IV and VI
    e. I and III, IV and V

13. Which of the following ions are aromatic? [see *McMurry*, **Section 15.6**]

        I   II   III

    a. I and II only
    b. II and III only
    c. III only
    d. I and III only
    e. all of the above

14. Which of the following are characteristics of aromatic systems in general? [see *McMurry*, Section 15.3, 4]
    I. Planar structure.
    II. Neutral.
    III. Carbons only in aromatic substructure.
    IV. Unhybridized p-orbitals on each aromatic system atom.
    a. I and II
    b. I, II and III
    c. I and IV
    d. II and III
    e. II, III, IV

15. What is the product formed when 1,2-diethylbenzene is treated with an aqueous solution of $KMnO_4$? [see *McMurry*, **Section 16.10**]
    a. Naphthalene
    b. *o*-benzenedicarboxylic acid
    c. No reaction occurs as the ring is inert.
    d. Benzoic acid
    e. Benzodiquinone

16. The non-aromatic reactant in electrophilic substitution reactions involving aromatic systems MUST be [see *McMurry*, **Section 16.2**]
    a. positively charged
    b. A Lewis Acid
    c. A Nucleophile
    d. negatively charged
    e. A Lewis Base

17. What is(are) the product(s) of the reaction of nitrobenzene with Cl_2 in the presence of $FeCl_3$? [see *McMurry*, **Section 16.2,5**]
    a. *p*-chloronitrobenzene
    b. *o*-chloronitrobenzene
    c. *m*-chloronitrobenzene
    d. a and c
    e. a and b

18. Which of the following statements is(are) true about electrophilic aromatic substitution (EAS) to nitrobenzene? [see *McMurry*, **Section 16.6**]
    I. Resonance stabilization acts to orient the new electrophile in the *ortho* position relative to the nitro group.
    II. Resonance stabilization acts to orient the new electrophile in the *meta* position relative to the nitro group.
    III. Resonance stabilization acts to orient the new electrophile in the *para* position relative to the nitro group.
    a. I
    b. II
    c. III
    d. I and III
    e. none

19. Benzene is reacted with chlorobenzene in the presence of $AlCl_3$. The product is [see *McMurry*, **Section 16.3**]
    a. dibenzene
    b. naphthalene
    c. anthacene
    d. graphite
    e. no reaction.

20. Rank the following from "most reactive" to "least reactive" with respect to electrophilic aromatic substitution. [see *McMurry*, **Section 16.5,6**]

    I. benzene   II. phenol   III. benzonitrile

    a. I > II > III
    b. II > I > III
    c. I > III > II
    d. III > II > I
    e. II > III > I

21. A chemist would choose to use $LiAlH_4$ (in ether) instead of $NaBH_4$ (in ether) for reductions in organic reactions because[see *McMurry*, **Section 17.8**]

    a. $LiAlH_4$ is stable in ether, whereas $NaBH_4$ is not.
    b. $NaBH_4$ is stable in ether, whereas $LiAlH_4$ is not.
    c. $LiAlH_4$ is a stronger reducing agent than $NaBH_4$.
    d. $LiAlH_4$ is unstable and potentially explosive if mishandled.

22. The molecule *p*-cyanophenol is a much stronger acid than phenol. Which of the following best explains this observation? [see *McMurry*, **Section 17.3**]
    a. aromatic stabilization of benzylic carbocations
    b. resonance stabilization of the resulting anion
    c. aromatic stabilization of benzylic carbanions
    d. resonance stabilization of the resulting cation
    e. hydrogen transfer of an aromatic proton

23. Alcohols have boiling points elevated from other molecules of similar size. Which of the following best explains this observation? [see *McMurry*, **Section 17.2**]
    a. van der Waals or dispersive interactions
    b. strong dipole/dipole interactions
    c. extensive hydrogen-bonding
    d. ordered packing of molecules
    e. most small alcohols are infinitely soluble in water

(Use the following for questions **24** and **25**.) The following product was prepared using the Williamson ether synthesis.

$$H_3C-O-C(CH_3)_2-CH_3$$

24. The reaction would proceed most readily to the correct product under which the following combinations of reagents / conditions ? [see *McMurry*, **Section 18.3**]
    a. $CH_3I + (CH_3)_2CHOH$ / PCC in CH_2Cl_2
    b. CH_3I / $NaBH_4$ in $(CH_3)_2CHOH$
    c. CH_3OH / 1) Na; 2) $(CH_3)_2CHI$ in ether
    d. $(CH_3)_2CHOH$ / 1) Na; 2) CH_3I in ether

25. The reaction occurs by which general type of mechanism? [see *McMurry*, **Section 18.3**]
    a. S_N2
    b. S_N1
    c. E2
    d. E1
    e. radical

Practice Exam #4 – McMurry, Chapters 17 - 24

1. Which of the following reagents would best be used to convert cyclohexanol into an aldehyde? [see *McMurry*, **Section 19.2**]
   a. PCC in CH_2Cl_2
   b. $Na_2Cr_2O_7$ in acid
   c. Tollens Reagent
   d. $LiAlH_4$ in ether
   e. None of these will carry out this reaction.

(Use the following for questions 2 - 5.) Carbonyl groups are important in organic synthesis because they are highly reactive in a variety of modes. Consider the basic carbonyl structure:

$$R_1 \underset{}{\overset{O}{\underset{\|}{C}}} R_2$$

2. The O end of the C=O bond: [see *McMurry*, **Section 19.6**]
   a. is vulnerable to attack by bases.
   b. is vulnerable to attack by acids.
   c. can be readily attacked by <u>either</u> acids or bases.
   d. is vulnerable to radical attack.
   e. is generally inert.

3. The O end of the C=O bond: [see *McMurry*, **Section 19.6**]
   a. is vulnerable to attack by bases.
   b. is vulnerable to attack by acids.
   c. can be readily attacked by <u>either</u> acids or bases.
   d. is vulnerable to radical attack.
   e. is generally inert.

4. Which of the following gives the most reactive compound? [see *McMurry*, **Section 19.5**]
   a. R1 = R2 = CH_3
   b. R1 = H, R2 = CH_3
   c. R1 = R2 = Ph
   d. R1 = H, R2 = Ph
   e. R1 = R2 = H

5. Which of the following best explains why correct answer above? [see *McMurry*, **Section 19.5**]
   a. Alkyl groups stabilize the charge on the carbonyl carbon.
   b. Resonance stabilization is important.
   c. The phenyl group(s) holds the system planar for best reaction.
   d. The site is less sterically hindered.
   e. The O can form a stabilizing intramolecular H-bond.

6. Rank the following acids in order of <u>increasing</u> acidity. [see *McMurry*, **Section 20.4**]
   I. propionic acid II. 3-chloropropionic acid III. Propanol IV. 2-chloropropionic acid
   a. I>II>III>IV
   b. IV>III>I>II
   c. IV>II>I>III
   d. IV>III>II>I
   e. III>I>II>IV

7. Carboxylic acids have higher boiling points than alkanes of similar molecular weights. Which of the following best explains this observation? [see McMurry, **Section 20.2**]

   a. Strong dipole/dipole interactions.
   b. Stronger than normal dispersive forces.
   c. General hydrogen bonding.
   d. H-bond dimers.
   e. Liquid crystalline packing.

8. There is a direct relationship between the deactivating strength of a substituent on the ring and the effect that substituent has on the acidity of benzoic acid. Which of the following best explains this observation? [see McMurry, **Section 20.5**]

   a. Deactivating groups are able to resonance stabilize the resulting anion.
   b. Deactivating strength is tied to steric interference.
   c. Deactivating groups are electron withdrawing and thus stabilize the resulting anion by inductive effects.
   d. a and b
   e. a and c

9. Addition of a metal hydroxide to a solution of acetic acid followed by evaporation produces which of the following? [see McMurry, **Section 20.3**]

   a. CO_2 gas and the purified metal left behind.
   b. A peroxide of the acetate and hydroxide anions.
   c. No reaction
   d. A chelated complex of two acetic acids per metal atom.
   e. A soluble metal acetate salt.

(Use the following for questions 10 – 14.) Select from the answer choices below the <u>product</u> of the reaction of an acid chloride (**RCOCl**) with the reagents/reactions (**R**) as noted in questions 10 - 14. Answers may be used more than once.

   a. a carboxylic acid
   b. an ester
   c. an amide
   d. an amine
   e. no new product

10. RCOCl + H_2O ⟶ ?

11. RCOCl + NH_3/H_2O ⟶ ?

12. RCOCl + ROH $\xrightarrow{\text{pyridine}}$?

13. RCOCl $\xrightarrow{\text{1. } H_2O,\ \text{2. } SOCl_2 \text{ in } CH_2Cl_2}$?

14. RCOCl $\xrightarrow{\text{1. } NH_3/H_2O,\ \text{2. } LiAlH_4 \text{ in ether}}$?

(Use the following molecule and labels for questions 15 – 16.)

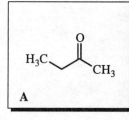

15. Which of the following statements is correct? [see McMurry, Section 22.1,5]
    a. Only protons labeled a, d, and e have enhanced acidity due to **resonance** effects.
    b. Only protons labeled a, b, c, and e have enhanced acidity due to **resonance** effects.
    c. Only protons labeled a, b, and c have enhanced acidity due to **resonance** effects.
    d. Only protons labeled a, b, and c have enhanced acidity due to **inductive** effects.
    e. Only protons labeled a, b, c, and d have enhanced acidity due to **inductive** effects.

16. Which proton is the MOST acidic? [see McMurry, Section 22.1,5]
    a.   b.   c.   d.   e.

Use the two molecules below for Question 17 and 18.

17. Which of the following statements is (are) true? [see McMurry, Section 22.1]
    a. A is more stable than B.
    b. B is more stable than A.
    c. Reduction of A by LiAlH$_4$ leads to B.
    d. A is an enolate anion.
    e. More than one of the above is true.

18. Which of the following statements is (are) true? [see McMurry, Section 22.1]
    a. A and B can be interconverted by a base catalyzed process.
    b. A and B can be interconverted by an acid catalyzed process
    c. The O atom in A acts as an electrophile in reactions.
    d. None of the above is true.
    e. More than one of the above is true.

19. Two mole equivalents of an <u>aldehyde</u> react in an alcoholic sodium ethoxide solution. The reaction name / product pair is: [see McMurry, Section 23.2]
    a. aldol condensation / β-keto ester
    b. Claisen condensation / β-keto ester
    c. aldol condensation / β-hydroxy aldehyde
    d. Claisen condensation /β-hydroxy aldehyde
    e. a mixture of four products

20. Two mole equivalents of an ester react in an alcoholic sodium ethoxide solution. The reaction name/product pair is: [see McMurry, **Section 23.8**]
    a. aldol condensation / β-keto ester
    b. Claisen condensation / β-keto ester
    c. aldol condensation / β-hydroxy aldehyde
    d. Claisen condensation / β-hydroxy aldehyde
    e. a mixture of four products

21. Certain single molecules can contain two aldehyde or two ester groups. If these molecules are mixed with sodium ethoxide in ethanol, what is the result? [see McMurry, **Section 23.7,10**]
    a. Claisen and/or aldol polymers are formed, as the two ends form links to other molecules.
    b. If large enough, the molecule will form a ring
    c. There are now eight possible products and all are formed.
    d. No reaction occurs under these conditions.

22. The Michael Reaction can be carried out with a variety of species, yielding a vast variety of products. Which of the following would result in the best reaction with an α,β-unsaturated ketone? [see McMurry, **Section 23.11**]
    a. β-keto ester
    b. Malonic ester
    c. Conjugated enone
    d. Nitrile

23. The Robinson Annulation Reaction results in which of the following? [see McMurry, **Section 23.13**]
    a. Stereochemically pure amino acids
    b. annulones
    c. polymers
    d. multiple fused-ring systems

24. Which of the following best defines the reactivity of amines? [see McMurry, **Section 24.4**]
    I. React as Lewis Acids.
    II. React as Lewis Bases.
    III. Usually accept protons.
    IV. Usually donate protons.

    a. I and III
    b. I and IV
    c. II and III
    d. II and IV

25. What is the product of the reaction shown below? [see McMurry, **Section 23.8**]

    PhNH$_2$ $\xrightarrow{\text{1. HNO}_2/\text{H}_2\text{SO}_4 \quad \text{2. H}_3\text{PO}_2}$

    a. benzene
    b. *p*-nitroaniline
    c. an amide
    d. a diazonium salt
    e. a mixture of products

Practice Exam #5 – "Alkanes and Cycloalkanes"

1. The most correct IUPAC name for the molecule shown on the right is:

   a. 1-aminodioxy-3,4-dimethylcyclopentane
   b. 1,2-dimethyl-4-nitrocyclopentane
   c. 1,2-dimethylcyclopentyl-4-nitrate
   d. 3,4-dimethyl-1-nitrocyclopentane
   e. 1-amino-2,3-dimethylcyclopentane

   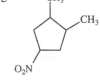

2. The most correct IUPAC name for the molecule shown on the right is:

   a. 1-chloromethyl-1,4-dichlorohexane
   b. 1-chloroethyl-1-(1,4-dichloro)hexane
   c. 1,2-dichloroethyl-3-chloropentane
   d. 1,2,5-trichloroheptane
   e. 3,6-dichlorohexyl-6-(1-chloro)methane

   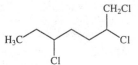

3. The most correct IUPAC name for the molecule shown on the right is:

   a. 5-ethyl-3-methyl-1-bromocyclohexane
   b. 1-bromo-(3-methyl-5-ethyl)cyclohexane
   c. 1-bromo-3-ethyl-5-methylcyclohexane
   d. 2-(3-bromo-5-methylcyclohexyl)ethane
   e. 3-ethyl-5-methyl-bromocyclohexane

   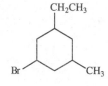

4. The most correct IUPAC name for the molecule shown on the right is:

   a. 1-(3-bromo-2-methylbutyl)cyclopropane
   b. 1-bromo-(3-methyl-5-ethyl)cyclohexane
   c. 3-bromo-1-cyclopropyl-2-methylbutane
   d. 2-(3-bromo-5-methylcyclohexyl)ethane
   e. 3-bromo-1-cyclopropyl-2-methylbutane

   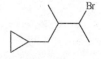

5. The most correct IUPAC name for the molecule shown on the right is:

   a. 1-chloro-2-methyl-4,4-dibromocyclohexane
   b. 4-chloro-1,1-dibromo-3-methylcyclohexane
   c. 1,1-dibromo-4-chloro-3-methylcyclohexane
   d. *trans*-1,1-dibromo-4-chloro-3-methylcyclohexane
   e. 1,1-dibromo-4-chloromethylcyclohexane

6. The most correct IUPAC name for the molecule shown on the right is:

   a. 3-(2-bromoethyl)-3-methylhexane
   b. 1-bromo-3-ethyl-3-methylhexane
   c. 1-bromo-3-propyl-3-methylpentane
   d. 3-(2-bromoethyl)-3-ethylpentane
   e. 2-(2-bromoethyl)-2-ethylpentane

7. The most correct IUPAC name for the molecule shown on the right is:

   a. *cis*-1-chloro-3-methylcyclohexane
   b. *cis*-1-chloromethylcyclohexane
   c. 1,3-diaxial-chloromethylcyclohexane
   d. *trans*-1-chloro-3-methylcyclohexane
   e. *cis*-1-chloro-*trans*-3-methylcyclohexane

   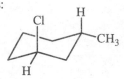

8. The most correct IUPAC name for the molecule shown on the right is:

   a. *trans*-1-chloro-3-(1-methylpropyl)cyclobutane
   b. *cis*-1-(3-chlorocyclobutyl)-1-methylpropane
   c. 1-(*cis*-3-chlorocyclobutyl)-1-methylpropane
   d. *trans*-1-(3-chlorocyclobutyl)-1-methylpropane
   e. *cis*-1-chloro-3-(1-methylpropyl)cyclobutane

   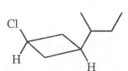

9. The two molecules shown on the right are:

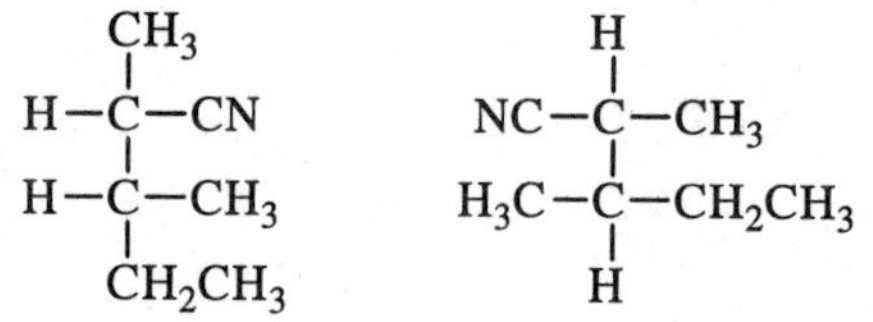

   a. identical
   b. constitutional isomers
   c. conformational isomers
   d. different compounds
   e. isomeric, but not constitutional isomers

10. Which statement is *not true* regarding the molecule on the right?

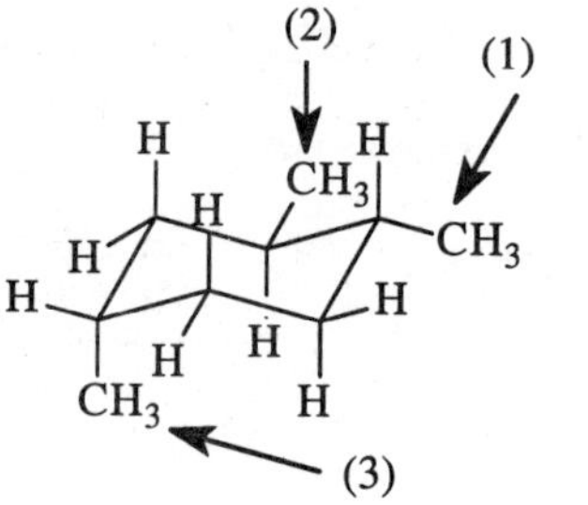

    a. methyl groups (1) & (2) are *trans* to each other
    b. methyl groups (2) & (3) are *trans* to each other
    c. methyl groups (1) & (3) are *cis* to each other
    d. in the most stable conformation, methyl (3) is equatorial
    e. if methyl group (3) is equatorial, methyl (2) will be axial

11. The two molecules shown on the right are:

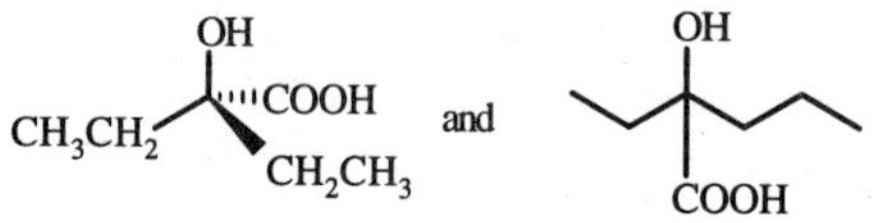

    a. .identical
    b. constitutional isomers
    c. conformational isomers
    d. rotamers, but not conformational isomers
    e. different chemical compounds

12. Which is *not true* regarding the molecule shown on the right?

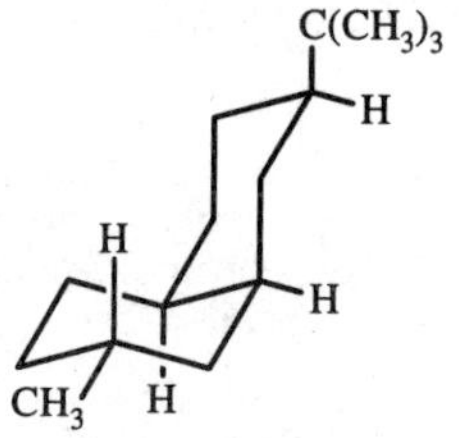

    a. the ring junction is *cis*
    b. the *tert*-butyl group is in an equatorial (stable) position
    c. the methyl group is *cis* to the carbons in the ring junction
    d. the compound cannot undergo ring inversion
    e. the *tert*-butyl group is *cis* to the carbons in the ring junction

13. The two compounds shown below are:
    a. identical   b. constitutional isomers   c. different, and not isomers   d. *cis-trans* isomers

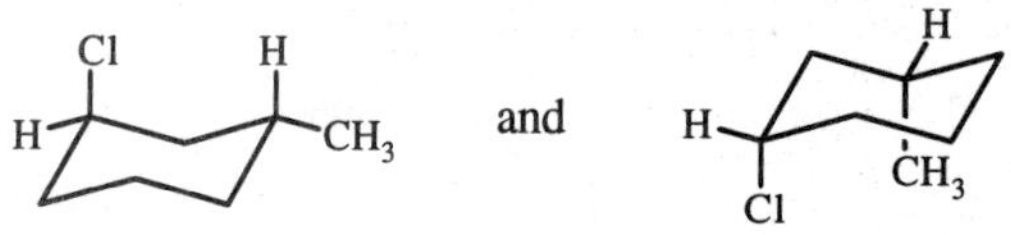

14. The two compounds shown below are:
    a. identical   b. constitutional isomers   c. different, and not isomers   d. *cis-trans* isomers

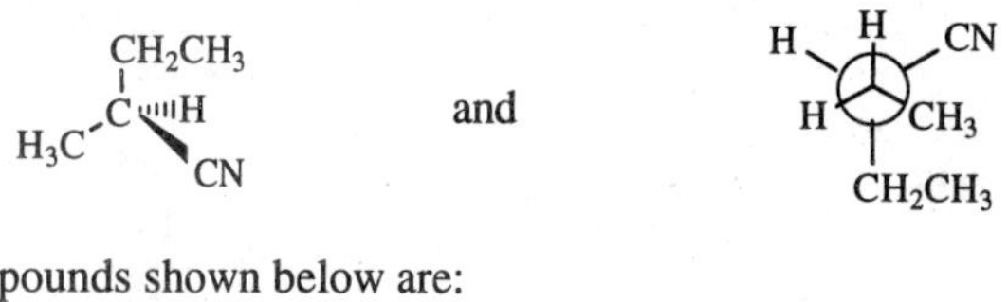

15. The two compounds shown below are:
    a. identical   b. constitutional isomers   c. different, and not isomers   d. *cis-trans* isomers

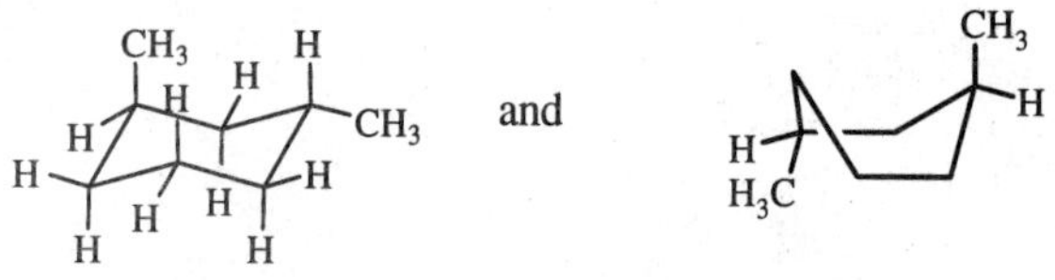

16. The two compounds shown below are:
   a. identical   b. constitutional isomers   c. different, and not isomers   d. *cis-trans* isomers

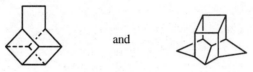

17. The two compounds shown below are:
   a. identical   b. constitutional isomers   c. different, and not isomers   d. *cis-trans* isomers

18. The two compounds shown below are:
   a. identical   b. constitutional isomers   c. different, and not isomers   d. *cis-trans* isomers

   CH₃CH₂(CHCH₃)CH₂CH₂Cl   and

19. The two compounds shown below are:
   a. identical   b. constitutional isomers   c. different, and not isomers   d. *cis-trans* isomers

20. Which of the molecules shown below is **Z**-1,2-dichloro-1-ethylcyclopentane?

21. Which of the molecules shown below is **E**-1,3-dimethylcyclohexane?

22. Which of the molecules shown below is 2,2-dimethylbutane?

23. Which of the molecules below have the emperical formula (CH$_2$)?
  a. a only **b.** b only **c.** c only **d.** d only **e.** a & b only **f.** a & d only

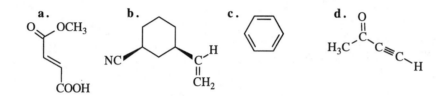

24. Which of the molecules below have three degrees of unsaturation?
  a. a only **b.** b only **c.** c only **d.** d only **e.** a & b only **f.** a & d only

25. Which of the molecules below are constitutional isomers?
  a. a only **b.** b only **c.** c only **d.** d only **e.** a & b only **f.** a, b & d only

Practice Exam #6 – "Spectroscopy"

1. In the mass spectrum, a compound with two bromine atoms will have:
   a. two molecular ions, in the ratio 1:1
   b. four molecular ions, in the ratio 1:1:1:1
   c. two molecular ions, in the ratio 2:1
   d. four molecular ions, in the ratio 1:2:2:1
   e. three molecular ions, in the ratio 1:2:1

2. In the ^{1}H NMR of diethylether (CH_3CH_2-O-CH_2CH_3) the signal from the CH_2 protons will appear as:
   a. a triplet
   b. a multiplet consisting of 9 peaks
   c. a multiplet consisting of seven peaks
   d. a multiplet consisting of 6 quartets
   e. a quartet

3. How many peaks will be present in the proton-decoupled ^{13}C NMR of the compound shown below:
   reminder: proton-decoupling destroys all splitting, reducing all absorbances to singlets
   a. 4
   b. 3
   c. 5
   d. 7
   e. 6

4. The base peak in the mass spectrum of benzaldehyde (Ph-C(=O)-H) is most likely to be due to:

   a. PhC(=O)H   b. Ph•+   c. Ph-C≡O+   d. tropylium+   e. cyclohexadienone•+

5. The major bands in the infrared spectrum of ethyl cyanoacetate will be:
   a. 3150 cm^{-1}; 2200 cm^{-1}; 1610 cm^{-1}
   b. 3450 cm^{-1}; 2930 cm^{-1}; 1760 cm^{-1}
   c. 2930 cm^{-1}; 2450 cm^{-1}; 1610 cm^{-1}
   d. 2930 cm^{-1}; 2200 cm^{-1}; 1750 cm^{-1}
   e. 3010 cm^{-1}; 2200 cm^{-1}; 1750 cm^{-1}

   NC-CH_2-C(=O)-O-CH_2-CH_3

6. Which of the following is **not true** regarding the spectra of acetophenone:
   a. in the ^{1}H NMR, the CH_3 group will appear as a singlet
   b. the ^{13}C NMR will have a singlet at ≈ 200 ppm
   c. the IR spectrum will display by a significant absorption at about 2400 cm^{-1}
   d. in the ^{1}H NMR, the CH_3 group will absorb at approximately δ 2.2
   e. in the mass spectrum, a significant peak will occur at m/e = 43

7. In the NMR, the term ppm (δ) refers to:
   a. the precessional path of the nuclear moment of inertia
   b. parts per million
   c. the concentration of the sample divided by 10^6
   d. the coupling constant, *J*, divided by the chemical shift of tetramethylsilane (Si(CH_3)$_4$)
   e. **b** and **c**

8. Which of the compounds shown below would be *most consistent* with the infrared spectrum shown below:

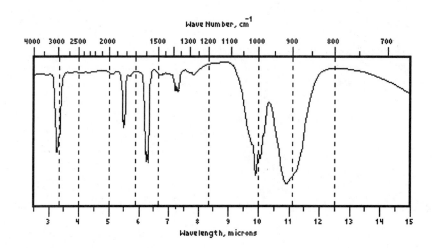

9. Which of the compounds shown below would be *most consistent* with the ^{1}H NMR spectrum shown below:

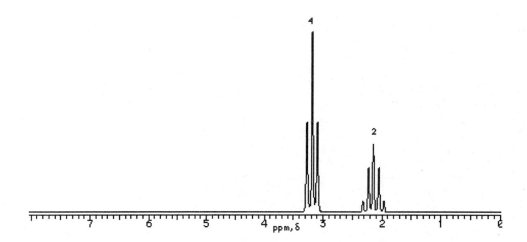

10. Which of the compounds shown below would be *most consistent* with the mass spectrum shown below:

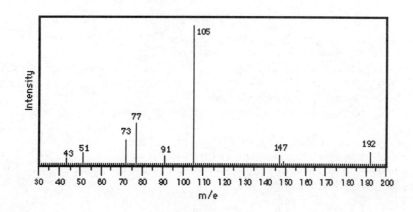

11. In the mass spectrum, the base peak:
    a. gives the molecular weight of the molecule
    b. represents the most stable Lewis base in the molecule
    c. represents the most reactive cation radical within the molecule
    d. is the most intense peak in the spectrum
    e. is formed by expulsion of the most stable cation radical

12. The ^{1}H NMR of 2-chloropropane will consist of:
    a. a doublet and a septet
    b. a multiplet consisting of ten peaks
    c. a singlet
    d. a quartet
    e. a multiplet consisting of five peaks

13. In the "aromatic" region of the ^{13}C NMR, a symmetrical 1,4-disubstituted aromatic compound will generally display, in a **proton-decoupled** spectrum:
    a. six peaks
    b. three peaks
    c. four peaks
    d. one singlet and one doublet
    e. two peaks

14. Which of the following is **true** regarding a **proton-decoupled** ^{13}C NMR:
    a. carbons bearing one hydrogen will be split into doublets
    b. the integration gives the number of hydrogens attached to the carbon giving rise to the absorption
    c. the signal intensity is not proportional to the number of carbons giving rise to the absorption
    d. splitting from adjacent ^{13}C atoms is suppressed
    e. splitting occurs only from hydrogens directly attached to the carbon giving rise to the absorption

15. The major bands in the infrared spectrum of ethyl propynoate will be:
    a. 3450 cm^{-1}; 2200 cm^{-1}; 1610 cm^{-1}
    b. 3100 cm^{-1}; 2930 cm^{-1}; 2200 cm^{-1}; 1760 cm^{-1}
    c. 2930 cm^{-1}; 2450 cm^{-1}; 1610 cm^{-1}
    d. 2930 cm^{-1}; 2200 cm^{-1}; 1710 cm^{-1}
    e. 3450 cm^{-1}; 3010 cm^{-1}; 2200 cm^{-1}; 1610 cm^{-1}

16. Which of the following is **true** regarding the spectra of 1-butyne:
    a. the ^{1}H NMR will consist of a quartet and a triplet
    b. the non-decoupled ^{13}C NMR will consist of two singlets
    c. the mass spectrum will be dominated by a peak at M-29
    d. the proton-decoupled ^{13}C NMR will consist of a triplet and a singlet
    e. the IR spectrum will have a significant absorption in the region 2200-2400 cm^{-1}

    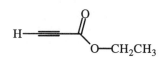

17. The ^{1}H NMR spectrum of 1,2-dimethoxyethane (CH$_3$–O–CH$_2$CH$_2$–O–CH$_3$) will consist of:
    a. one singlet, δ 3.5, area 6, and two triplets, δ 3-4, area two each
    b. two singlets in the region δ 3-4
    c. one singlet, δ 3.5, area 6, and one triplet, δ 4.0, area 4
    d. two singlets, δ 3-4, area 3 each, and two triplets, also δ 3-4, area two each
    e. one singlet, δ 3.5, area 6, and one multiplet (5 peaks), δ 4.0, area 4

18. In the ^{1}H NMR, the most highly **shielded** protons (low δ values) which are typically encountered are:
    a. carboxylic acid protons
    b. *tert*-butyl groups
    c. aromatic protons
    d. alkyne protons
    e. the methyl groups of tetramethyl silane

19. Which of the following is **true** regarding NMR spectroscopy:
    a. in a magnetic field, nuclei are split *equally* into high and low spin states
    b. nuclei with non-integer spins give rise to NMR signals
    c. adjacent ^{13}C atoms do not magnetically couple with each other
    d. decoupling is a technique commonly used to enhance the chemical shift of selected atoms
    e. in NMR, ppm is used as an abbreviation for *Please Pass the Margarine*

20. Which of the following is **not true** regarding infrared spectroscopy:
    a. the carbonyl stretch is highly assymmetric and is therefore a major peak in the IR spectrum
    b. hydroxyl groups give rise to broad absorbance bands due to hydrogen bonding
    c. the "fingerprint region" of the spectrum consists of complex stretching, bending and overtone bands, but is unique for a given molecule
    d. stretching vibrations from *symmetrical* bonds are not observed in the IR spectrum
    e. higher energy stretching bands occur at *lower* wave numbers on the cm^{-1} scale

21. The ^{1}H NMR spectrum of cyclohexane (in which ring inversion is very fast) will consist of:

   a. one doublet and one quartet
   b. one singlet
   c. six pairs of doublets and six quartets
   d. one singlet and one doublet
   e. one singlet and one triplet

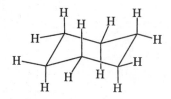

22. For each of the following disubstituted-benzene regioisomers, indicate the number of peaks you expect **in the aromatic region** of the **proton-decoupled** ^{13}C NMR:

   Compound   **Number of Peaks** *(Aromatic Region Only!)*

   a. methylbenzene (toluene)   _____

   b. 1,3-dimethylbenzene   _____

   c. 1,4-dimethylbenzene   _____

   d. 1,3,5-trimethylbenzene   _____

   e. hexamethylbenzene   _____

Cumulative Exam #1 – McMurry, Chapters 1 - 24

1. How many electrons can occupy the d-subshell of an atom? [see *McMurry*, **Section 1.3**]
    a. 2
    b. 10
    c. 8
    d. 6
    e. 14

2. Water, H_2O, has [see *McMurry*, **Section 1.6**]
    a. 4 bonding pair of electrons
    b. 3 bonding pair and 1 lone pair of electrons
    c. 2 bonding pair and 2 lone pair of electrons
    d. 2 sigma (s) bonds and two pi (p) bonds

3. Which of the following statements is true? [see *McMurry*, **Section 2.1**]
    a. Covalent bonds are characterized by equal sharing of electrons between atoms.
    b. Ionic bonds buildup electron density between atoms.
    c. Ionic bonds are MUCH stronger than covalent bonds.
    d. Covalent bonds are directionally oriented in space, whereas ionic bonds are not.
    e. Ionic bonds occur between non-metals.

4. Which of the following has the most polar X-H bonds?
    a. CH_4
    b. NH_3
    c. H_2O
    d. HF
    e. HCl

(Use the following information for questions 5-6.) Consider the set of alkanes with the molecular formula C_5H_{12}, usually referred to as a group as the "pentanes".

5. How many constitutional isomers does C_5H_{12} have?
    a. 1
    b. 2
    c. 3
    d. 4
    e. 5

6. How many of the constitutional isomers of C_5H_{12} are branched? [see *McMurry*, **Section 3.2,3**]
    a. 1
    b. 2
    c. 3
    d. 4
    e. 5

7. Using the IUPAC naming system, how many of the C_5H_{12} constitutional isomers will have a name ending in "pentane"? [see *McMurry*, **Section 3.2,3**]
    a. 1
    b. 2
    c. 3
    d. 4
    e. all of them

8. The least stable conformation of 1,2-dichloroethane is [see McMurry, Section 4.1]

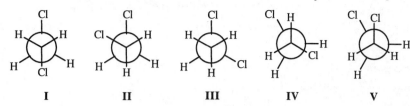

I   II   III   IV   V

   a. I
   b. II
   c. III
   d. IV
   e. V

9. Which of the following C_6H_{12} isomers would have the highest heat of combustion? [see McMurry, Section 4.5,6]
   a. propylcyclopropane
   b. ethylcyclobutane
   c. methylcyclopentane
   d. cyclohexane
   e. Since they all have the same formula, they are all the same.

(Use the following information for questions 10-11.) Again, considering the polar reaction we used in class for an example of reaction mechanisms and the following reaction profile:

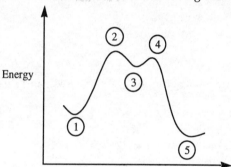

10. Which point(s) represent transition state(s)? [see McMurry, Section 5.9]
    a. 2 only
    b. 3 only
    c. 4 only
    d 2 and 4
    e. 2, 3, and 4

11. Which point(s) represent an intermediate? [see McMurry, Section 5.9,10]
    a. 2 only
    b. 3 only
    c. 4 only
    d 2 and 4
    e. 2, 3, and 4

12. Twist ethylene (II below) is less stable than regular ethylene (I below) because [see McMurry, Section 6.4]

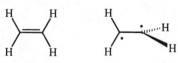

Form I   Form II

   a. the staggered arrangement of the H atoms decreases stability.
   b. the activation energy for transition to the twist form is higher than to the regular form.
   c. the C–C bond is shorter in I than in II.
   d. dipole moments no longer cancel one another out.
   e. orbital overlap, and hence bonding, is reduced.

13. The reaction in the previous question is an example of [see McMurry, Section 7.7]
    a. hydroboration
    b. oxidation
    c. *anti* addition
    d. *syn* addition
    e. none of these is correct

14. Addition of HBr to propene produces bromopropane in good yield in the presence of peroxides. Which of the following statements is correct? [see McMurry, Section 7.10]
    I. The addition is anti-Markovnikov.
    II. The addition is Markovnikov.
    III. The reaction is a radical chain reaction.
    IV. The reaction is a radical polar reaction.
    a. I only
    b. II only
    c. I and III
    d. I and IV
    e. II and III

15. What characteristic of alkynes would make it difficult to prepare cyclohexyne? [see McMurry, Section 8.1]
    a. requirement for linearity at the triple bond
    b. large electron density between carbons in a triple bond
    c. short C-C distance in triple bonds
    d. the C-C triple bond must be internal to the chain
    e. all of these

16. Which of the following are unusual about the acetylide anion and serves to make it a useful reactant? [see McMurry, Section 8.9]
    a. A H cleaved a C is unique in reactivity.
    b. A carbocation (C electrophile) is produced.
    c. A carbanion (C nucleophile) is produced.
    d. None of these is particularly significant.
    e. Given differing reaction conditions, all of these are true.

17. Which of the following molecules has the S stereochemical orientation? [see McMurry, Section 9.6]

    a. [structure: central C with H up, H (dashed) left, H₃C down-left, Cl right]

    b. [structure: central C with CH₃ up, H (dashed) left, H down-left, Cl right]

    c. [structure: central C with CH₃ up, Br (dashed) left, H down-left, Cl right]

    d. [structure: central C with Cl up, Br (dashed) left, H down-left, CH₃ right]

18. Which of the following is a relatively insignificant factor affecting the magnitude of specific optical rotation? [see *McMurry*, **Section 9.4**]
    a. concentration of the optically active substance
    b. purity of the sample
    c. temperature of the measurement
    d. length of the sample tube
    e. all are significant

19. What is the product of the reaction of cyclohexene with N-bromosuccinimide (NBS)? [see *McMurry*, **Section 10.5**]
    a. bromocyclohexane
    b. 2-bromocyclohexene
    c. 1,2-dibromocyclohexane
    d. hexane
    e. cyclohexane

20. What role does NBS serve in the reaction in reaction above? [see *McMurry*, **Section 10.5**]
    a. Produces a highly reactive bromide anion.
    b. Produces a Br• radical.
    c. Produces Br_2 for addition to the double bond.
    d. Provides and oxidant to cleave the double bond.
    e. Produces H_2 for addition to the double bond.

21. Which of the following characteristics do S_N1 reactions possess? [see *McMurry*, **Section 11.6**]
    I. unimolecular kinetics
    II. two step process
    III. do NOT completely invert stereochemical configuration
    a. I and II only
    b. II and III only
    c. II only
    d. none of these
    e. all of these

22. Which of the following is the BEST leaving group for S_N2 reactions? [see *McMurry*, **Section 11.5**]
    a. I⁻
    b. F⁻
    c. OAc⁻
    d. MeO⁻
    e. H_2N^-

23. If the following substrate is chiral, which of following reaction mechanisms would result in a racemic mixture of products? [see *McMurry*, **Section 11.5**]

    a. S_N1
    b. S_N2
    c. E1
    d. E2
    e. More than one of the above.

24. Electromagnetic radiation interacts with molecules at a variety of levels. Which of the following correctly lists the type of interaction with a molecule (in order) for infrared, UV-Visible, and NMR experiments? [see McMurry, **Section 12.5**]
    a. vibrational, nuclear spin, electronic transitions
    b. vibrational, electronic transitions, nuclear spin
    c. electronic transitions, vibrational, nuclear spin
    d. electronic transitions, nuclear spin, vibrational
    e. nuclear spin, electronic transitions, vibrational

25. How many ¹H NMR signals (unsplit by spin coupling) would cis-1,2-dichlorocyclopropane give? [see McMurry, **Section 13.9**]

    cis -1,2-dichlorocyclopropane

    a. one
    b. two
    c. three
    d. four
    e. five

26. How many (unsplit) ¹³C NMR signals would cis-1,2-dichlorocyclopropane give? [see McMurry, **Section 13.3**]
    a. one
    b. two
    c. three
    d. four
    e. five

27. Which of the following statements is (are) true? [see McMurry, **Section 13.9**]
    a. Aromatic protons are downfield of alkene protons.
    b. Aromatic protons are upfield of alkene protons.
    c. TMS is downfield of all other molecules.
    d. Electron withdrawing groups result in more shielded protons.
    e. More than one of the above is true.

28. The 1,4 monoaddition product(s) for HCl reacting with 2,4-hexadiene is(are) [see McMurry, **Section 14.5**]
    a. No reaction.
    b. 5-chloro-2-hexene
    c. 4-chloro-2-hexene
    d. 2-chloro-3-hexene
    e. more than one of the above

29. What is the product when benzene reacts with H₂ over Ni? [see McMurry, **Section 15.3**]
    a. 1,3-cyclohexadiene
    b. cyclohexene
    c. cyclohexane
    d. n-hexane
    e. No reaction occurs as the ring is inert.

30. Which of the following conjugated cyclic polyenes is aromatic? [see *McMurry*, **Section 15.5,8**]
    I. C_7H_7
    II. C_8H_8
    III. $C_{10}H_{10}$
    a. I and II only
    b. II and III only
    c. III only
    d. I
    e. all of the above

31. A student in lab class runs a nitration reaction using two substrates. The first is chlorobenzene and the second is toluene. Which of the following statements is TRUE? [see *McMurry*, **Section 16.5**]
    a. The reactions proceed at the same rate.
    b. The reaction with chlorobenzene proceeds more rapidly.
    c. The reaction with toluene proceeds more rapidly.
    d. It is not possible to differentiate between the two in terms of rate.

32. Which of the following statements is(are) true about electrophilic addition to benzene? [see *McMurry*, **Section 16.2**]
    I. One of the sets of _ electrons acts as a nucleophile.
    II. One of the sets of _ electrons acts as an electrophile.
    III. Powerful electrophiles are required for reaction to occur.
    a. I and II
    b. II and III
    c. I and III
    d. all of these

33. Which of the following statements is(are) true about electrophilic aromatic substitution (EAS) to toluene? [see *McMurry*, **Section 16.6**]
    I. Certain resonance forms make the ring <u>more reactive</u> than benzene.
    II. Certain resonance forms make the ring <u>less reactive</u> than benzene.
    III. Inductive effects make the ring <u>more reactive</u> than benzene.
    a. I
    b. II
    c. III
    d. none of these

34. Which of the following pairs of reagents can be used to prepare 2-methyl-3-pentanol? [see *McMurry*, **Section 17.6**]

    a. acetaldehyde (H_3C-CHO) and $(CH_3)_3C-MgBr$

    b. acetaldehyde (H_3C-CHO) and $CH_3CH(MgBr)CH_2CH_3$

    c. formaldehyde ($H-CHO$) and $H_3C-C(MgBr)(CH_3)CH_3$... (actually $(CH_3)_2C(MgBr)CH_3$)

    d. H_3C-CH_2-CHO and $(CH_3)_2CH-MgBr$

35. Ethers share which of the following characteristics? [see McMurry, **Section 18.2,5**]
    I. Generally chemically inert.
    II. Vulnerable to acid-catalyzed cleavage.
    III. Vulnerable to base-catalyzed cleavage.
    IV. Generally insoluble in water, even when small.
    a. I and II
    b. II and III
    c. I and IV
    d. II, III, and IV
    e. I, II, and IV

36. Epoxides (oxiranes) are highly reactive species. Which of the following statements <u>best</u> accounts for this observation? [see McMurry, **Section 18.7**]
    a. The carbocation is stabilized.
    b. The resulting carbanion is stabilized.
    c. A stable, inert ether is formed on reaction.
    d. The C–O bond is strengthened by adjacent groups.
    e. The three-membered ring is highly strained.

37. Nucleophilic addition reactions to carbonyl groups can be catalyzed by either bases or acids. Acid catalyzes these reactions by [see McMurry, **Section 19.6**]
    a. making the C of the carbonyl group more electrophilic.
    b. making the C of the carbonyl group more nucleophilic.
    c. shifting the equilibrium of the reaction.
    d. being a powerful nucleophile itself.
    e. none of these.

38. Reaction of a primary amine with an aldehyde or ketone produces [see McMurry, **Section 19.9**]
    a. an amide
    b. a secondary amine
    c. a nitrile
    d. an imine
    e. none of these.

39. Which of the following is characteristic of the IR spectrum of carboxylic acids? [see McMurry, **Section 20.9**]
    a. A broad strong peak between 3300 and 3500 cm^{-1}.
    b. A broad strong peak between 2500 and 3300 cm^{-1}.
    c. A strong narrow peak around 1750 cm^{-1}.
    d. a <u>and</u> b
    e. b <u>and</u> c

40. The following molecule is classified as a(n) [see McMurry, **Section 21.1**]

    a. amide.
    b. carboxylic acid.
    c. acid anhydride.
    d. lactone.
    e. acid halide.

41. Which of the following carboxylic acid derivatives is the most reactive towards nucleophilic acyl substitution. [see *McMurry*, **Section 21.2**]
    a. amides
    b. acid chlorides
    c. acid anhydrides
    d. nitriles
    e. esters

    a. The N resonance stabilizes the C atom so that it acts like a carbonyl center.
    b. There is back-donation of electron density from the N lone pairs.
    c. They possess three bonds between C and an electronegative element.
    d. They are sterically hindered in the same fashion.
    e. c <u>and</u> d

42. The molecule below exists in equilibrium between the two forms shown. Which of the following statements <u>best</u> describes the equilibrium position for this system? [see *McMurry*, **Section 22.1**]

    a. Equilibrium to the <u>left</u> is preferred to vast excess.
    b. Equilibrium to the <u>right</u> is preferred to vast excess.
    c. Equilibrium to the <u>left</u> is unusually high for this general type of system due to resonance stabilization.
    d. Equilibrium to the <u>right</u> is unusually high for this general type of system due to resonance destabilization.
    e. Transformation between the forms above can be accomplished only by acid catalysis. The transformation reaction does NOT occur in basic solution.

43. Which of the following is true about enolate anions? [see *McMurry*, **Section 22.5,6**]
    I. Prepared by the addition of base to a carbonyl with an a-H.
    II. Prepared by the addition of acid to a carbonyl with an a-H.
    III. More reactive than an enol form of the original carbonyl.
    IV. More likely to react as an electrophile than a nucleophile.
    a. I, III, and IV
    b. II, III, and IV
    c. I and III
    d. I and IV
    e. II and III

44. What is the product when an aldol is dehydrated? [see *McMurry*, **Section 23.4**]
    a. an acid anhydride
    b. malonate diester
    c. b-hydroxyaldehyde
    d. conjugated enone
    e. b-ketoester

45. Which of the following best describes what happens in an aldol condensation reaction? [see *McMurry*, **Section 23.2**]
    a. The acid catalyst activates on aldehyde so that it can act as an electrophile and attack an unactivated molecule.
    b. The base abstracts an a hydrogen producing an electrophile that reacts with an unactivated aldehyde.
    c. The base abstracts an a hydrogen producing an nucleophile that reacts with an unactivated aldehyde.
    d. Two moles of aldehyde are consumed in an endothermic reaction, caused water to condense into a vapor and form droplets on the inside of the reaction vessel.
    e. The reaction initially forms and *enol* moiety, which rearranges to the *keto* form.

46. Arenediazonium salts are prepared by reaction of [see *McMurry*, **Section 24.7**]
    a. nitrobenzene with H_2SO_4 and HNO_2
    b. nitrobenzene with sodium ethoxide in alcohol
    c. aniline with sodium ethoxide in alcohol
    d. aniline with H_2SO_4 and HNO_2
    e. None of these.

(Use the following reaction sequence for questions **47-49**. Assume that product from the previous reaction is present for the next step.) [see *McMurry*, **Section 24.6**]

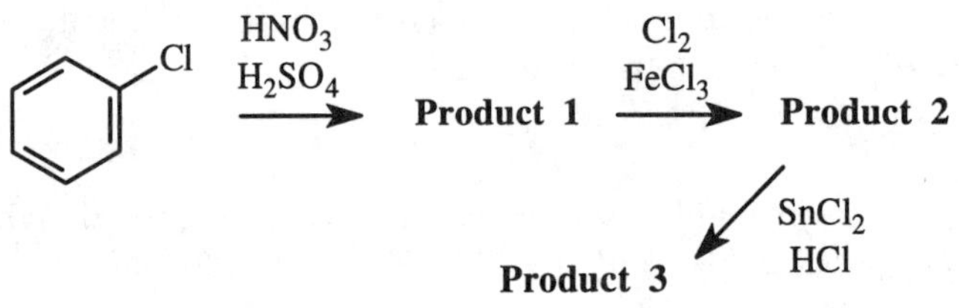

$$\text{PhCl} \xrightarrow[H_2SO_4]{HNO_3} \text{Product 1} \xrightarrow[FeCl_3]{Cl_2} \text{Product 2} \xrightarrow[HCl]{SnCl_2} \text{Product 3}$$

47. Which of the following is **Product 1**?
    a. aniline
    b. arenediazonium salt
    c. *p*-chloroaniline
    d. *m*-chloroaniline
    e. *p*-nitrochlorobenzene

48. Which of the following is **Product 2**?
    a. p-chloroaniline
    b. chlorobenzene
    c. 2,4-dichloroaniline
    d. 3,4-dichloraniline
    e. 3,4-dichloronitrobenzene

49. Which of the following is **Product 3**?
    a. aniline
    b. benzene
    c. 1,3-dichlorobenzene
    d. 1,2-dichlorobenzene
    e. 3,4-dichloroaniline

50. Reaction of an arenediazonium salt with which of the following sets of reagents would produce nitrobenzene? [see *McMurry*, **Section 24.7**]
    a. nitric acid
    b. nitrous acid
    c. hyponitrous acid
    d. N_2 gas bubbled through the solution.
    e. None of these.

Cumulative Exam #2 – McMurry, Chapters 1 - 24

Use the following molecule for questions 1 - 2. [see *McMurry*, Section 1.9]

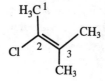

1. Carbon #2
   a. is sp³ hybridized
   b. is sp² hybridized
   c. is sp hybridized
   d. has three p bonds and one s bond
   e. is a tetrahedral center

2. Carbon #1
   a. is sp³ hybridized
   b. is sp² hybridized
   c. is sp hybridized
   d. has three p bonds and one s bond
   e. is a trigonal planar center

   Given:
Acid	pK_a
HF	3.2
H_2O	15.7
HCN	9.2

3. Which acid is strongest? [see *McMurry*, Section 2.8]

   a. HF   b. HCN   c. H_2O

4. Which base is strongest? [see *McMurry*, Section 2.8]

   a. HF   b. HCN   c. H_2O

5. Given the following reaction, which is the conjugate base according to Brønsted-Lowry Acid/Base Theory? [see *McMurry*, Section 2.7]

   $$H_3C-C(=O)OH \quad + \quad {}^{\ominus}OH \quad \rightleftharpoons \quad H_2O \quad + \quad H_3C-C(=O)O^{\ominus}$$
   $$\text{I} \qquad\qquad \text{II} \qquad\qquad \text{III} \qquad\qquad \text{IV}$$

   a. I   b. II   c. III   d. IV

6. An alkane, C_6H_{14}, reacts with Cl_2 in the presence of UV light to produce FIVE constitutional isomers with the formula $C_6H_{13}Cl$. The structure of the parent alkane is: [see *McMurry*, Section 3.5]

   a. $CH_3CH_2CH_2CH_2CH_2CH_3$

   b. $CH_3CHCH_2CH_2CH_3$
      $\quad\;\; |$
      $\quad\;\; CH_3$

   c. $CH_3CH_2CHCH_2CH_3$
      $\qquad\quad |$
      $\qquad\quad CH_3$

   d. $CH_3\underset{\underset{CH_3}{|}}{\overset{\overset{CH_3}{|}}{C}}CH_2CH_3$

   e. $CH_3CH\underset{\underset{CH_3}{|}}{\overset{\overset{CH_3}{|}}{C}H}CH_3$

7. The folded conformation that unsubstituted cyclic alkanes (C ≤ 6) takes on is caused by [see *McMurry*, **Section 4.3**]

   I. strained bond angles
   II. destabilizing eclipsing interactions
   III. destabilizing 1,3-diaxial interactions

   a. I and II
   b. I and III
   c. II and III
   d. all of these important
   e. none of these important

8. Baeyer correctly predicted small cycloalkane rings would be highly strained and reactive. He also <u>incorrectly</u> predicted that larger rings would be impossible to synthesize due to instability. What did his model leave out? [see *McMurry*, **Section 4.4**]
   a. He did not understand reduced eclipsing interactions.
   b. He assumed that the molecules were flat.
   c. He neglected steric strain.
   d. He neglected orbital expansion.
   e. More than one of the above.

9. In reference to the conformers defined by the C2 to C3 bond axis in butane and pentane, the energy difference between the *anti* and *gauche* conformers for pentane is larger than it is for butane. Which of the following best explains this? [see *McMurry*, **Section 4.3**]
   a. Enhanced eclipsing interactions.
   b. Orbitals are off-center.
   c. Enhanced steric strain.
   d. None of the above.
   e. More than one of the above.

(Use the following information for questions 10-12.) Again, considering the polar reaction we used in class for an example of reaction mechanisms and the following reaction profile:

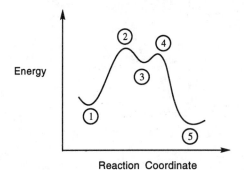

10. The energy of the reaction is the difference between points [see *McMurry*, **Section 5.7,9**]

    a. 2 and 1
    b. 3 and 1
    c. 5 and 1
    d. 4 and 1
    e. 4 and 3

11. The energy of the rate-determining step is the difference between points [see *McMurry*, **Section 5.7,9**]

    a. 2 and 1
    b. 3 and 1
    c. 5 and 1
    d. 4 and 1
    e. 4 and 3

12. As drawn, the diagram shows that the overall reaction is [see McMurry, **Section 5.7,9**]

    a. exothermic
    b. endothermic
    c. sterically hindered
    d. polar
    e. radical

13. Which is the proper name for the following compound? [see McMurry, **Section 6.5**]

    a. 2-pentene
    b. *cis* - 2-pentene
    c. *trans* -2-pentene
    d. E -2-pentene
    e. Z - 2-pentene

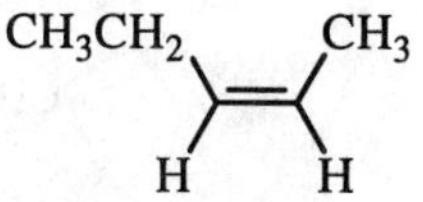

The following general reaction will be used for Questions 14 – 15. Different reagents and sets of conditions will be supplied. Choose the correct product from the choices listed. [see McMurry, **Section 7.8**]

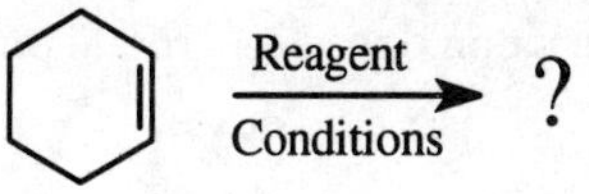

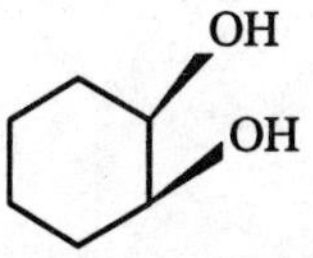

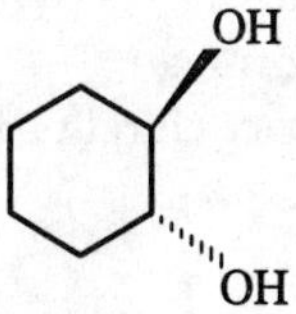

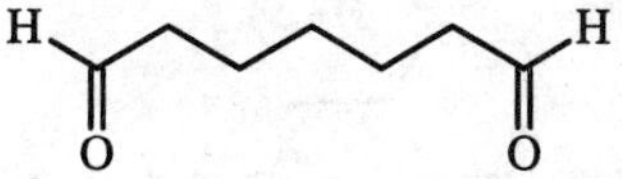

  e. No reaction.

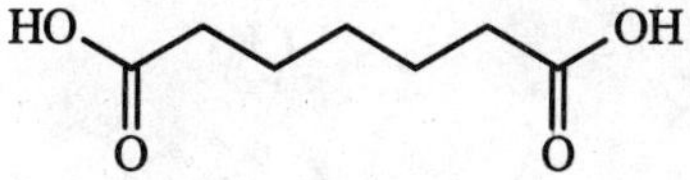

14. *Reagent:* $KMnO_4$ *Conditions:* aqueous acid

15. *Reagent:* 1. O_3; 2. Zn, HOAc

16. If Lindlar's Catalyst is used to hydrogenate 2-pentyne, which of the following products will be produced? [see McMurry, **Section 8.6**]

    a. *n*-pentane
    b. 1,3-pentadiene
    c. 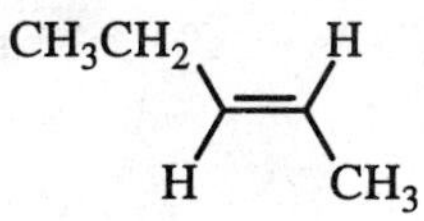
    d.

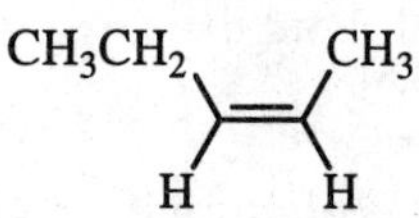

17. Propyne is reacted with mercuric sulfate in aqueous acid. Which of the following products is isolated from the reaction mixture? [see McMurry, Section 8.5]

    a. (acetone structure)   b. (propen-2-ol structure)

    c. (2-methyl-2,3-diol structure with H₃C, CH₃, OH, HO)   d. (structure with H₃C, H, OH, CH₂OH)

18. Consider the following reaction: [see McMurry, Section 9.15]

    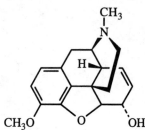

    The resulting solution from the above reaction does NOT rotate plane-polarized light. The correct explanation for this is that

    a. none of the product(s) possess a chiral center.
    b. the carbocation intermediate is pyramidal in geometry.
    c. a racemic mixture is formed.
    d. achiral reactants produce chiral products.
    e. no reaction occurs.

19. How many stereogenic centers are present in the following molecule. [see McMurry, Section 9.9]

    (morphine-like structure)

    a. 3   b. 4   c. 5   d. 6   e. more than 6

20. Which of the following best accounts for the stability of the allyl radical? [see McMurry, Section 10.6]

    a. Steric hindrance in the primary alkyl radical is at a minimum
    b. The balance between R groups and H's is ideal in the secondary radical.
    c. Tertiary alkyl groups are the most stable.
    d. Resonance effects stabilize this form.

21. Which of the following are unusual about the Grignard reagent and serves to make it a useful reactant? [see McMurry, Section 10.8]
    a. The chloride anion is stable.
    b. A carbocation (C electrophile) is produced.
    c. A carbanion (C nucleophile) is produced.
    d. The metal has empty d-orbitals that promote reactivity.
    e. Given differing reaction conditions, all of these are true.

22. What would be the predominant product from the following reaction, and which mechanism would be preferred? [see McMurry, **Section 11.11**]

$$(CH_3)_2CH-X \xrightarrow{\text{Strong Base}}$$

   a. propene, E1
   b. propene, E2
   c. 2-propanol, S_N2
   d. 1-propanol, S_N1
   e. 2,2-propanediol, oxidation

23. An INCREASE in which of the three effects noted below leads a particular subtrate to react *via* the S_N1 pathway rather than the S_N2 pathway? [see McMurry, **Section 11.6**]

   I. steric hindrance
   II. leaving group stability
   III. carbocation stability

   a. I
   b. II
   c. III
   d. None of these has the desired effect.
   e. More than one of the above.

24. A compound of the formula C_2H_5ON has a strong IR peak at 1700 cm^{-1} and a medium strength narrow peak at 3300 cm^{-1}. It is likely a(n) [see McMurry, **Section 12.9**]

   a. carboxylic acid
   b. amide
   c. alcohol
   d. enol
   e. diazonium salt

25. Peaks are observed in the IR spectra for a compound at 1750, 2850, 2900, 2925, and 3400 wavenumbers. Which of the following is the most likely structure for this compound. [see McMurry, **Table 12.1**]

   a. aniline (C$_6$H$_5$NH$_2$)
   b. H$_2$N-CH$_2$CH$_2$-O-CH$_3$
   c. H-C(COOCH$_2$CH$_3$)$_2$-H (diethyl malonate-type, CH$_2$(COOCH$_2$CH$_3$)$_2$)
   d. H$_3$C-C(=O)-NH$_2$
   e. None of the above.

26. Which of the following analytical techniques used in organic chemistry is primarily used to identify functional groups? [see McMurry, **Section 12.6,9**]
   a. Mass spectroscopy
   b. IR
   c. elemental analysis
   d. NMR
   e. UV-Visible spectroscopy

27. How many ¹H NMR signals would you expect from *p*-dimethoxybenzene? [see *McMurry*, **Section 13.9**]

    a. 2
    b. 3
    c. 4
    d. 5
    e. >5

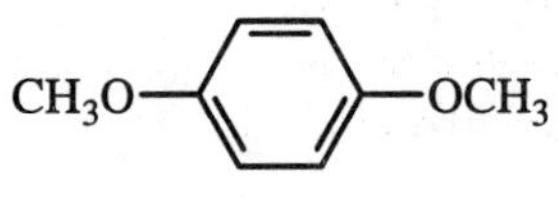

    $CH_3O-\langle\text{benzene}\rangle-OCH_3$

    para-dimethoxybenzene

28. How many ¹³C NMR signals would you expect from *p*-dimethoxybenzene? [see *McMurry*, **Section 13.3**]

    a. 2
    b. 3
    c. 4
    d. 5
    e. >5

29. Which of the following correctly describes the NMR experiment? [see *McMurry*, **Section 13.1**]
    a. NMR light is passed through the sample and an absorbance spectrum is collected.
    b. A solid sample is rotated in a radio frequency field and magnetic signal is noted.
    c. A radio frequency field is applied and magnetic field is tuned.
    d. A magnetic field is applied and radio frequency radiation is tuned.
    e. A magnetic field is applied and microwave radiation is tuned.

30. On monoaddition of HCl to 1,3-cyclohexadiene, multiple products are formed. Select the products from the choices provided below. [see *McMurry*, **Section 14.5**]
    a. 3-chlorocyclohexene
    b. 4-chlorocyclohexene
    c. 2-chlorocyclohexene
    d. a and b only
    e. a, b, and c

31. Match the 3 steroid structures shown below to the following λ_{max} values respectively: 257 nm, 304 nm, 356 nm. [see *McMurry*, **Section 14.12**]

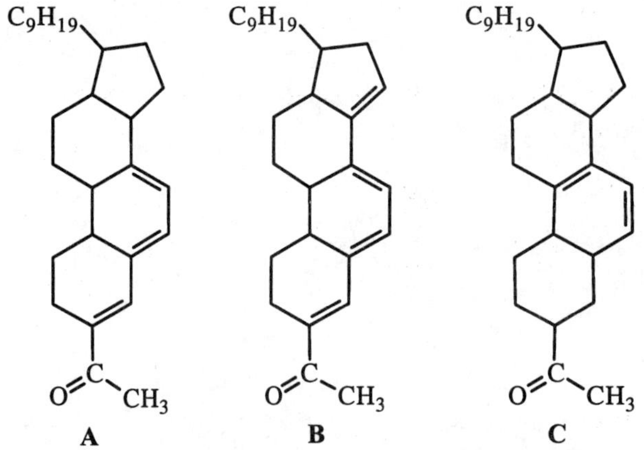

    A   B   C

    a. A, B, C
    b. A, C, B
    c. B, A, C
    d. C, A, B
    e. C, B, A

32. Which of the following are characteristics of benzene that differentiate it from other organic compounds? [see McMurry, Section 15.1,3]
   I. Benzene undergoes substitution rather that addition.
   II. Benzene is easily hydrogenated.
   III. Benzene has no characteristic aroma.
   a. I and II only
   b. II and III only
   c. III only
   d. I
   e. all of the above

33. What rule can be used to determine in a species is aromatic? [see McMurry, Section 15.5,8]
   a. Hund's Rule
   b. Frosts's Circle
   c. Huckel's 4n+2
   d. neutral, flat and conjugated
   e. Le Chatelier's Principle

34. *Ortho/para*-directing groups for aromatic electrophilic substitution <u>always</u> have[see McMurry, Section 16.5,6]
   a. a negative charge
   b. activating effect on ring
   c. a lone-pair of electrons
   d. a positive charge
   e. none of the above

(Use the following information for questions 35 and 36.) A Friedel-Crafts acylation is carried out on phenol:

35. Which of the following sets of reagents are used for this procedure? [see McMurry, Section 16.4]
   a. $AlCl_3 / Cl_2$
   b. $KMnO_4 / H^+$
   c. HNO_3 / H_2SO_4
   d. $AlCl_3 / RCOOH$
   e. $AlCl_3 / RCOCl$

36. Which of the following is the correct product for this reaction? [see McMurry, Section 16.5]

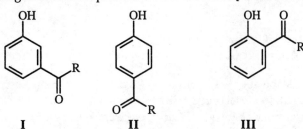

   a. I and II only
   b. II and III only
   c. I and III only
   d. none of the above
   e. all of the above

37. Which of the following compounds is the <u>most</u> oxidized C compound? [see McMurry, Section 17.8]
   a. primary alkane
   b. alcohol
   c. aldehyde
   d. carboxylic acid

38. Which of the following is true of the IR spectra of compounds containing an –OH group. [see *McMurry*, **Section 17.12**]
    a. The position O-H stretch can be used to differentiate between phenols, alkyl alcohols, and carboxylic acids
    b. the C–O stretching frequency is altered by H-bonding
    c. H-bonding has no effect on the -OH stretching frequency
    d. the alcoholic proton is shifted downfield from alkyl protons.
    e. the shape of a peak in the spectrum can be altered by dilution.

39. Mild oxidation of thiols lead to[see *McMurry*, **Section 18.11**]
    a. sulfides
    b. thioacetamides
    c. sulfonic acids
    d. sulfonates
    e. disulfides

40. Nucleophilic addition reactions to carbonyl groups can be catalyzed by either bases or acids. Base catalyzes these reactions by[see *McMurry*, **Section 19.6**]
    a. making the C of the carbonyl group more electrophilic.
    b. making the C of the carbonyl group more nucleophilic.
    c. shifting the equilibrium of the reaction.
    d. being a powerful nucleophile itself.
    e. none of these.

41. Which of the following is the most predominant product from the reaction of cyclohexanone with excess methanol in the presence of HCl? [see *McMurry*, **Section 19.11**]
    a. an alcohol
    b. an aldehyde
    c. a cleaved ring with aldehyde on one end and ketone on the other
    d. an acetal
    e. a hemiacetal

42. For most carboxylic acids, the pK_a is in the range[see *McMurry*, **Section 20.3**]
    a. 1-2
    b. 4-5
    c. 6-7
    d. 10-11
    e. 15-16

43. Why are nitriles grouped with carboxylic acid derivatives based on reactivity? [see *McMurry*, **Section 21.8**]
    a. The N resonance stabilizes the C atom so that it acts like a carbonyl center.
    b. There is back-donation of electron density from the N lone pairs.
    c. They possess three bonds between C and an electronegative element.
    d. They are sterically hindered in the same fashion.
    e. c and d

44. Which of the following statements is (are) true? [see *McMurry*, **Section 22.5,6**]
    a. Enols are more reactive than enolate anions.
    b. Enolates are more reactive than enols.
    c. Enols are easily isolated in basic solutions.
    d. Enolates are easily isolated in acidic solutions.
    e. More than one of the above is true.

45. An aldehyde is placed in a solution of sodium hydroxide and Cl_2 is added. The solution is then neutralized with acid. Which of the following products is observed? [see McMurry, **Section 22.7**]
   a. $CHCl_3$
   b. alcohol
   c. acid anhydride
   d. acid halide
   e. alkyl halide

(Use the following information for questions 46 and 47.) Two different aldehydes reacting via the aldol condensation usually form <u>four</u> products.

46. Which of the following best explains this observation? [see McMurry, **Section 23.6**]
   a. One aldol is formed, which can rearrange into the other products upon heating.
   b. As the reaction proceeds, some of the intermediates are esters, so some Claisen products are also formed.
   c. Aldols reversibly rearrange into conjugated enones, so significant amounts of both enone and aldol are present in most cases.
   d. Each of the aldehydes will act both as electrophile and nucleophile giving four different combinations.

47. Some aldehyde combinations do NOT form four products. Which of the following aldehydes would lead to only ONE product in combination with formaldehyde? [see McMurry, **Section 23.6**]

   a. H_3C-CHO
   b. CH_3CH_2-CHO
   c. $(CH_3)_2CH-CHO$
   d. $(CH_3)_3C-CHO$

(Use the following compounds for Questions 48 and 49.)

I: PhCH$_2$NH$_2$ II: PhCH$_2$NHCH$_3$ III: PhCH$_2$N(CH$_3$)$_2$ IV: PhNH$_2$

48. Which of these is a quaternary amine? [see McMurry, **Section 24.1**]

   a. I   b. II   c. III   d. IV   e. None of these.

49. Which is the LEAST basic? [see McMurry, **Section 24.5**]

   a. I   b. II   c. III   d. IV

50. Benzenediazonium bisulfate salts are primarily used [see McMurry, **Section 24.8**]
   a. to produce mixed nitro/amine groups.
   b. to couple two nitroarene systems together.
   c. to make a nitro group attached to a ring reactive.
   d. to replace an arylamine group with another functional group.
   e. to cleave benzene rings.